Principles of
STRUCTURAL DESIGN

Wood, Steel, and Concrete

Principles of
STRUCTURAL DESIGN

Wood, Steel, and Concrete

RAM S. GUPTA

CRC Press
Taylor & Francis Group
Boca Raton   London   New York

CRC Press is an imprint of the
Taylor & Francis Group, an **informa** business

CRC Press
Taylor & Francis Group
6000 Broken Sound Parkway NW, Suite 300
Boca Raton, FL 33487-2742

© 2011 by Taylor and Francis Group, LLC
CRC Press is an imprint of Taylor & Francis Group, an Informa business

No claim to original U.S. Government works

Printed in the United States of America on acid-free paper
10 9 8 7 6 5 4 3 2

International Standard Book Number: 978-1-4200-7339-3 (Hardback)

Library of Congress Cataloging-in-Publication Data

Gupta, Ram S.
    Principles of structural design : wood, steel, and concrete / author, Ram S. Gupta.
        p. cm.
    "A CRC title."
    Includes bibliographical references and index.
    ISBN 978-1-4200-7339-3 (hardcover : alk. paper)
    1. Structural design. 2. Building materials. I. Title.

TA633.G87 2011
624.1'771--dc22   2010006796

Visit the Taylor & Francis Web site at
http://www.taylorandfrancis.com

and the CRC Press Web site at
http://www.crcpress.com

Contents

PART I Design Loads

PART II   Wood Structures

PART III   *Steel Structures*

PART IV   Reinforced Concrete Structures

Preface

This book fills in a gap that exists for a textbook that provides a comprehensive design course in wood, steel, and concrete. It presents up-to-date information and practices of structural design just at a right level for architecture and construction management majors. It is also suitable for civil engineering and general engineering undergraduate programs, where the curriculum includes a joint coursework in structural design with wood, steel, and concrete.

The book is at the elementary level and it has a code connected design focus. In a structural system, there are three main elements: the tension, compression, and bending members based on the stress that they are subjected to, singularly or in combination. The book presents these elements systematically following the latest load resistance factor design (LRFD) approach, and uses several fully solved examples. If one is interested in designing elements according to the codes for each of the construction materials, this book is an excellent resource.

The book is divided into 4 parts, spanning 16 chapters. Part I (Chapters 1 through 5) provides a detailed coverage of loads, load combinations, and the specific code requirements for different types of loads. The LRFD philosophy and the unified approach to design have also been explained in this part.

Part II (Chapters 6 through 8) covers sawn lumber, structural glued laminated timber, and structural composite lumber; the last of these, which commonly includes laminated veneer lumber, is finding an increased application in wood structures. First, the conceptual designs of tension, compression, and bending members are reviewed, and then the effects of the column and beam stabilities and the combined forces are discussed.

The frame is an essential component of steel structures. Part III (Chapters 9 through 13) covers the design of individual tension, compression, and bending members. Additionally it provides a theoretical background and designs of braced and unbraced frames with fully solved examples. Open-web steel joists and joist girders, though separate from the American Institute of Steel Construction, have been included since they form a common type of flooring system for steel-frame buildings.

Connections are a subject of special interest because of their weak links in structures and neglect by engineers. Separate detailed chapters on wood (Chapter 8) and steel (Chapter 13) connections present the design of the common types of connecting elements for the two materials.

In concrete, there is no direct tension member and shear is handled differently. Part IV (Chapters 14 through 16) covers reinforced beams and slabs, shear and torsion, compression and combined compression and flexure in relation to basic concrete structures.

This is a self-contained book. Instead of making references to codes, manuals, and other sources of data, a voluminous amount of material on section properties, section dimensions, specifications, reference design values, load tables, and other design aids that would be needed for the design of structures have been included in the book.

During the preparation of the manuscript, help came from many quarters. My wife, Saroj Gupta, and daughters, Sukirti Gupta and Sudipti Gupta, typed and edited the chapters. The senior students of my structural design class provided very helpful input to the book; Ignacio Alvarez prepared the revised illustrations, Andrew Dahlman, Ryan Goodwin, and George Schork reviewed the end-of-chapter problems. Joseph Clements, David Fausel, and other staff at CRC Press provided valuable support that led to the completion of the book. I extend my sincere thanks to these individuals and to my colleagues at Roger Williams University who provided a helping hand from time to time.

Author

Ram S. Gupta holds a master of engineering degree from IIT, Roorkee, India, and a PhD from Polytechnic University, New York. He is a registered professional engineer in Rhode Island and Massachusetts.

Dr. Gupta has 40 years of experience working on projects in the United States, Australia, India, and Liberia (West Africa), and is currently working as a professor of engineering at Roger Williams University (RWU), Bristol, Rhode Island. He has been a full-time faculty member at RWU since 1981. He was a rotary scholar professor at Kathmandu University, Dhulikhel, Nepal, and a Fulbright scholar at the Indian Institute of Technology, Kanpur, India.

Dr. Gupta is president of Delta Engineers, Inc., an Rhode Island-based consulting company, specializing in structural and water resource disciplines.

Besides contributing to a very large number of research papers, he has authored two very successful books: *Hydrology and Hydraulic Systems*, 3rd edition (Waveland Press, Long Grove, IL, 2008), *Introduction to Environmental Engineering and Science*, 2nd edition (ABS Consulting, Rockville, MD, 2004), and *Principles of Structural Design: Wood, Steel, and Concrete* (Taylor & Francis, Boca Raton, FL, 2010).

Part I

Design Loads

1 Design Criteria

CLASSIFICATION OF BUILDINGS

Buildings and other structures are classified based on the nature of occupancy according to Table 1.1. The occupancy categories range from I to IV where occupancy category I represents buildings and other structures that pose no danger to human life in the event of failure and the occupancy category IV represents all essential facilities. Each structure is assigned the highest applicable occupancy category. An assignment of more than one occupancy category to the same structure based on the use and loading conditions is permitted.

BUILDING CODES

To safeguard public safety and welfare, town and cities across the United States follow certain codes for design and construction of buildings and other structures. Until recently, towns and cities modeled their codes based on the following three regional codes, which are revised normally at 3 year intervals:

1. The BOCA* National Building Code
2. The Uniform Building Code
3. The Standard Building Code

The International Codes Council was created in 1994 for the purpose of unifying these codes into a single set of standards. The council included the representatives from the three regional code organizations. The end result was the preparation of the *International Building Code* (IBC), which was first published in 2000, with a second revision in 2003 and a third revision in 2006. Now, practically all local and state authorities follow the IBC. For the specifications of loads to which the structures should be designed, the IBC makes a direct reference to the American Society of Civil Engineers' publication *Minimum Design Loads for Buildings and Other Structures* commonly referred to as the ASCE 7-05.

STANDARD UNIT LOADS

The primary loads on a structure are dead loads due to weight of the structural components and live loads due to structural occupancy and usage. The other common loads are snow loads, wind loads, and seismic loads. Some specific loads to which a structure could additionally be subjected to comprise of soil loads, hydrostatic force, flood loads, rain loads, and ice loads (atmospheric icing). The ASCE 7-05 specifies the standard unit loads that should be adopted for each category of loading. These have been described in Chapters 2 through 5 for the main categories of loads.

---

* Building Officials and Code Administrators.

TABLE 1.1
Occupancy Category of Buildings and Other Structures

Nature of Occupancy	Category
Agriculture, temporary structures, storage	I
All buildings and structures except classified as I, III, and IV	II
Buildings and other structures that can cause a substantial economic impact and/or mass disruption of day-to-day civil lives, including the following:	III
More than 300 people congregation	
Day care with more than 150	
School with more than 250 and college with more than 500	
Resident health care with 50 or more	
Jail	
Power generation, water treatment, wastewater treatment, telecommunication centers	
Essential facilities, including the following:	IV
Hospitals	
Fire, police, ambulance	
Emergency shelters	
Facilities need in emergency	

Source: Courtesy of American Society of Civil Engineers, Reston, VA.

TRIBUTARY AREA

Since the standard unit load in the ASCE 7-05 is for a unit area, it needs to be multiplied by the effective area of the structural element on which it acts to ascertain the total load. In certain cases, the ASCE 7-05 specifies the concentrated load, then its location needs to be considered for the maximum effect. In a parallel framing system shown in Figure 1.1, beam CD receives the load from the floor that extends half way to the next beam ($B/2$) on each side, as shown by the hatched area. Thus, the tributary area of the beam is $B \times L$ and the load, $W = w \times B \times L$, where w is the unit standard load. Exterior beam AB receives the load from one side only extending half way to the next beam. Hence the tributary area is $1/2B \times L$.

Suppose we consider a strip of 1 ft width as shown in Figure 1.1. The area of the strip is ($1 \times B$). The load of the strip is $w \times B$, which represents the uniform load per running ft (or meter) of the beam.

The girder is point loaded at the locations of beams by the beam reactions. However, if the beams are closely spaced, the girder could be considered to bear uniform load from the tributary area of $1/2B \times L$.

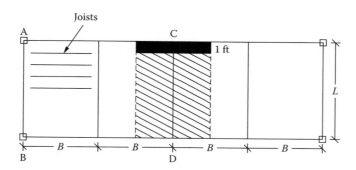

FIGURE 1.1   Parallel framing system.

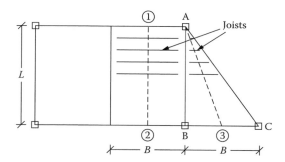

FIGURE 1.2 Triangular loaded frame.

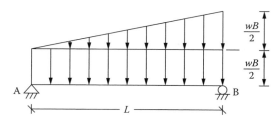

FIGURE 1.3 Load distribution on beam AB.

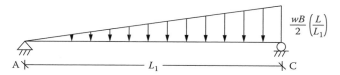

FIGURE 1.4 Load distribution on beam AC.

In Figure 1.2, beam AB supports a rectangular load from an area A, B, 1, 2, the load is $wBL/2$ and also a triangular load from an area A, B, 3 the load is $(1/2)w(B/2)L$ or $wBL/4$.

This has a distribution as shown in Figure 1.3. Beam AC supports the triangular load from area A, C, 3 which is $wBL/4$. However, the loading on the beam is not straightforward because the length of the beam is not L but $L_1 = \left(\sqrt{L^2 + B^2}\right)$. The triangular loading will be as shown in Figure 1.4 to represent the total load (the area under the load diagram) of $wBL/4$.

The framing of a floor system can be arranged in more than one manner. The tributary area and the loading pattern on the framing elements will be different for different framing systems, as shown in Figures 1.5 and 1.6.

Example 1.1

In Figure 1.2, the span L is 30 ft, the spacing B is 10 ft. The distributed standard unit load on the floor is 60 lb/ft². Determine the tributary area and show the loading on beams AB and AC.

Solution
Beam AB

1. Rectangular tributary area/ft beam length $= 1 \times 5 = 5$ ft²/ft
2. Uniform load/ft $=$ (standard unit load $\times$ tributary area) $= (60\,\text{lb/ft}^2)\,(5\,\text{ft}^2/\text{ft}) = 300\,\text{lb/ft}$

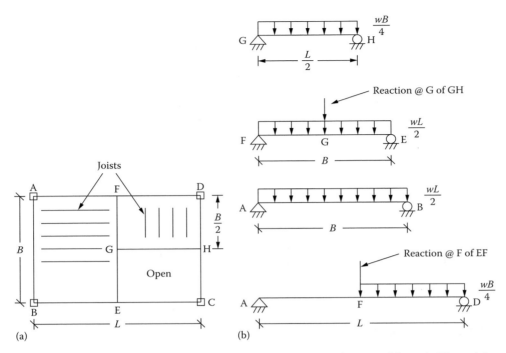

FIGURE 1.5   (a) A framing arrangement. (b) Distribution of loads on elements of frame in Figure 1.5.

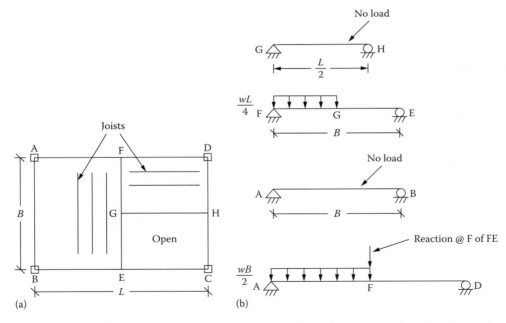

FIGURE 1.6   (a) An alternative framing arrangement. (b) Distribution of loads on frame in Figure 1.6.

   3. Triangular tributary area (total) = 1/2 (5) (30) = 75 ft²
   4. Total load of triangular area = 60 × 75 = 4500 lb
   5. Area of triangular load diagram = 1/2wL
   6. Equating items (4) and (5): 1/2wL = 4500 or w = 300 lb/ft
   7. Loading is shown in Figure 1.7

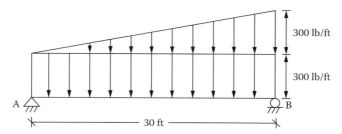

FIGURE 1.7   Distribution of loads on beam AB of Example 1.1.

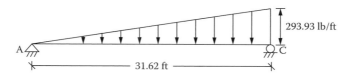

FIGURE 1.8   Distribution of loads on beam AC of Example 1.1.

Beam AC

1. Tributary area $= 75 \, ft^2$
2. Total load $= 60 \times 75 = 4500 \, lb$
3. Length of beam AC, $L = \left(\sqrt{30^2 + 10^2} \right) = 31.62 \, ft$
4. Area of triangular load diagram $= 1/2 wL = 1/2 w \, (31.62)$
5. Equating (2) and (4): $1/2 w \, (31.62) = 4500$ or $w = 293.93 \, lb/ft$
6. The loading is shown in Figure 1.8

WORKING STRESS DESIGN, STRENGTH DESIGN, AND UNIFIED DESIGN OF STRUCTURES

There are two approaches to design: the traditional approach and comparatively a newer approach. The distinction between them can be understood from the stress–strain diagram. The stress–strain diagram with labels for a ductile material is shown in Figure 1.9. The diagram for a brittle material is similar except that there is only one hump indicating both the yield and ultimate strength point, and the graph at the beginning is not really (it is close to) a straight line.

The allowable stress is the ultimate strength divided by a factor of safety. It falls on the straight line portion within the elastic range. In the allowable stress design (ASD) or working stress design (WSD) method, the design is carried out so that when the computed design load, known as the *service load*, is applied on a structure, the actual stress created does not exceed the allowable stress limit. Since the allowable stress is well within the ultimate strength, the structure is safe. This method is also known as the *elastic design approach*.

In the other method, known variously as the *strength design*, the *limit design*, or the *load resistance factor design* (LRFD), the design is carried out at the ultimate strength level. Since we do not want the structure to fail, the design load value is magnified by a certain factor known as the *load factor*. Since the structure at ultimate level is designed for loads higher than the actual loads, it does not fail. In the strength design, the strength of the material is taken to be the ultimate strength, and a resistance factor (of less than 1) is applied to the ultimate strength to account for the uncertainties associated with determination of the ultimate strength.

The LRFD method is more efficient than the ASD method. In ASD method, a single factor of safety is applied to arrive at the design stress level. In LRFD, different load factors are applied

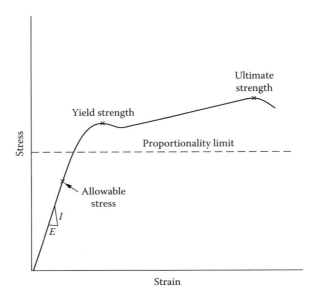

FIGURE 1.9   Stress–strain relation of a ductile material.

depending upon the reliability to which the different loads could be computed. Moreover, the resistance factors are applied to account for the uncertainties associated with the strength values.

The American Concrete Institute (ACI) was the first regulatory agency to adopt the (ultimate) strength design approach in early 1970 because concrete does not behave as an elastic material and it does not display the linear stress–strain relationship at any stage. The American Institute of Steel Construction (AISC) adopted the LRFD specifications in the beginning of 1990. On the other hand, the American Forest and Paper Association included the LRFD provisions only recently in the 2005 edition of the *National Design Specification for Wood Construction*.

The *AISC Manual 2005* has proposed a unified approach wherein they have combined the ASD and the LRFD methods together in a single documentation. The principle of unification is as follows.

The nominal strength of a material is a basic quantity that corresponds to the ultimate strength of the material. In terms of the force, the nominal (force) strength is equal to the yield or ultimate strength (stress) times the sectional area of the member. In terms of the moment, the nominal (moment) strength is equal to the ultimate strength times the section modulus of the member. Thus,

$$P_n = F_y A \tag{1.1}$$

$$M_n = F_y S \tag{1.2}$$

where
   A is the area of cross section
   S is the section modulus

In the ASD approach, the nominal strength of a material is divided by a factor of safety to convert it to the allowable strength. Thus,

$$\text{Allowable (force) strength} = \frac{P_n}{\Omega} \tag{1.3}$$

$$\text{Allowable (moment) strength} = \frac{M_n}{\Omega} \tag{1.4}$$

where Ω is the factor of safety.

For a safe design, the load or moment applied on the member should not exceed the allowable strength. Thus, the basis of the ASD design is as follows:

$$P_a \leq \frac{P_n}{\Omega} \tag{1.5}$$

and

$$M_a \leq \frac{M_n}{\Omega} \tag{1.6}$$

where

P_a is the service design load combination

M_a is the moment due to service design load application

Using Equations 1.5 or 1.6, the required cross-sectional area or the section modulus of the member can be determined.

The common ASD procedure works at the stress level. The service (applied) load, P_a, is divided by the sectional area, A, or the service moment, M_a, is divided by the section modulus, S, to obtain the applied or the created stress due to the loading, σ_a. Thus, the cross-sectional area and the section modulus are not used on the strength side but on the load side in the usual procedure. It is the ultimate or yield strength (stress) that is divided by the factor of safety to obtain the permissible stress, σ_p. To safeguard the design, it is ensured that the applied stress σ_a does not exceed the permissible stress σ_p.

For the purpose of unification of the ASD and LRFD approaches, the above procedure considers the strength in terms of the force or the moment. In the LRFD approach, the nominal strengths are the same as given by Equations 1.1 and 1.2. The design strength are given by

$$\text{Design (force) strength} = \phi P_n \tag{1.7}$$

$$\text{Design (moment) strength} = \phi M_n \tag{1.8}$$

where ϕ is the resistance factor.

The basis of design is

$$P_u \leq \phi P_n \tag{1.9}$$

$$M_u \leq \phi M_n \tag{1.10}$$

where

P_u is the factored design loads

M_u is the maximum moment due to factored design loads

From the above relations, the required area or the section modulus can be determined, which are the parts of P_n and M_n in Equations 1.1 and 1.2.

A link between the ASD and the LRFD approaches can be made as follows:
From the ASD Equation 1.5, at the upper limit

$$P_n = \Omega P_a \qquad (1.11)$$

Considering only the dead load and live load, $P_a = D + L$. Thus,

$$P_n = \Omega(D + L) \qquad (1.12)$$

From the LRFD Equation 1.9 at the upper limit

$$P_n = \frac{P_u}{\phi} \qquad (1.13)$$

Considering only the factored dead load and live load, $P_u = 1.2D + 1.6L$. Thus,

$$P_n = \frac{(1.2D + 1.6L)}{\phi} \qquad (1.14)$$

Equating Equations 1.12 and 1.14

$$\frac{(1.2D + 1.6L)}{\phi} = \Omega(D + L) \qquad (1.15)$$

or

$$\Omega = \frac{1(1.2D + 1.6L)}{\phi(D + L)} \qquad (1.16)$$

The factor of safety, Ω, has been computed as a function of the resistance factor, ϕ, for various selected live-to-dead load ratios in Table 1.2.

The 2005 AISC Specifications has used the relation $\Omega = 1.5/\phi$ throughout the manual to connect the ASD and the LRFD approaches together. Wood and concrete structures are relatively heavier, i.e., L/D ratio is less than 3 and the factor of safety Ω tends to be lower than $1.5/\phi$, but a value of 1.5 could be reasonably used for those structures as well, because the variation of the factor is not significant. This book uses the LRFD basis of design for all structures.

ELASTIC AND PLASTIC DESIGNS

The underlined concept in the preceding section was that a limiting state is reached when the stress level at any point in a member approaches the yield strength value of the material and the corresponding load is the design capacity of the member.

Let us revisit the stress–strain diagram for a ductile material like steel. The initial portion of the stress–strain curve of Figure 1.9 has been drawn again in Figure 1.10 to a greatly enlarged horizontal scale. The yield point F_y is a very important property of structural steel. After an initial yield, a steel element elongates in the plastic range without any appreciable change in stress level. This elongation is a measure of the ductility and serves a useful purpose in steel design. The strain and stress diagrams for a rectangular beam due to an increasing loading are shown in Figures 1.11 and 1.12.

TABLE 1.2
Ω as a Function of ϕ for Various L/D Ratios

L/D Ratio (Select)	Ω from Equation 1.16
1	$1.4/\phi$
2	$1.47/\phi$
3	$1.5/\phi$
4	$1.52/\phi$

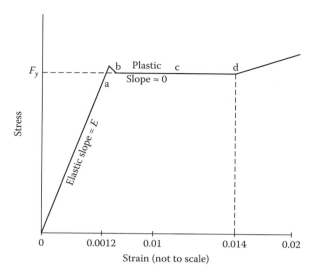

FIGURE 1.10 Initial portion of stress–strain relation for ductile material.

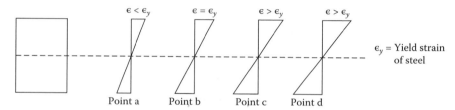

FIGURE 1.11 Strain variation in a rectangular section.

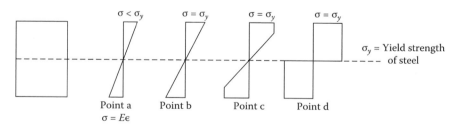

FIGURE 1.12 Stress variation in a rectangular section.

Beyond the yield strain at point b, as a load increases, the strain continues to rise in the plastic range and the stress at yield level extends from the outer fibers into the section. At point d, the entire section has achieved the yield stress level and no more stress capacity is available to develop. This is known as the *fully plastic state* and the moment capacity as the *full plastic moment*. The full moment is the ultimate capacity of a section. Beyond that a structure will collapse. When a full moment capacity is reached, we say that a *plastic hinge* has formed. In a statically determinate structure, the formation of one plastic hinge will lead to the collapse mechanism. Two or more plastic hinges are required in a statically indeterminate structure for a collapse mechanism. In general, for a complete collapse mechanism,

$$n = r + 1 \qquad (1.17)$$

where

n is the number of plastic hinges

r is the degree of indeterminacy

ELASTIC MOMENT CAPACITY

As stated earlier, commonly the structures are designed for elastic moment capacity, i.e., the failure load is based on the stress reaching a yield level at any point. Consider that on a rectangular beam of Figure 1.10, at position b when the strain has reached to the yield level, a full elastic moment M_E acts. This is shown in Figure 1.13.

Total compression force:

$$C = \frac{1}{2}\sigma_y A_c = \frac{1}{2}\sigma_y \frac{bd}{2} \tag{a}$$

Total tensile force:

$$T = \frac{1}{2}\sigma_y A_t = \frac{1}{2}\sigma_y \frac{bd}{2} \tag{b}$$

These act at the centroids of the stress diagram in Figure 1.13.

$$M_E = \text{force} \times \text{moment arm}$$

$$M_E = \left(\frac{1\sigma_y}{2}\frac{bd}{2}\right) \times \left(\frac{2d}{3}\right) \tag{c}$$

$$M_E = \sigma_y \frac{bd^2}{6} \tag{1.18}$$

It should be noted that $bd^2/6 = S$, the section modulus and the above relation is given by $M = \sigma_y S$. In terms of the moment of inertia, this relation is $M = \sigma_y I/C$. In the case of a nonsymmetrical section, the neutral axis is not in the center and there are two different values of c and, accordingly, two different section moduli. The smaller M_E is used for the moment capacity.

PLASTIC MOMENT CAPACITY

Consider a full plastic moment M_p acting on the rectangular beam section at the stress level d of Figure 1.10. This is shown in Figure 1.14.

Total compression force:

$$C = \sigma_y A_c = \sigma_y \frac{bd}{2} \tag{a}$$

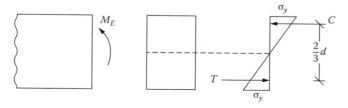

FIGURE 1.13 Full elastic moment acting on a rectangular section.

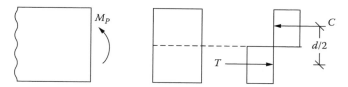

FIGURE 1.14   Full plastic moment acting on a rectangular section.

Total tensile force:

$$T = \sigma_y A_t = \sigma_y \frac{bd}{2} \tag{b}$$

$$M_p = \text{force} \times \text{moment arm}$$

$$= \sigma_y \frac{bd}{2} \times \frac{d}{2} \tag{c}$$

or

$$M_p = \sigma_y \frac{bd^2}{4} \tag{1.19}$$

This is given by

$$M_p = \sigma_y Z \tag{1.20}$$

where Z is called the *plastic section modulus*. For a rectangle, the plastic section modulus is 1.5 times of the (elastic) section modulus and the plastic moment capacity (M_p) is 1.5 times the elastic moment capacity (M_E). The ratio between the full plastic and full elastic moment of a section is called the *shape factor*. In other words, for the same design moment value, the section is smaller according to the plastic design.

The plastic analysis is based on the collapse load mechanism and requires knowledge of how a structure behaves when the stress exceeds the elastic limit. The plastic principles are used in the design of steel structures.

Example 1.2

For a steel beam section shown in Figure 1.15, determine the (a) elastic moment capacity, (b) plastic moment capacity, and (c) shape factor. The yield strength is 210 MPa.

Solution

    a. Elastic moment capacity
       1. Refer to Figure 1.15a
       2. $C = T = \dfrac{1}{2}(210 \times 10^6)(0.05 \times 0.075) = 393.75 \times 10^3 \, \text{N}$
       3. $M_E = (393.75 \times 10^3) \times 0.1 = 39.38 \times 10^3 \, \text{N-m}$
    b. Plastic moment capacity
       1. Refer to Figure 1.15b
       2. $C = T = (210 \times 10^6)(0.05 \times 0.075) = 787.5 \times 10^3 \, \text{N}$
       3. $M_p = (787.5 \times 10^3) \times 0.075 = 59.06 \times 10^3 \, \text{N-m}$

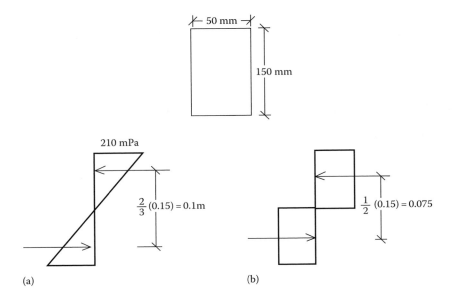

FIGURE 1.15   Beam section. (a) Elastic moment capacity. (b) Plastic moment capacity.

c. Shape factor

$$SF = \frac{M_P}{M_E} = \frac{59.06 \times 10^3}{39.38 \times 10^3} = 1.5$$

Example 1.3

The design moment for a rectangular beam is 40 kN-m. The yield strength of the material is 200 MPa. Design a section having the width-to-depth ratio of 0.5 according to the (a) elastic theory, (b) plastic theory.

Solution

a. Elastic theory
 1. $M_E = \sigma_y S$
    or

$$S = \frac{M_E}{\sigma_y} = \frac{40 \times 10^3}{200 \times 10^6} = 0.2 \times 10^{-3}\, m^3$$

 2. $\frac{1}{6}bd^2 = 0.2 \times 10^{-3}$

$$\frac{1}{6}(0.5d)(d^2) = 0.2 \times 10^{-3}$$

    or

    $d = 0.134$ m

    and

    $b = 0.076$ m

b. Plastic theory

   1. $M_p = \sigma_y Z$

     or

$$Z = \frac{M_p}{\sigma_y} = \frac{40 \times 10^3}{200 \times 10^6} = 0.2 \times 10^{-3}\, \text{m}^3$$

   2. $\frac{1}{4} bd^2 = 0.2 \times 10^{-3}\, \text{m}^3$

$$\frac{1}{4}(0.5d)(d^2) = 0.2 \times 10^{-3}\, \text{m}^3$$

   or

   $d = 0.117\, \text{m}$

   and

   $b = 0.058\, \text{m}$

THE COMBINATION OF LOADS

Various types of loads that act on a structure were described in the "Standard Unit Loads" section. For designing a structure, its elements or foundation, the loads are considered to act in the following combinations with the load factors as indicated in order to produce the most unfavorable effect on the structure or its element. The dead load, roof live load, floor live load, and snow load are gravity loads that act vertically downward. Wind load and seismic load have the vertical as well as the lateral components. The vertically acting roof live load, live load, wind load (simplified approach), and snow load are considered to be acting on the horizontal projection of any inclined surface. However, the dead load and the vertical component of the earthquake load act over the entire inclined length of the member.

For the LRFD, the ASCE 7-05 has recommended the following seven combinations with respect to common types of loads:

1. $1.4D$ (1.21)

2. $1.2D + 1.6L + 0.5(L_r \text{ or } S)$ (1.22)

3. $1.2D + 1.6(L_r \text{ or } S) + fL \text{ or } 0.8W$ (1.23)

4. $1.2D + 1.6W + fL + 0.5(L_r \text{ or } S)$ (1.24)

5. $1.2D + E_v + E_h + fL + 0.2S$ (1.25)

6. $0.9D + 1.6W$ (1.26)

7. $0.9D - E_v + E_h$ (1.27)

where

   D is the dead load

   L is the live load

L_r is the roof live load
S is the snow load
W is the wind load
E_h is the horizontal earthquake load
E_v is the vertical earthquake load
$f = 0.5$ for all occupancies when the unit live load does not exceed 100 psf except for garage and public assembly places and value of 1 is for 100 psf load and for any load on garage and public place

For other special loads like fluid load, flood load, rain load, and earth pressure, a reference is made to Chapter 2 of the ASCE 7-05.

Example 1.4

A simply supported roof beam receives loads from the following sources taking into account the respective tributary areas. Determine the loading diagram for the beam according to the ASCE 7-05 combinations.

1. Dead load (1.2 k/ft acting on roof slope 10°)
2. Roof live load (0.24 k/ft)
3. Snow load (1 k/ft)
4. Wind load at roof level (15 k)
5. Earthquake load at roof level (25 k)
6. Vertical earthquake load (0.2 k/ft)

Solution

1. The dead load and the vertical earthquake load since related with the dead load act on the entire member length. The other vertical forces act on the horizontal projection, according to the code.
2. Adjusted dead load on horizontal projection = 1.2/cos 10° = 1.22 k/ft
3. Adjusted vertical earthquake load on horizontal project = 0.2/cos 10° = 0.20 k/ft
4. Equation 1.21: $W_u = 1.4D = 1.4(1.22) = 1.71$ k/ft
5. Equation 1.22: $W_u = 1.2D + 1.6L + 0.5$ (L_r or S)
   This combination is shown in Table 1.3.
6. Equation 1.23: $W_u = 1.2D + 1.6(L_r$ or $S) + (0.5L$ or $0.8W)$
   This combination is shown in Table 1.4.

TABLE 1.3
Dead, Live, and Snow loads

Source	D, k/ft	L, k/ft	L_r or S, k/ft	Combined Value	Diagram
Load	1.22	—	1		1.964 k/ft
Load factor	1.2		0.5		
Factored vertical load	1.464		0.5	1.964 k/ft	
Factored horizontal load	—	—	—		

7. Equation 1.24: $W_u = 1.2D + 1.6W + 0.5L + 0.5(L_r \text{ or } S)$
   This combination is shown in Table 1.5.
8. Equation 1.25: $W_u = 1.2D + E_v + E_h + 0.5L + 0.2S$
   This combination is shown in Table 1.6.
9. Equation 1.26: $W_u = 0.9D + 1.6W$
   This combination is shown in Table 1.7.
10. Equation 1.27: $W_u = 0.9D + E_h - E_v$
    This combination is shown in Table 1.8.

Items 4, 5, and 9 can be eliminated as they are less than the other combinations. Items 6, 7, and 8 should be evaluated for the maximum effect and item 10 for the least effect.

TABLE 1.4
Dead, Live, Snow, and Wind loads

Source	D, k/ft	L, k/ft	S, k/ft	W, k	Combined Value	Diagram
Load	1.22	—	1	15		3.06 k/ft
Load factor	1.2		1.6	0.8		
Factored vertical load	1.464		1.6		3.06 k/ft	12 k
Factored horizontal load				12	12 k	L

TABLE 1.5
Dead, Live, Snow, and Wind loads

Source	D, k/ft	L, k/ft	S, k/ft	W, k	Combined Value	Diagram
Load	1.22	—	1	15		1.964 k/ft
Load factor	1.2		0.5	1.6		
Factored vertical load	1.464		0.5		1.964 k/ft	24 k
Factored horizontal load				24	24 k	L

TABLE 1.6
Dead, Live, Snow, and Earthquake Loads

Source	D, k/ft	L, k/ft	S, k/ft	E_v, k/ft	E_h, k	Combined Value	Diagram
Load	1.22	—	1	0.2	25		1.864 k/ft
Load factor	1.2		0.2	1	1		25 k
Factored vertical load	1.464		0.2	0.2		1.864	L
Factored horizontal load					25	25	

TABLE 1.7
Dead and Wind Loads

Source	D, k/ft	W, k	Combined Value	Diagram
Load	1.22	15		1.1 k/ft
Load factor	0.9	1.6		
Factored vertical load	1.1		1.1 k/ft	
Factored horizontal load		24	24 k	

TABLE 1.8
Dead and Earthquake Loads

Source	D, k/ft	E_v, k/ft	E_h, k	Combined Value	Diagram
Load	1.22	(−) 0.2	25		0.9 k/ft
Load factor	0.9	1	1		
Factored vertical load	1.1	(−) 0.2		0.9 k/ft	
Factored horizontal load			25	25 k	

PROBLEMS

Note: In Problems 1.1 through 1.6, the loads given are the factored loads.

1.1 A floor framing plan is shown in Figure P1.1. The standard unit load on the floor is 60 lb/ft². Determine the design uniform load/ft on the joists and the interior beam.

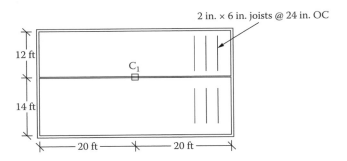

FIGURE P1.1 Floor framing plan.

1.2 In Figure 1.5, length, $L = 50$ ft and width, $B = 30$ ft. For a floor loading of 100 lb/ft², determine the design loads on beams GH, EF, and AB.

1.3 In Figure 1.6, length, $L = 50$ ft and width, $B = 30$ ft and the loading is 100 lb/ft², determine the design loads on beams GH, EF, and AB.

1.4 An open well is framed so that beams CE and DE sit on beam AB, as shown in Figure P1.4. Determine the design load for beam CE and girder AB. The combined unit of dead and live loads is 80 lb/ft².

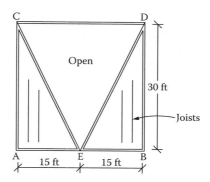

FIGURE P1.4 An open well frame.

1.5 A roof is framed as shown on Figure P1.5. The load on the roof is $3\,kN/m^2$. Determine the design load distribution on the ridge beam.

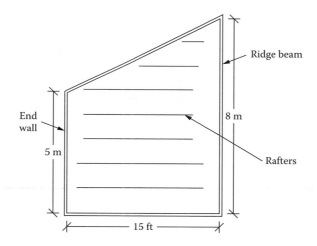

FIGURE P1.5 Roof frame.

1.6 Determine the size of a square wood column C_1 of Problem 1.1 shown in Figure P1.1. Use a resistance factor of 0.8 and assume no slenderness effect. The yield strength of wood in compression is 4000 psi.

1.7 The service dead and live loads acting on a round tensile member of steel are 10 and 20 k, respectively. The resistance factor is 0.9. Determine the diameter of the member. The yield strength of steel is 36 ksi.

1.8 A steel beam spanning 30 ft is subjected to a service dead load of 400 lb/ft and a service live load of 1000 lb/ft. What is the size of a rectangular beam if the depth is twice the width? The resistance factor is 0.9. The yield strength of steel is 50 ksi.

1.9 Design the interior beam of Problem 1.1 in Figure P1.1. The resistance factor is 0.9. The depth is three times of the width. The yield strength of wood is 4000 psi.

1.10 For a steel beam section shown in Figure P1.10, determine the (1) elastic moment capacity, (2) plastic moment capacity, and (3) shape factor. The yield strength is 50 ksi.

FIGURE P1.10 Rectangular beam section.

1.11 For a steel beam section shown in Figure P1.11, determine the (1) elastic moment capacity, (2) plastic moment capacity, and (3) shape factor. The yield strength is 210 MPa.

[*Hint*: For the elastic moment capacity, use the relation $M_E = \sigma_y I/C$. For the plastic capacity, find the compression (or tensile) forces separately for web and flange of the section and apply these at the centroid of the web and flange, respectively.]

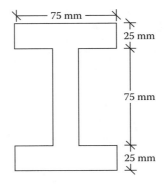

FIGURE P1.11   An *I*-beam section.

1.12 For a circular wood section as shown in Figure P1.12, determine the (1) elastic moment capacity, (2) plastic moment capacity, and (3) shape factor. The yield strength is 2000 psi.

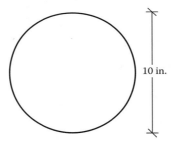

FIGURE P1.12   A circular wood section.

1.13 For the asymmetric section shown in Figure P1.13, determine the plastic moment capacity. The plastic neutral axis (where $C = T$) is at 20 mm above the base. The yield strength is 275 MPa.

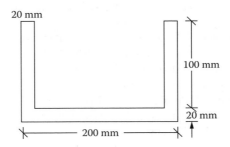

FIGURE P1.13   An asymmetric section.

1.14 The design moment capacity of a rectangular beam section is 2000 ft-lb. The material's strength is 10,000 psi. Design a section of width-to-depth ratio of 0.6 according to the (1) elastic theory, (2) plastic theory.

1.15 For Problem 1.14, design a circular section.

1.16 The following vertical loads are applied on a structural member. Determine the critical vertical load in psf for all the ASCE 7-05 combinations.

1. Dead load (on a 15° inclined member)	10 psf
2. Roof live load	20 psf
3. Wind load (vertical component)	15 psf
4. Snow load	30 psf
5. Earthquake load (vertical only)	2 psf

1.17 A floor beam supports the following loads. Determine the load diagrams for the various loads combinations.

1. Dead load	1.15 k/ft
2. Live load	1.85 k/ft
3. Wind load (horizontal)	15 k
4. Earthquake load (horizontal)	20 k
5. Earthquake load (vertical)	0.3 k

1.18 A simply supported floor beam is subject to the loads as shown in Figure P1.18. Determine the loading diagrams for load combinations according to Equations 1.22, 1.24, and 1.25.

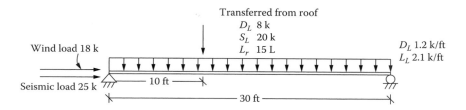

FIGURE P1.18 Loads on a beam.

2 Primary Loads: Dead Loads and Live Loads

DEAD LOADS

Dead loads are due to the weight of all materials that constitute a structural member. This also includes the weight of fixed equipment that are built into the structure such as piping, ducts, air conditioning, and heating equipment. The specific or unit weights of materials are available from different sources. The dead loads are, however, expressed in terms of uniform loads on a unit area (e.g., pounds per square ft). The weights are converted to dead loads taking into account the tributary area of a member. For example, a beam section weighting 4.5 lb/ft when spaced 16 in. (1.33 ft) on center will have a uniform dead load of 4.5/1.33 = 3.38 psf. If the same beam section is spaced 18 in. (1.5 ft) on center, the uniform dead load will be 4.5/1.5 = 3.5 psf. The spacing of beam section may not be known to begin with, as this might be an objective of the design.

Moreover, the estimation of dead load of a member requires knowledge as to what items and materials constitute that member. For example, a wood roof comprises of the roof covering, sheathing, framing, insulation, and ceiling.

It is expeditious to assume a reasonable dead load for the structural member, only to be revised when found grossly out of order.

The dead load of a building of light frame construction is about 10 lb/ft^2 for a flooring or roofing system without the plastered ceilings, and 20 lb/ft^2 with the plastered ceiling. For concrete flooring system, each 1 in. thick slab has a uniform load of about 12 psf; 36 psf for 3 in. slab. To this at least 10 psf should be added for the supporting system. Dead loads are gravity forces that act vertically downward. On a sloped roof the dead load acts over the entire inclined length of the member.

Example 2.1

The framing of a roof consists of the following:

Asphalt shingles (2 psf), 3/4 in. plywood (2.5 psf), 2 × 8 framing @ 12 in. on center (2.5 psf), fiberglass 0.5 in. insulation (1 psf), and plastered ceiling (10 psf). Determine the roof dead load. Make provisions for reroofing (3 psf).

Solution

	psf
Shingles	2
Plywood	2.5
Framing	2.5
Insulation	1
Ceiling	10
Reroofing	3
Roof dead load	21

LIVE LOADS

Live loads also act vertically down like dead loads but are distinct from the latter as they are not an integral part of the structural element. Roof live loads, L_r, are associated with maintenance of a roof by workers, equipment, and material. They are treated separately from the other types of live loads, L, that are imposed by the use and occupancy of the structure. The ASCE 7-05 specifies the minimum uniformly distributed load or the concentrated load that should be used as a live load for the intended purpose. Both, the roof live load and the floor live load are subjected to a reduction when they act on a large tributary area since it is less likely that the entire large area will be loaded to the same magnitude of high unit load. This reduction is not allowed when an added measure of safety is desired for important structures.

FLOOR LIVE LOADS

The floor live load is estimated by the equation

$$L = kL_o \tag{2.1}$$

where
    L_o is the basic design live load (see the "Basic Design Live Load, L_o" section below)
    k is the area reduction factor (see the "Effective Area Reduction Factor" section below)

BASIC DESIGN LIVE LOAD, L_o

The ASCE 7-05 provides a comprehensive table for basic design loads arranged by occupancy and use of a structure. This has been consolidated under important categories in Table 2.1.

To generalize, the basic design live loads are as follows.

Above the ceiling storage areas: 20 psf; one or two family sleeping area: 30 psf; normal use rooms: 40 psf; special use rooms (office, operating, reading, fixed sheet arena): 50–60 psf; public assembly places: 100 psf; lobbies, corridors, platforms, and stadium*: 100 psf for first floor, 80 psf for other floors; light industrial uses: 125 psf; and heavy industrial uses: 250 psf.

EFFECTIVE AREA REDUCTION FACTOR

The members that have more than 400 ft² of influential area are subject to a reduction of the standard live loads. The influence area is defined as the tributary area, A_T, multiplied by an element factor, K_{LL}, as listed in Table 2.2.

The following cases are excluded from the live load reduction:

1. Heavy live load that exceeds 100 psf
2. Passenger car garages
3. Public assembly areas

Except the above three items, for all other cases the reduction factor is given by

$$k = \left(0.25 + \frac{15}{\sqrt{K_{LL}A_T}} \right) \tag{2.2}$$

---

* In addition to vertical loads, horizontal swaying forces as follows are applied to each row of sheets: 24 lb per linear ft of seat in the direction to each row of sheets, and 10 lb per linear ft of sheet in the direction perpendicular to each row of sheets. Both horizontal forces need not be applied simultaneously.

TABLE 2.1
Summarized Basic Design Live Loads

Category	Uniform Load, psf
Residential	
Storage area	20
Sleeping area (dwelling)	30
Living area, stairs (dwelling)	40
Hotel room	40
Garage	40
Office	50
Computer room/facility	100
School classroom	40
Hospital	
Patient room	40
Operation room/lab	60
Library	
Reading room	60
Stacking room	150
Industrial manufacturing/warehouse	
Light	125
Heavy	250
Corridor/lobby	
First floor	100
Above first floor	80
Public places[a]	100

[a] Balcony, ballroom, fire escape, gymnasium, public stairs/exits, restaurant, stadium, store, terrace, theatre, yard, etc.

TABLE 2.2
Live Load Element Factor K_{LL}

Structure Element	K_{LL}
Interior columns	4
Exterior columns without cantilever slabs	4
Edge columns with cantilever slabs	3
Corner columns with cantilever slabs	2
Edge beams without cantilever slabs	2
Interior beams	2
All other members not identified including	1
Edge beams with cantilever slabs	
Cantilever beams	
One-way slabs	
Two-way slabs	
Members without provisions for continuous shear transfer normal to their span	

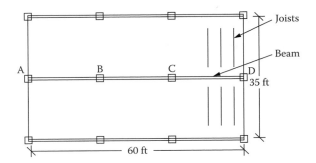

FIGURE 2.1   Floor framing plan.

As long as the following limits are observed, Equation 2.2 can be applied to any area. However with the limits imposed, the factor k becomes effective when $K_{LL}A_T > 400$ as stated earlier:

1. k factor should not be more than 1.
2. k factor should not be less than 0.5 for members supporting one floor and 0.4 for members supporting more than one floor.

Example 2.2

The first floor framing plan of a single family dwelling is shown in Figure 2.1. Determine the magnitude of the live load on the interior column C.

Solution

1. From Table 2.1, $L_o = 40\,\text{psf}$
2. Tributary area $A_T = 20 \times 17.5 = 350\,\text{ft}^2$
3. From Table 2.2, $K_{LL} = 4$
4. $K_{LL}A_T = 4 \times 350 = 1400$
5. From Equation 2.2,

$$k = \left(0.25 + \frac{15}{\sqrt{K_{LL}A_T}}\right)$$

$$= \left(0.25 + \frac{15}{\sqrt{1400}}\right) = 0.65$$

6. From Equation 2.1, $L = kL_o = 0.65\,(40) = 26\,\text{psf}$.

OTHER PROVISIONS FOR FLOOR LIVE LOADS

Besides the uniformly distributed live loads, the ASCE 7-05 also indicates the concentrated live loads in certain cases that are assumed to be distributed over an area of 2.5 ft × 2.5 ft. The maximum effect of either the uniformly distributed or concentrated load has to be considered. In most cases, the uniformly distributed load controls.

The buildings where partitions are likely to be erected, a uniform partition live load is provided in addition to the basic design loads. The minimum partition load is 15 psf. The partition live loads are not subjected to reduction for large effective area.

Live loads include an allowance for an ordinary impact. However, where unusual vibrations and impact forces are involved, the live loads should be increased. The moving loads shall be increased by an impact factor as follows: (1) elevator, 100%; (2) light shaft or motor-driven machine, 20%; (3) reciprocating machinery, 50%; and (4) hangers for floor or balcony, 33%. Including these effects

$$\text{Total LL/unit area} = \text{unit LL} \ (1 + \text{IF}) + \text{PL} \ \{\text{min 15 psf}\} \tag{2.3}$$

where
   LL is the live load
   IF is the impact factor
   PL is the partition load

ROOF LIVE LOADS, L_r

Roof live loads happen for a short time during the roofing or reroofing process. In the load combinations, either the roof live load L_r or the snow load S is included, since both of these are not likely to occur simultaneously.

The standard roof live load for ordinary flat, sloped, or curved roofs is 20 psf. This can be reduced to a minimum value of 12 psf based on the tributary area being larger than 200 ft² and/or the roof slope being more than 18.4°. When less than 20 psf of the roof live loads are applied to a continuous beam structure, the reduced roof live load is applied to adjacent spans or to alternate spans, whichever produces the greatest unfavorable effect.

The roof live load is estimated by

$$L_r = R_1 R_2 L_o \tag{2.4}$$

where
   L_r is the reduced roof live load on a horizontally projected surface
   L_o is the basic design load for ordinary roof = 20 psf
   R_1 is the tributary area reduction factor, see the "Tributary Area Reduction Factor, R_1" section below
   R_2 is the slope reduction factor, see the "Slope Reduction Factor" section below

Tributary Area Reduction Factor, R_1

This is given by

$$R_1 = 1.2 - 0.001 A_T \tag{2.5}$$

where A_T is the horizontal projection of roof tributary area in ft².
   This is subject to the following limitations:

   1. R_1 should not exceed 1
   2. R_1 should not be less than 0.6

Slope Reduction Factor

This is given by

$$R_2 = 1.2 - 0.6 \tan \theta \tag{2.6}$$

where θ is the roof slope angle.

1. R_2 should not exceed 1
2. R_2 should not be less than 0.6

Example 2.3

The horizontal projection of a roof framing plan of a building is similar to Figure 2.1. The roof pitch is 7 on 12. Determine the roof live load acting on column C.

Solution

1. $L_o = 20\,\text{psf}$
2. $A_T = 20 \times 17.5 = 350\,\text{ft}^2$
3. From Equation 2.5: $R_1 = 1.2 - 0.001\,(350) = 0.85$
4. Pitch of 7 on 12, $\tan\theta = 7/12$ or $\theta = 30.256°$
5. From Equation 2.6: $R_2 = 1.2 - 0.6\tan 30.256° = 0.85$
6. From Equation 2.4: $L_r = (0.85)\,(0.85)\,(20) = 14.45\,\text{psf} > 12\,\text{psf}$ OK

The above computations are for an ordinary roof. Special purpose roofs such as roof gardens have loads up to 100 psf. These are permitted to be reduced according to the floor live load reduction as discussed in the "Floor Live Loads" section.

PROBLEMS

2.1 A floor framing consists of the following:
Hardwood floor (4 psf), 1 in. plywood (3 psf), 2 in. × 12 in. framing @ 4 in. on center (2.6 psf), ceiling supports (0.5 psf), gypsum wallboard ceiling (5 psf).
Determine the floor dead load.

2.2 In Problem 2.1, the floor covering is replaced by a 1 in. concrete slab and the framing by 2 in. × 12 in. at 3 in. on center. Determine the floor dead load.
[*Hint*: Weight of concrete/unit area = 1 ft × 1 ft × 1/12 ft × 150.]

2.3 For the floor framing plan of Example 2.2, determine the design live load on the interior beam BC.

2.4 An interior steel column of an office building supports loads from two floors. The column-to-column distance among all columns in the floor plan is 40 ft. Determine the design live load on the column.

2.5 The framing plan of a light industrial building is shown in Figure P2.5. Determine the live load on column A.

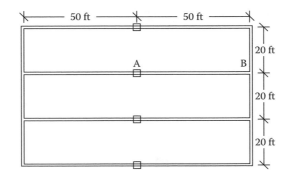

FIGURE P2.5 Framing plan of a building.

2.6 Determine the live load on the slab resting on column A of Problem 2.5.

2.7 The building in Problem 2.5 includes partitioning of the floor and it is equipped with a reciprocating machine that induces vibrations on the floor. Determine the design live load on beam AB.

2.8 Determine the roof live load acting on the end column D of a roofing plan shown in Figure P2.8.

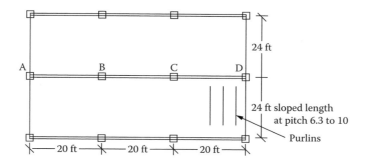

FIGURE P2.8   Roofing plan.

2.9 Determine the roof live load on purlins of Figure P2.8 if they are 4 ft apart.

2.10 A roof framing section is shown in Figure P2.10. The length of the building is 40 ft. The ridge beam has supports at two ends and at mid-length. Determine roof live load on the ridge beam.

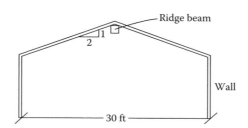

FIGURE P2.10   Side elevation of a building.

2.11 Determine the load on the walls due to roof live load of Problem 2.10.

2.12 An interior column supports loads from a roof garden. The tributary area to the column is 250 ft². Determine the roof live load. Assume a basic roof garden load of 100 psf.

3 Snow Loads

INTRODUCTION

Snow is a controlling roof load in about half of all the states in the United States. It is a cause of frequent and costly structural problems. Snow loads have the following components:

1. Balanced snow load
2. Extra load due to rain on snow
3. Partial loading of the balanced snow load
4. Unbalanced snow load due to a drift on one roof
5. Unbalanced load due to a drift from an upper to a lower roof
6. Sliding snow load

Snow loads are assumed to act on the horizontal projection of the roof surface.

BALANCED SNOW LOAD

This is the basic snow load to which a structure is subjected to. The procedure to determine the balanced snow load is as follows:

1. Determine the ground snow load, p_g, from the snow load map in the ASCE 7-05, reproduced in Figure 3.1.
2. Convert the ground snow load to flat roof snow load (roof slope ≤5°), p_f, with consideration given to the (1) roof exposure, (2) roof thermal condition, and (3) occupancy category of the structure.
3. Apply a roof slope factor to the flat roof snow load to determine the sloped (balanced) roof snow load.
4. Combining the above steps, the sloped roof snow load is calculated from

$$p_s = 0.7 C_s C_e C_t I p_g \tag{3.1}$$

where
p_g is the 50 year ground snow load from Figure 3.1
I is the importance factor (see the "Importance Factor" section)
C_t is the thermal factor (see the "Thermal Factor, C_t" section)
C_e is the exposure factor (see the "Exposure Factor, C_e" section)
C_s is the roof slope factor (see the "Roof Slope Factor, C_s" section)

The slope of a roof is defined as a *low slope* if (1) it is a mono-slope roof having a slope of less than 15°; (2) it is a hip and gable roof having a slope of less than (i) 2.4° or (ii) $(70/W) + 0.5$ whichever is higher, where W is the horizontal distance from eave to ridge of the roof in feet; and (3) it is a curved roof having the vertical angle from the eave to crown of less than 10°.

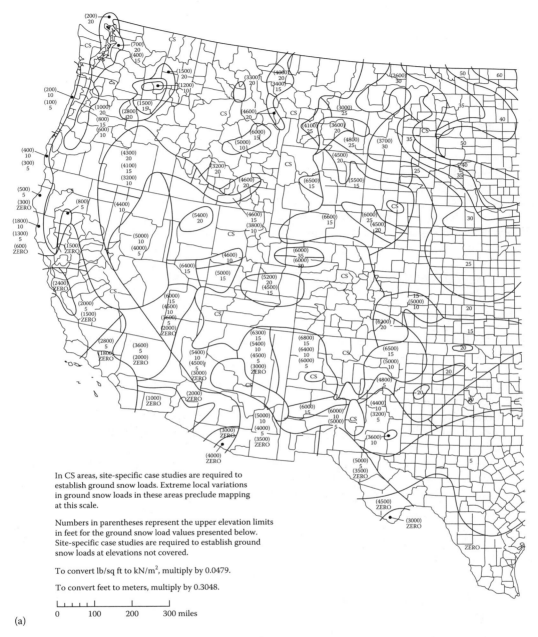

In CS areas, site-specific case studies are required to establish ground snow loads. Extreme local variations in ground snow loads in these areas preclude mapping at this scale.

Numbers in parentheses represent the upper elevation limits in feet for the ground snow load values presented below. Site-specific case studies are required to establish ground snow loads at elevations not covered.

To convert lb/sq ft to kN/m², multiply by 0.0479.

To convert feet to meters, multiply by 0.3048.

```
|  |  |  |  |  |  |  |  |  |  |  |
0        100      200     300 miles
```

(a)

FIGURE 3.1   Ground snow loads, p_g, for the United States (lb/ft²). (Courtesy of American Society of Civil Engineers, Reston, VA.)

For a low sloped roof, the magnitude of p_s from Equation 3.1 should not be less than the following values:

1. When the ground snow load, p_g, is 20 lb/ft² or less, the minimum roof snow load for the low-slope roof is given by

$$p_s = C_s I p_g \tag{3.2}$$

(b)

FIGURE 3.1 (continued)

2. When p_g exceeds $20\,\text{lb/ft}^2$, the minimum roof snow load for the low-slope roof is given by

$$p_s = 20C_s I$$

(3.3)

IMPORTANCE FACTOR

Depending upon the occupancy category identified in the "Classification of Buildings" section in Chapter 1, the importance factor is determined from Table 3.1.

TABLE 3.1
Importance Factor for Snow Load

Category of Occupancy	Importance Factor
I. Structures of low hazard to human life	0.8
II. Standard structures	1.0
III. High occupancy structures	1.1
IV. Essential structures	1.2

Source: Courtesy of American Society of Civil Engineers, Reston, VA.

THERMAL FACTOR, C_t

The factors are given in Table 3.2. The intent is to account for the heat loss through the roof and its effect on snow accumulation. For modern, well-insulated, energy-efficient construction with eave and ridge vents, the common C_t value is 1.1.

EXPOSURE FACTOR, C_e

The factors, as given in Table 3.3, are a function of the surface roughness (terrain type) and the location of the structure within the terrain (sheltered to fully exposed).

It should be noted that Exposure A representing centers of large cities where over half the buildings are in excess of 70 ft is not recognized separately in the ASCE 7-05. This type of terrain is included into Exposure B.

The sheltered areas correspond to the roofs that are surrounded on all sides by the obstructions that are within a distance of $10h_o$, where h_o is the height of the obstruction above the roof level. Fully exposed roofs have no obstruction within $10h_o$ on all sides including no large roof top equipment

TABLE 3.2
Thermal Factor for Snow Load

Thermal Condition	C_t
Green house with interior temperature of at least 50°	0.85
Well above freezing (warm or hot) roofs	1.0
Just above freezing or well-insulated, ventilated roofs	1.1
Below freezing structure	1.2

TABLE 3.3
Exposure Factor for Snow Load

Terrain	Fully Exposed	Partially Exposed	Sheltered
B. Urban, suburban wooded	0.9	1.0	1.2
C. Open, flat open grasslands, water surfaces in hurricane-prone regions	0.9	1.0	1.1
D. Open, smooth mud and salt flats, water surfaces in non-hurricane-prone regions	0.8	0.9	1.0
Above the tree line in mountainous region	0.7	0.8	—
Alaska: treeless	0.7	0.8	—

TABLE 3.4
Roof Slope Factor, C_s

Thermal Factor	Unobstructed Slippery Surface $R \geq 30\,\text{ft}^2$ F/Btu for unventilated and $R \geq 20\,\text{ft}^2$ F/Btu for ventilated	Other Surfaces
Warm roofs ($C_t \leq 1$)	$\theta = 0° - 5°\ C_s = 1$	$\theta = 0° - 30°\ C_s = 1$
	$\theta = 5° - 70°\ C_s = 1 - \dfrac{\theta - 5°}{65°}$	$\theta = 30° - 70°\ C_s = 1 - \dfrac{\theta - 30°}{40°}$
	$\theta > 70°\ C_s = 0$	$\theta > 70°\ C_s = 0$
Cold roofs ($C_t = 1.1$)	$\theta = 0° - 10°\ C_s = 1$	$\theta = 0° - 37.5°\ C_s = 1$
	$\theta = 10° - 70°\ C_s = 1 - \dfrac{\theta - 10°}{60°}$	$\theta = 37.5° - 70°\ C_s = 1 - \dfrac{\theta - 37.5°}{32.5°}$
	$\theta > 70°\ C_s = 0$	$\theta > 70°\ C_s = 0$
Cold roofs ($C_t = 1.2$)	$\theta = 0° - 15°\ C_s = 1$	$\theta = 0° - 45°\ C_s = 1$
	$\theta = 15° - 70°\ C_s = 1 - \dfrac{\theta - 15°}{55°}$	$\theta = 45° - 70°\ C_s = 1 - \dfrac{\theta - 45°}{25°}$
	$\theta > 70°\ C_s = 0$	$\theta > 70°\ C_s = 0$

Note: θ is the slope of roof.

or tall parapet walls. The partially exposed roofs represent structures that are not sheltered or fully exposed. The partial exposure is a most common exposure condition.

ROOF SLOPE FACTOR, C_s

This factor decreases as the roof slope increases. Also, the factor is smaller for the slippery roofs and for the warm roof surfaces.

The ASCE 7-05 provides the graphs of C_s versus roof slope for three separate thermal factor, C_t values, i.e., C_t of ≤ 1.0 (warm roofs), C_t of 1.1 (cold well-insulated and ventilated roofs), and C_t of 1.2 (cold roofs). On the graph for each value of the thermal factor, two curves are shown. The dashed line is for an unobstructed slippery surface and the solid line is for other surfaces. The dashed line of unobstructed slippery surfaces has smaller C_s values.

The unobstructed surface has been defined as a roof on which no object exists that will prevent snow from sliding and that a sufficient space is available below the eaves where the sliding snow can accumulate. The slippery surface includes metal, slate, glass, and membranes. For warm roof case ($C_t \leq 1$), to qualify as an unobstructed slippery surface there is a further requirement with respect to the R (the thermal resistance) value. The values of C_s can be expressed mathematically, as given in Table 3.4. It will be seen that for a non-slippery surface like asphalt shingles, which is a common case, the C_s factor is relevant only for roofs having a slope larger than 30°; for slopes larger than 70°, $C_s = 0$.

RAIN-ON-SNOW SURCHARGE

An extra load of $5\,\text{lb/ft}^2$ has to be added due to rain on snow for locations where the following two conditions apply: (1) the ground snow load, p_g, is $\leq 20\,\text{lb/ft}^2$ and (2) the roof slope is less than $W/50$; W being the horizontal eave-to-ridge roof distance. This extra load is only for the balanced snow load case and should not be used in the partial, drift, and sliding load cases.

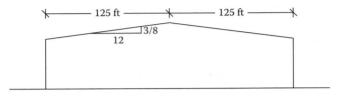

FIGURE 3.2   Low slope roof.

Example 3.1

Determine the balanced load for an unheated building of ordinary construction shown in Figure 3.2 in suburban area having trees obstruction less than $10h_o$. The ground snow load is 20 psf.

Solution

A. Parameters
   1. $p_g = 20$ psf
   2. Unheated roof $C_t = 1.20$
   3. Ordinary building $I = 1.0$
   4. Suburban area (terrain B), sheltered, Exposure factor, $C_e = 1.2$
   5. Roof angle, $\tan\theta = \dfrac{3/8}{12} = 0.0313; \theta = 1.8°$
   6. $\dfrac{70}{W} + 0.5 = \dfrac{70}{125} + 0.5 = 1.06$
   7. $\theta < 2.4°$, it is a low slope, the minimum load equation applies
   8. $\dfrac{W}{50} = \dfrac{125}{50} = 2.5$
   9. $\theta < 2.5°$, and $p_g = 20$ psf, rain-on-snow surcharge $= 5$ lb/ft^2
   10. From Table 3.4, $C_s = 1.0$
B. Snow loads
   1. From Equation 3.1

   $$p_s = 0.7C_sC_eC_tIp_g$$
   $$= 0.7(1)(1.2)(1.2)(1)(20) = 20.16 \, \text{lb/ft}^2$$

   2. Minimum snow load, from Equation 3.2

   $$p_s = (1)(1)(20) = 20 \, \text{lb/ft}^2$$

   3. Add rain-on-snow surcharge

   $$p_b = 20.16 + 5 = 25.16 \, \text{lb/ft}^2$$

Example 3.2

Determine the balanced snow load for an essential facility in Seattle, Washington, having a roof eave to ridge width of 100 ft and a height of 25 ft. It is a warm roof structure.

Solution

A. Parameters
   1. Seattle, Washington, $p_g = 20$ psf
   2. Warm roof $C_t = 1.00$
   3. Essential facility, $I = 1.2$
   4. Category B, urban area, partially exposed (default), Exposure factor, $C_e = 1.00$

5. Roof slope, $\tan\theta = \dfrac{25}{100} = 0.25; \theta = 14°$
6. It is not a low slope, the minimum snow equation is not applicable
7. θ is not less than $W/50$, there is no rain-on-snow surcharge
8. For a warm roof, other structures, from Table 3.4. $C_s = 1$

B. Snow loads

$$p_s = 0.7C_sC_eC_tIp_g$$

$$= 0.7(1)(1)(1)(1.2)(20) = 16.8\,\text{lb/ft}^2$$

PARTIAL LOADING OF THE BALANCED SNOW LOAD

The partial loads are different from the unbalanced loads. In unbalanced loads, snow is removed from one portion and is deposited on another portion. In the case of partial loading, snow is removed from one portion via scour or melting but it is not added to another portion. The intent is that in a continuous span structure, a reduction in snow loading on one span might induce the heavier stresses in some other portion than those occur with the entire structure being loaded. The provision requires that a selected span or spans should be loaded with one-half of the balanced snow load and the remaining spans with the full balanced snow load. This should be evaluated for various alternatives to assess the greatest effect on the structural member.

UNBALANCED SNOW LOAD DUE TO DRIFT

The unbalanced loading condition results from the drifting process when a blowing wind depletes snow from the upwind direction to pile it up in the downward direction. There are two drift cases:

1. Across the ridge drift when snow is removed from the windward panel of roof and is deposited on the leeward side of the same roof.
2. Drift from an upper roof on to a lower roof when two different levels are involved.

The unbalanced snow loading for hip and gable roofs is discussed below. For curved, saw tooth, and dome roofs, a reference is made to the ASCE 7-05.

ACROSS THE RIDGE SNOW DRIFT ON A ROOF

Across the ridge drift, the balanced load and unbalanced load should be analyzed separately. For the drift to occur on any roof, it should neither be a low-slope roof nor a steep roof. Thus, the following two conditions should be satisfied for across the ridge unbalanced snow loading:

1. The roof slope should be equal or larger than both (1) 2.4° and (2) $(70/W) + 0.5$, where W is the horizontal eave-to-ridge distance in feet.
2. The roof slope should be less than 70°.

When the above two conditions are satisfied, the unbalanced load distribution is expressed in two different ways:

1. For narrow roofs ($W \le 20\,\text{ft}$) of simple structural systems like the prismatic wood rafters or light gauge roof rafters spanning from eave to ridge, the windward side is taken as free of snow, and on the leeward side the total drift is represented by a uniform load from eave to ridge as follows. (Note this is the total load and is not an addition to the balanced load.)

$$p_u = Ip_g \tag{3.4}$$

2. For wide roofs ($W > 20$ ft) of any structures as well as the narrow roofs of other than the simple structures stated above, the drift is triangular in shape but is represented by a more user friendly rectangular surcharge over the balanced load.

On the windward side a uniform load of $0.3p_s$ is applied, where p_s is the balanced snow load from the "Balanced Snow Load" section. On the leeward side, a rectangular load is placed adjacent to the ridge, on top of the balanced load, p_s, as follows:

$$\text{Uniform load, } p_u = \frac{h_d \gamma}{\sqrt{s}} \tag{3.5}$$

$$\text{Horizontal extent from ridge, } L = \frac{8h_d\sqrt{s}}{3} \tag{3.6}$$

where
$\dfrac{1}{s}$ is the roof slope

γ is the unit weight of snow in lb/ft^3, given by

$$\gamma = 0.13p_g + 14 \le 30 \text{ lb/ft}^3 \tag{3.7}$$

h_d is the height of drift in feet on the leeward roof, given by

$$h_d = 0.43(W)^{1/3}(p_g + 10)^{1/4} - 1.5 \tag{3.8}$$

W is the horizontal distance from eave to ridge of roof in feet

If $W < 25$ ft, use $W = 25$ ft.

Example 3.3

Determine the unbalanced drift snow load for Example 3.1.

Solution

1. Roof slope, $\theta = 1.8°$
2. Since roof slope $<2.4°$, there is no unbalanced drift snow load

Example 3.4

Determine the unbalanced drift snow load for Example 3.2.

Solution

A. On leeward side
   1. Roof slope, $\theta = 14°$, it is not a low-slope roof
   2. $W > 20$ ft, it is a wide roof

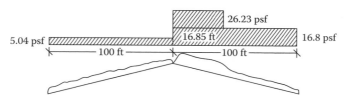

FIGURE 3.3 Drift snow load on a roof.

3. $p_g = 20\,\text{psf}$ and $p_s = 16.8\,\text{lb/ft}^2$ (from Example 3.2)

4. $\text{slope} = \dfrac{1}{s} = \dfrac{3}{12}$ or $s = 4$

5. $h_d = 0.43(W)^{1/3}(p_g + 10)^{1/4} - 1.5$

$\quad\quad = 0.43(100)^{1/3}(20 + 10)^{1/4} - 1.5 = 3.16\,\text{ft}$

6. Unit weight of snow

$\quad \gamma = 0.13p_g + 14 \le 30$

$\quad\quad = 0.13(20) + 14 = 16.6\,\text{lb/ft}^3$

7. $p_u = \dfrac{h_d \gamma}{\sqrt{s}}$

$\quad\quad = \dfrac{(3.16)(16.6)}{\sqrt{4}} = 26.23\,\text{lb/ft}^2$

8. Horizontal extent, $L = \dfrac{8h_d\sqrt{s}}{3} = \dfrac{8(3.16)\sqrt{4}}{3} = 16.85\,\text{ft}$

B. On windward side

9. $p_u = 0.3\ p_s = 0.3\ (16.8) = 5.04\,\text{psf}$

10. This is sketched in Figure 3.3

SNOW DRIFT FROM A HIGHER TO A LOWER ROOF

In a higher–lower level roofs combination, the drift from the higher roof accumulates on the lower roof. This drift is a surcharge that is superimposed on the balanced snow roof load of the lower roof. The drift accumulation, when the higher roof is on the windward side, is shown in Figure 3.4. This is known as the *leeward snow drift.*

When the higher roof is on the leeward side, the drift accumulation, known as the *windward snow drift*, is more complex. It starts as a quadrilateral shape because of the wind vortex and ends up in a triangular shape, as shown in Figure 3.5.

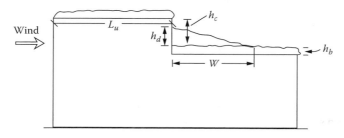

FIGURE 3.4 Leeward snow drift.

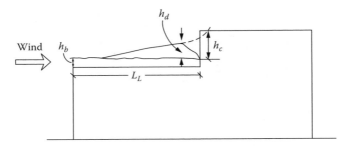

FIGURE 3.5   Windward snow drift.

LEEWARD SNOW DRIFT

In Figure 3.4, if h_c/h_b is less than 0.2, the drift load is not applied, where h_b is the balanced snow depth determined by dividing the balanced snow load, p_s, by unit load of snow, γ, computed by Equation 3.7. The term h_c represents the difference of elevation between high and low roofs subtracted by h_b, as shown in Figure 3.4.

The drift is represented by a triangle, as shown in Figure 3.6:

$$h_d = 0.43(L_u)^{1/3}(p_g + 10)^{1/4} - 1.5 \tag{3.9}$$

where L_u is the horizontal length of the roof upwind of the drift, as shown in Figure 3.4.

The corresponding maximum snow load is

$$p_d = \gamma h_d \tag{3.10}$$

The width of the snow load (base of the triangle) has the following value for two different cases:

1. For $h_d \leq h_c$

$$w = 4 h_d \tag{3.11}$$

2. For $h_d > h_c$

$$w = \frac{4 h_d^2}{h_c} \tag{3.12}$$

but w should not be greater than $8 h_c$.

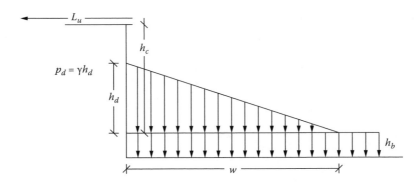

FIGURE 3.6   Configuration of snow drift.

Although in Equation 3.12 w is computed by the value of h_d from Equation 3.9 which is higher than h_c, and since the drift height cannot exceed the upper roof level, the height of the drift itself is subsequently changed as follows:

$$h_d = h_c \tag{3.13}$$

If width, w, is more than the lower roof length, L_L, then the drift shall be truncated at the end of the roof and not reduced to zero there.

WINDWARD SNOW DRIFT

In Figure 3.5, if h_c/h_b is less than 0.2, the drift load is not applied. The drift is given by a triangle similar to the one shown in Figure 3.6. However, the value of h_d is replaced by the following:

$$h_d = 0.75[0.43(L_L)^{1/3}(p_g + 10)^{1/4} - 1.5] \tag{3.14}$$

where L_L is the lower roof length as shown in Figure 3.5.

Equations 3.11 and 3.12 apply to windward width also.

The larger of the leeward and windward heights from the "Leeward Snow Drift" and the "Windward Snow Drift" sections is used in the design.

Example 3.5

A two-story residential building has an attached garage, as shown in Figure 3.7. The residential part is heated and has well-insulated, ventilated roof, whereas the garage is unheated. Both roofs of 4 on 12 slope have metal surfaces consisting of the purlins spanning eave to ridge.

The site is a forested area in a small clearing among huge trees. The ground snow load is 40 psf. Determine the snow load on the lower roof.

Solution

1. The upper roof is subjected to the balanced snow load and the unbalanced across the ridge drift load due to wind in transverse direction.

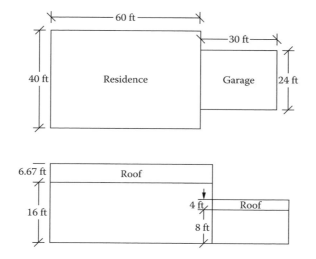

FIGURE 3.7   Higher–lower roof drift.

2. The lower roof is subjected to the balanced snow load, the unbalanced across the ridge load due to transverse directional wind, and the drift load from upper to lower roof due to longitudinal direction wind. Only the lower roof is analyzed here.
3. For the lower roof, the balanced load
   a. Unheated roof, $C_t = 1.2$
   b. Residential facility, $I = 1.0$
   c. Terrain B, sheltered, $C_e = 1.2$
   d. 4 on 12 slope, $\theta = 18.43°$
   e. For slippery unobstructed surface at $C_t = 1.2$, from Table 3.4

   $$C_s = 1 - \frac{(\theta - 15)}{55} = 1 - \frac{(18.43 - 15)}{55} = 0.94$$

   f. $p_s = 0.7 C_s C_e C_t I p_g$
   $$= 0.7(0.94)(1.2)(1.2)(1)(40) = 37.90 \text{ lb/ft}^2$$

4. For the lower roof, across the ridge unbalanced load
   a. $W = 12 < 20$ ft, roof rafter system, the simple case applies
   b. Windward side no snow load
   c. Leeward side

   $$p_u = I p_g = 1(40) = 40 \text{ psf}$$

5. For lower roof, upper–lower roof unbalanced load
   a. From Equation 3.7

   $$\gamma = 0.13 p_g + 14 = 0.13(40) + 14 = 19.2 \text{ lb/ft}^3$$

   b. $h_b = \dfrac{p_s}{\gamma} = \dfrac{37.9}{19.2} = 1.97 \text{ ft}$
   c. $h_c = (22.67 - 12) - 1.97 = 8.7 \text{ ft}$

   $$\frac{h_c}{h_b} = \frac{8.7}{1.97} = 4.4 > 0.2 \text{ drift load to be considered}$$

   d. Leeward drift
      From Equation 3.9

   $$h_d = 0.43(L_u)^{1/3}(p_g + 10)^{1/4} - 1.5$$
   $$= 0.43(60)^{1/3}(40 + 10)^{1/4} - 1.5 = 2.97 \text{ ft}$$

   Since $h_c > h_d$, $h_d = 2.97$ ft
   e. $p_d = \gamma h_d = (19.2)(2.97) = 57.03 \text{ lb/ft}^2$
   f. From Equation 3.11

   $$w = 4 h_d = 4(2.97) = 11.88 \text{ ft}$$

   g. Windward drift

   $$h_d = 0.75 \left[0.43(L_t)^{1/3}(p_g + 10)^{1/4} - 1.5 \right]$$
   $$= 0.75 \left[0.43(30)^{1/3}(40 + 10)^{1/4} - 1/5 \right]$$
   $$= 1.54 \text{ ft} < 2.97 \text{ ft, Leeward controls}$$

6. Figure 3.8 presents the three loading cases for the lower roof.

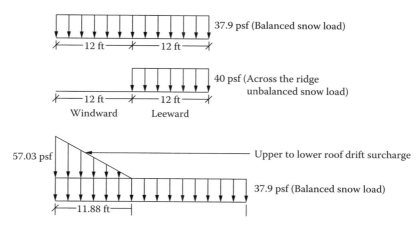

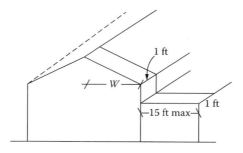

FIGURE 3.8 Loading on lower roof.

SLIDING SNOW LOAD ON LOWER ROOF

A sliding snow load from an upper to a lower roof is superimposed on the balanced snow load. The sliding load (plus the balanced load) and the lower roof drift load (plus the balanced load) are considered as two separate cases and the higher one is used. One basic difference between a slide and a drift is that in the former case, snow slides off the upper roof along the slope by the action of gravity and the lower roof should be in front of the sloping surface to capture this load. In the latter case, wind carries the snow downstream and thus the drift can take place lengthwise perpendicular to the roof slope, as in Example 3.5.

The sliding snow load is applied to the lower roof when the upper slippery roof has a slope of more than $\theta = 2.4°$ (1/4 on 12) or when the non-slippery upper roof has the slope greater than $9.5°$ (2 on 12).

With reference to Figure 3.9, the total sliding load per unit distance (length) of eave is taken as $0.4 p_f W$, which is uniformly distributed over a maximum lower roof width of 15 ft. If the width of the lower roof is less than 15 ft, the sliding load is reduced proportionately. The effect is that it is equivalent to distribution over a 15 ft width.

Thus,

$$S_L = \frac{0.4 p_f W}{15} \ \text{lb/ft}^2 \tag{3.15}$$

where
  p_f is the flat upper roof snow load (psf)
  W is the horizontal distance from ridge to eave of the upper roof

FIGURE 3.9 Sliding snow load.

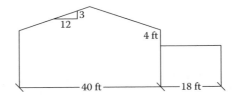

FIGURE 3.10   Sliding snow load on a flat roof.

Example 3.6

Determine the sliding snow load on an unheated flat roof garage attached to a residence, as shown in Figure 3.10, It is in an suburban area with scattered trees. The $p_g = 20$ psf. Assume that the upper roof flat snow load is 18 psf.

Solution

A. Balanced load on garage
   1. Unheated roof, $C_t = 1.2$
   2. Normal usage, $I = 1$
   3. Terrain B, partial exposure, $C_e = 1$
   4. Flat roof, $C_s = 1$
   5. Since $p_g = 20$ and $\theta = 0$, the minimum load applies
   6. Since $p_g = 20$ and $\theta < W/50$, rain-on-snow surcharge applies, but not included in the sliding load case
   7. $p_s = 0.7 C_s C_e C_t I p_g$
      $= 0.7(1)(1)(1.2)(1)(20) = 16.8 \ \text{lb/ft}^2$
   8. Minimum snow load

   $$p_{min} = C_s I p_g$$
   $$= (1)(1)20 = 20 \text{psf} \leftarrow \text{Controls}$$

   9. Rain-on-snow surcharge = 5 psf
B. Sliding snow load
   1. Upper roof slope $\theta = 14° > 9.5°$, sliding applies
   2. $p_f = 18 \ \text{lb/ft}^2$ (given)
   3. $S_L = \dfrac{0.4 p_f W}{15} = \dfrac{(0.4)(18)(20)}{15} = 9.6$ psf distributed over 15 ft length

Figure 3.11 presents the loading cases for garage.

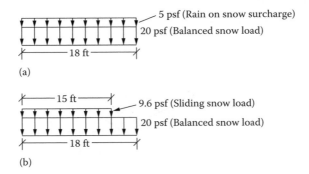

FIGURE 3.11   Loading on lower roof. (a) Balanced load and (b) balanced plus sliding snow load.

PROBLEMS

3.1 Determine the balanced snow load on a residential structure shown in Figure P3.1 in a suburban area. The roof is well insulated and ventilated. There are few trees behind the building to create obstruction. The ground snow load is 20 lb/ft².

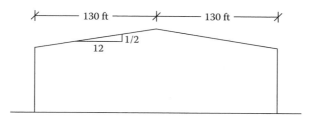

FIGURE P3.1 Suburban residence.

3.2 Solve Problem 3.1 except that the eave-to-ridge distance is 30 ft.

3.3 It is a heated warm roof structure in an urban area surrounded by obstructions from all sides. The eve to ridge distance is 25 ft and height is 7 ft. The ground snow load is 30 psf. Determine the balanced snow load.

3.4 The roof of a high occupancy structure is insulated and well ventilated in a fully open countryside. The eve to ridge distance is 20 ft and height is 4 ft. The ground snow load is 25 psf. Determine the balanced snow load.

3.5 Determine the unbalanced load for Problem 3.1.

3.6 Determine the unbalanced load for Problem 3.2.

3.7 Determine the unbalanced snow load for Problem 3.3.

3.8 Determine the unbalanced snow load for Problem 3.4.

3.9 Determine snow load on the lower roof of a building where ground snow load is 30 lb/ft². The elevation difference between the roofs is 5 ft. The higher roof is 70 ft wide and 100 ft long. It is a heated and unventilated office building. The lower roof is 60 ft wide and 100 ft long. It is an unheated storage area. Both roofs have 5 on 12 slope of metallic surfaces without any obstructions. The building is located in an open country with no obstructions. The building is laid out lengthwise, as shown in Figure P3.9.

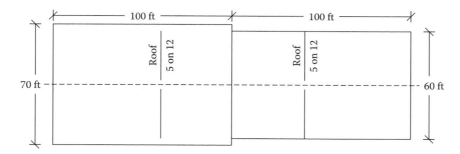

FIGURE P3.9 Different level roofs lengthwise.

3.10 Solve Problem 3.9 except that the roofs elevation difference is 3 ft.

3.11 Solve Problem 3.9 when the building is laid out side by side, as shown in Figure P3.11. The lowest roof is flat.

3.12 Solve Problem 3.11 for the sliding snow load.

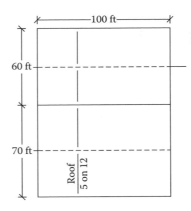

FIGURE P3.11 Different level roofs side by side.

3.13 Determine the snow load due to sliding effect for a heated storage area attached to an office building with a well-ventilated/insulated roof in an urban area in Rhode Island having scattered obstructions, as shown in Figure P3.13.

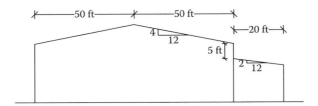

FIGURE P3.13 Urban building sliding snow.

3.14 Determine the sliding load for an unheated garage attached to a cooled roof of a residence shown in Figure P3.14 in a partially exposed suburban area. The ground snow load is 15 lb/ft².

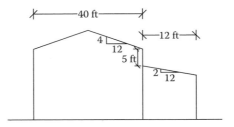

FIGURE P3.14 Suburban residence sliding snow.

4 Wind Loads

INTRODUCTION

The ASCE 7-05 prescribes three procedures to determine design wind loads for buildings and other structures:

Method 1: Simplified procedure
Method 2: Analytical procedure
Method 3: Wind tunnel procedure

In these procedures, two separate categories have been identified for wind load provisions:

1. Main wind force–resisting system (MWFRS): MWFRS represents the entire structure comprising of an assemblage of the structural elements constituting the structure that can sustain wind from more than one surface.
2. Components and cladding (C and C): These are the individual elements that face wind directly and pass on the loads to the MWFRS.

The broad distinction is apparent. The entire lateral force–resisting system (LFRS) as a unit that transfers loads to the foundation belongs to the first category. In the second category, the cladding comprises of wall and roof coverings like sheathing and finish material, curtain walls, exterior windows, and doors. The components include fasteners, purlins, girts, and roof decking.

However, there are certain elements like trusses and studs that are part of the MWFRS but could also be treated as individual components.

The C and C loads are higher than MWFRS loads since they are developed to represent the peak gusts over small areas that result from localized funneling and turbulence of wind.

An interpretation has been made that while using MWFRS, the combined interactions of the axial and bending stresses due to the vertical loading together with the lateral loading should be used. But in the application of C and C, either the axial or the bending stress should be considered individually. They are not combined together since the interaction of loads from multiple surfaces is not intended to be used in C and C.

THE SIMPLIFIED PROCEDURE FOR MWFRS

The simplified procedure is used for the regular-shaped diaphragm buildings with flat, gabled, or hipped roofs. The building height should not exceed 60 ft and also the height should not exceed the least horizontal dimension. Any special wind effects or torsion loading also prevents the use of this method. However, the simplified procedure can be applied to one- and two-story buildings in most locations.

The combined windward and leeward net wind pressure, p_s, is determined by the following simplified equation:

$$p_s = \lambda K_{zt} I p_{s30} \tag{4.1}$$

where
I is the importance factor, Table 4.1
λ is the adjustment factor for structure height and exposure, Tables 4.2 and 4.3
K_{zt} is the topographic factor
p_{s30} is the simplified standard design wind pressure, Table 4.4

TABLE 4.1
Importance Factor for Wind Load

Occupancy Category	Non-Hurricane-Prone Regions or Hurricane-Prone Regions of Velocity ≤100 mph and Alaska	Hurricane-Prone Regions of Velocity >100 mph
I. Low hazard to human life structures	0.87	0.77
II. Normal structures	1.00	1.00
III. High occupancy structures	1.15	1.15
IV. Essential structures	1.15	1.15

TABLE 4.2
Exposure Category for Wind Load

Surface Roughness	Exposure Category
Urban and suburban areas, wooded areas, closely spaced dwellings	B
Open terrain-flat open country, grasslands, water surfaces in hurricane-prone regions	C
Smooth mud and salt flats, water surfaces in non-hurricane-prone regions, ice	D

TABLE 4.3
Adjustment Factor for Height and Exposure

Mean Roof Height (ft)	Exposure		
	B	C	D
15	1.00	1.21	1.47
20	1.00	1.29	1.55
25	1.00	1.35	1.61
30	1.00	1.40	1.66
35	1.05	1.45	1.70
40	1.09	1.49	1.74
45	1.12	1.53	1.78
50	1.16	1.56	1.81
55	1.19	1.59	1.84
60	1.22	1.62	1.87

The pressure p_s is the pressure that acts horizontally on the vertical and vertically on the horizontal projection of the structure surface. It represents the net pressure that algebraically sums up the external and internal pressures acting on a building surface. Furthermore in the case of MWFRS, for the horizontal pressures that act on the building envelope, the p_s combines the windward and leeward pressures.

The plus and minus signs signify the pressures acting toward and away, respectively, from the projected surfaces.

TABLE 4.4
Simplified Wind Pressures, p_{s30} (psf)

Basic Wind Speed (mph)	Roof Angle (°)	Load Case	Horizontal Pressures				Vertical Pressures				Overhangs	
			A	B	C	D	E	F	G	H	E_{OH}	G_{OH}
85	0–5	1	11.5	−5.9	7.6	−3.5	−13.8	−7.8	−9.6	−6.1	−19.3	−15.1
	10	1	12.9	−5.4	8.6	−3.1	−13.8	−8.4	−9.6	−6.5	−19.3	−15.1
	15	1	14.4	−4.8	9.6	−2.7	−13.8	−9.0	−9.6	−6.9	−19.3	−15.1
	20	1	15.9	−4.2	10.6	−2.3	−13.8	−9.6	−9.6	−7.3	−19.3	−15.1
	25	1	14.4	2.3	10.4	2.4	−6.4	−8.7	−4.6	−7.0	−11.9	−10.1
		2	—	—	—	—	−2.4	−4.7	−0.7	−3.0	—	—
	30–45	1	12.9	8.8	10.2	7.0	1.0	−7.8	0.3	−6.7	−4.5	−5.2
		2	12.9	8.8	10.2	7.0	5.0	−3.9	4.3	−2.8	−4.5	−5.2
90	0–5	1	12.8	−6.7	8.5	−4.0	−15.4	−8.8	−10.7	−6.8	−21.6	−16.9
	10	1	14.5	−6.0	9.6	−3.5	−15.4	−9.4	−10.7	−7.2	−21.6	−16.9
	15	1	16.1	−5.4	10.7	−3.0	−15.4	−10.1	−10.7	−7.7	−21.6	−16.9
	20	1	17.8	−4.7	11.9	−2.6	−15.4	−10.7	−10.7	−8.1	−21.6	−16.9
	25	1	16.1	2.6	11.7	2.7	−7.2	−9.8	−5.2	−7.8	−13.3	−11.4
		2	—	—	—	—	−2.7	−5.3	−0.7	−3.4	—	—
	30–45	1	14.4	9.9	11.5	7.9	1.1	−8.8	0.4	−7.5	−5.1	−5.8
		2	14.4	9.9	11.5	7.9	5.6	−4.3	4.8	−3.1	−5.1	−5.8
100	0–5	1	15.9	−8.2	10.5	−4.9	−19.1	−10.8	−13.3	−8.4	−26.7	−20.9
	10	1	17.9	−7.4	11.9	−4.3	−19.1	−11.6	−13.3	−8.9	−26.7	−20.9
	15	1	19.9	−6.6	13.3	−3.8	−19.1	−12.4	−13.3	−9.5	−26.7	−20.9
	20	1	22.0	−5.8	14.6	−3.2	−19.1	−13.3	−13.3	−10.1	−26.7	−20.9
	25	1	19.9	3.2	14.4	3.3	−8.8	−12.0	−6.4	−9.7	−16.5	−14.0
		2	—	—	—	—	−3.4	−6.6	−0.9	−4.2	—	—
	30–45	1	17.8	12.2	14.2	9.8	1.4	−10.8	0.5	−9.3	−6.3	−7.2
		2	17.8	12.2	14.2	9.8	6.9	−5.3	5.9	−3.8	−6.3	−7.2
105	0–5	1	17.5	−9.0	11.6	−5.4	−21.1	−11.9	−14.7	−9.3	−29.4	−23.0
	10	1	19.7	−8.2	13.1	−4.7	−21.1	−12.8	−14.7	−9.8	−29.4	−23.0

(continued)

TABLE 4.4 (continued)
Simplified Wind Pressures, p_{s30} (psf)

Basic Wind Speed (mph)	Roof Angle (°)	Load Case	Zones									
			Horizontal Pressures				Vertical Pressures				Overhangs	
			A	B	C	D	E	F	G	H	E_{OH}	G_{OH}
	15	1	21.9	−7.3	14.7	−4.2	−21.1	−13.7	−14.7	−10.5	−29.4	−23.0
	20	1	24.3	−8.4	16.1	−3.5	−21.1	−14.7	−14.7	−11.1	−29.4	−23.0
	25	1	21.9	3.5	15.9	3.5	−9.7	−13.2	−7.1	−10.7	−18.2	−15.4
		2	—	—	—	—	−3.7	−7.3	−1.0	−4.6	—	—
	30–45	1	19.6	13.5	15.7	10.8	1.5	−11.9	0.6	−10.3	−6.9	−7.9
		2	19.6	13.5	15.7	10.8	7.6	−5.8	6.5	−4.2	−6.9	−7.9
110	0–5	1	19.2	−10.0	12.7	−5.9	−23.1	−13.1	−16.0	−10.1	−32.3	−25.3
	10	1	21.6	−9.0	14.4	−5.2	−23.1	−14.1	−16.0	−10.8	−32.3	−25.3
	15	1	24.1	−8.0	16.0	−4.6	−23.1	−15.1	−16.0	−11.5	−32.3	−25.3
	20	1	26.6	−7.0	17.7	−3.9	−23.1	−16.0	−16.0	−12.2	−32.3	−25.3
	25	1	24.1	3.9	17.4	4.0	−10.7	−14.6	−7.7	−11.7	−19.9	−17.0
		2	—	—	—	—	−4.1	−7.9	−1.1	−5.1	—	—
	30–45	1	21.6	14.8	17.2	11.8	1.7	−13.1	0.6	−11.3	−7.6	−8.7
		2	21.6	14.8	17.2	11.8	8.3	−6.5	7.2	−4.6	−7.6	−8.7
120	0–5	1	22.8	−11.9	15.1	−7.0	−27.4	−15.6	−19.1	−12.1	−38.4	−30.1
	10	1	25.8	−10.7	17.1	−6.2	−27.4	−16.8	−19.1	−12.9	−38.4	−30.1
	15	1	28.7	−9.5	19.1	−5.4	−27.4	−17.9	−19.1	−13.7	−38.4	−30.1
	20	1	31.6	−8.3	21.1	−4.6	−27.4	−19.1	−19.1	−14.5	−38.4	−30.1
	25	1	28.6	4.6	20.7	4.7	−12.7	−17.3	−9.2	−13.9	−23.7	−20.2
		2	—	—	—	—	−4.8	−9.4	−1.3	−6.0	—	—
	30–45	1	25.7	17.6	20.4	14.0	2.0	−15.6	0.7	−13.4	−9.0	−10.3
		2	25.7	17.6	20.4	14.0	9.9	−7.7	8.6	−5.5	−9.0	−10.3
125	0–5	1	24.7	−12.9	16.4	−7.6	−29.7	−16.9	−20.7	−13.1	−41.7	−32.7
	10	1	28.0	−11.6	18.6	−6.7	−29.7	−18.2	−20.7	−14.0	−41.7	−32.7
	15	1	31.1	−10.3	20.7	−5.9	−29.7	−19.4	−20.7	−14.9	−41.7	−32.7
	20	1	34.3	−9.0	22.9	−5.0	−29.7	−20.7	−20.7	−15.7	−41.7	−32.7
	25	1	31.0	5.0	22.5	5.1	−13.8	−18.8	−10.0	−15.1	−25.7	−21.9

Wind	Angle	Case										
130	30–45	2	—	—	—	—	-5.2	-10.2	-1.4	-6.5	—	-11.2
	30–45	1	27.9	19.1	22.1	15.2	2.2	-16.9	0.8	-14.5	-9.8	-11.2
	0–5	2	27.9	19.1	22.1	15.2	10.7	-8.4	9.3	-6.0	-9.8	-11.2
	0–5	1	26.8	-13.9	17.8	-8.2	-32.2	-18.3	-22.4	-14.2	-45.1	-35.3
	10	1	30.2	-12.5	20.1	-7.3	-32.2	-19.7	-22.4	-15.1	-45.1	-35.3
	15	1	33.7	-11.2	22.4	-6.4	-32.2	-21.0	-22.4	-16.1	-45.1	-35.3
	20	1	37.1	-9.8	24.7	-5.4	-32.2	-22.4	-22.4	-17.0	-45.1	-35.3
	25	1	33.6	5.4	24.3	5.5	-14.9	-20.4	-10.8	-16.4	-27.8	-23.7
140	30–45	2	—	—	—	—	-5.7	-11.1	-1.5	-7.1	—	-12.1
	30–45	1	30.1	20.6	24.0	16.5	2.3	-18.3	0.8	-15.7	-10.6	-12.1
	0–5	2	30.1	20.6	24.0	16.5	11.6	-9.0	10.0	-6.4	-10.6	-12.1
	0–5	1	31.1	-16.1	20.6	-9.6	-37.3	-21.2	-26.0	-16.4	-52.3	-40.9
	10	1	35.1	-14.5	23.3	-8.5	-37.3	-22.8	-26.0	-17.5	-52.3	-40.9
	15	1	39.0	-12.9	26.0	-7.4	-37.3	-24.4	-26.0	-18.6	-52.3	-40.9
	20	1	43.0	-11.4	28.7	-6.3	-37.3	-26.0	-26.0	-19.7	-52.3	-40.9
	25	1	39.0	6.3	28.2	6.4	-17.3	-23.6	-12.5	-19.0	-32.3	-27.5
145	30–45	2	—	—	—	—	-6.6	-12.8	-1.8	-8.2	—	-14.0
	30–45	1	35.0	23.9	27.8	19.1	2.7	-21.2	0.9	-18.2	-12.3	-14.0
	0–5	2	35.0	23.9	27.8	19.1	13.4	-10.5	11.7	-7.5	-12.3	-14.0
	0–5	1	33.4	-17.3	22.1	-10.3	-40.0	-22.7	-27.9	-17.6	-56.1	-43.9
	10	1	37.7	-15.6	25.0	-9.1	-40.0	-24.5	-27.9	-18.8	-56.1	-43.9
	15	1	41.8	-13.8	27.9	-7.9	-40.0	-26.2	-27.9	-20.0	-56.1	-43.9
	20	1	46.1	-12.2	30.8	-6.8	-40.0	-27.9	-27.9	-21.1	-56.1	-43.9
	25	1	41.8	6.8	30.3	6.9	-18.6	-25.3	-13.4	-20.4	-34.6	-29.5
150	30–45	2	—	—	—	—	-7.1	-13.7	-1.9	-8.8	—	-15.0
	30–45	1	37.5	25.6	29.8	20.5	2.9	-22.7	1.0	-19.5	-13.2	-15.0
	0–5	2	35.7	25.6	29.8	20.5	14.4	-11.3	12.6	-8.0	-13.2	-15.0
	0–5	1	35.7	-18.5	23.7	-11.0	-42.9	-24.4	-29.8	-18.9	-60.0	-47.0
	10	1	40.2	-16.7	26.8	-9.7	-42.9	-26.2	-29.8	-20.1	-60.0	-47.0
	15	1	44.8	-14.9	29.8	-8.5	-42.9	-28.0	-29.8	-21.4	-60.0	-47.0

(continued)

TABLE 4.4 (continued)
Simplified Wind Pressures, p_{s30} (psf)

Basic Wind Speed (mph)	Roof Angle (°)	Load Case	Zones									
			Horizontal Pressures				Vertical Pressures				Overhangs	
			A	B	C	D	E	F	G	H	E_{OH}	G_{OH}
	20	1	49.4	−13.0	32.9	−7.2	−42.9	−29.8	−29.8	−22.6	−60.0	−47.0
	25	1	44.8	7.2	32.4	7.4	−19.9	−27.1	−14.4	−21.8	−37.0	−31.6
		2	—	—	—	—	−7.5	−14.7	−2.1	−9.4	—	—
	30–45	1	40.1	27.4	31.9	22.0	3.1	−24.4	1.0	−20.9	−14.1	−16.1
		2	40.1	27.4	31.9	22.0	15.4	−12.0	13.4	−8.6	−14.1	−16.1
170	0–5	1	45.8	−23.8	30.4	−14.1	−55.1	−31.3	−38.3	−24.2	−77.1	−60.4
	10	1	51.7	−21.4	34.4	−12.5	−55.1	−33.6	−38.3	−25.8	−77.1	−60.4
	15	1	57.6	−19.1	38.3	−10.9	−55.1	−36.0	−38.3	−27.5	−77.1	−60.4
	20	1	63.4	−16.7	42.3	−9.3	−55.1	−38.3	−38.3	−29.1	−77.1	−60.4
	25	1	57.5	9.3	41.6	9.5	−25.6	−34.8	−18.5	−28.0	−47.6	−40.5
		2	—	—	—	—	−9.7	−18.9	−2.6	−12.1	—	—
	30–45	1	51.5	35.2	41.0	28.2	4.0	−31.3	1.3	−26.9	−18.1	−20.7
		2	51.5	35.2	41.0	28.2	19.8	−15.4	17.2	−11.0	−18.1	−20.7

Source: Courtesy of American Society of Civil Engineers, Reston, VA.

The following are the steps of the procedure:

1. Determine the basic wind speed, V, for the location from Figure 4.1, reproduced from the ASCE 7-05.
2. Determine the importance factor, I, based on the occupancy category from Table 4.1.
3. Determine the upwind exposure category depending upon the surface roughness that prevails in the upwind direction of the structure, as indicated in Table 4.2.
4. Determine the height and exposure adjustment coefficient λ from Table 4.3.
5. The topographic factor, K_{zt}, has to be applied to a structure that is located on an isolated hill of at least 60 ft height for exposure B, and of at least 15 ft height for exposure C and D and that it should be unobstructed by any similar hill for at least a distance of 100 times the height of the hill or two miles, whichever is higher. The factor is assessed by the three multipliers that are presented in Figure 6-4 of the ASCE 7-05. For usual cases $K_{zt} = 1$.
6. Determine p_{s30} from Table 4.4, reproduced from the ASCE 7-05.

HORIZONTAL PRESSURE ZONES FOR MWFRS

The horizontal pressures acting on the vertical plane are separated into the following four pressure zones, as shown in Figure 4.2:

A: End zone of wall
B: End zone of (vertical projection) roof
C: Interior zone of wall
D: Interior zone of (vertical projection) roof

The dimension of the end zones A and B are taken equal to $2a$, where the value of a is the smaller of the following two values:

1. 0.1 times the least horizontal dimension
2. 0.4 times the roof height, h

The height, h, is the mean height of roof from the ground. For roof angle <10°, it is the height to the eave.

If the pressure in zone B or D is negative, treat it to be zero in computing the total horizontal force.

For wind in longitudinal direction (wind acting on width), the zones B and D do not exist, as in Figure 4.2.

VERTICAL PRESSURE ZONES FOR MWFRS

The vertical pressures on the roof are likewise separated into the following four zones:

E: End zone of (horizontal projection) windward roof
F: End zone of (horizontal projection) leeward roof
G: Interior zone of (horizontal projection) windward roof
H: Interior zone of (horizontal projection) leeward roof

Where the end zones E and G fall on a roof overhang, the pressure values under the columns E_{OH} and G_{OH} in wind Table 4.4 are used for windward side. For leeward side, the basic values are used.

For transverse wind direction, the dimension of the end zones E and F is taken to be the horizontal distance from edge to ridge, as shown in Figure 4.3. For longitudinal wind direction, the roof angle$=0$ is used and the zones E and F are to the mid-length of the structure, as shown in Figure 4.3.

Minimum Pressure for MWFRS

The minimum wind load computed for MWFRS should not be less than $10\,\text{lb/ft}^2$ multiplied by the vertically projected building area. It is equivalent to the pressure p_s of $10\,\text{lb/ft}^2$ on zones A, B, C, and D and the pressure of zero on zones E, F, G, and H.

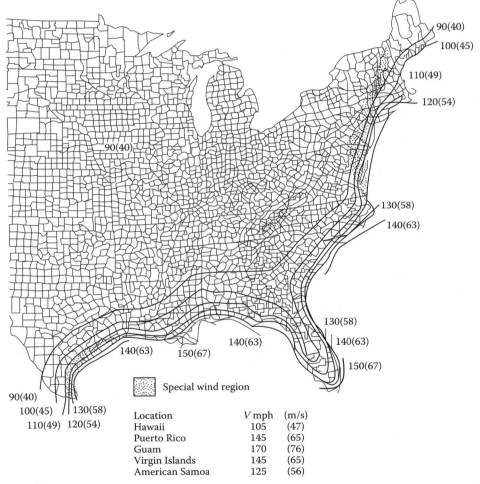

Location	V mph	(m/s)
Hawaii	105	(47)
Puerto Rico	145	(65)
Guam	170	(76)
Virgin Islands	145	(65)
American Samoa	125	(56)

Notes:

1. Values are nominal design 3 s gust wind speeds in miles per hour (m/s) at 33 ft (10 m) above ground for Exposure C category.

2. Linear interpolation between wind contours is permitted.

3. Islands and coastal areas outside the last contour shall use the last wind speed contour of the coastal area.

4. Mountainous terrain, gorges, ocean promontories, and special wind regions shall be examined for unusual wind conditions.

(a)

FIGURE 4.1   Basic wind speed. (Courtesy of American Society of Civil Engineers, Reston, VA.)

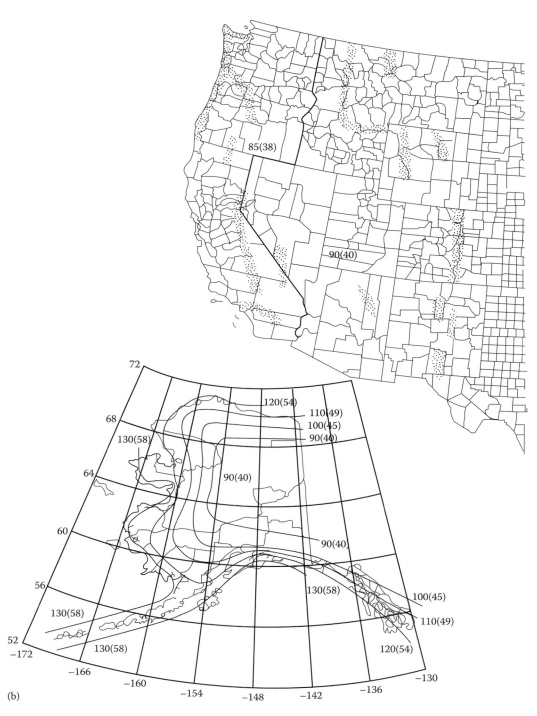

(b)

FIGURE 4.1 (continued)

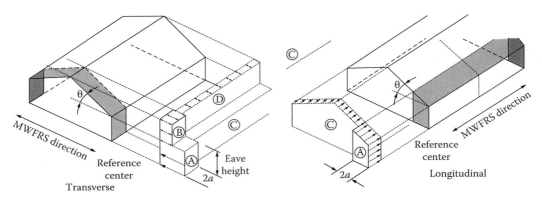

FIGURE 4.2 Horizontal pressure zones.

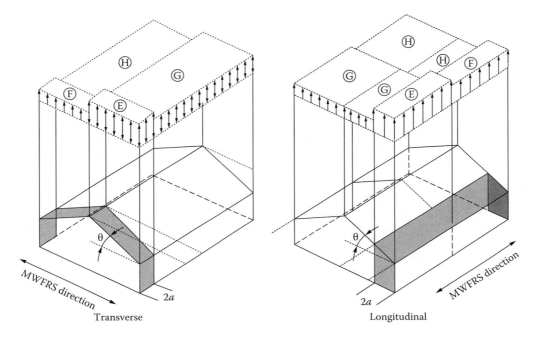

FIGURE 4.3 Vertical pressure zones.

Example 4.1

A two-story essential facility shown in Figure 4.4 is an enclosed wood frame building located in Seattle, Washington. Determine the design wind pressures for MWFRS in both principal directions of the building and the forces acting on the transverse section of the building. The wall studs and roof rafters are 16 in. on center. $K_{zt}=1.0$.

Solution

I. Design parameters
  1. Roof slope, $\theta=14°$
  2. $h_{mean}=22 + 6.25/2 =25.13$ ft
  3. End zone dimension, a: smaller of
      a. $0.4\,h_{mean}=0.4\,(25.13) =10$ ft
      b. 0.1width$=0.1(50)=5$ft ← controls
         Length of end zone$=2a=10$ ft

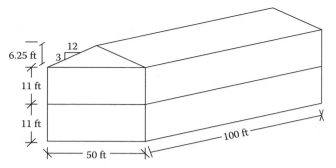

FIGURE 4.4 Two-story framed building.

4. Basic wind speed, $V = 85$ mph
5. Importance factor for essential facility $= 1.15$
6. Exposure category $= B$
7. λ from Table 4.3 up to 30 ft $= 1.0$
8. $K_{zt} = 1.00$ (given)
9. $p_s = \lambda K_{zt} I p_{s30}$
   $p_s = (1.0)(1.0)(1.15) p_{s30} = 1.15 p_{s30}$

II. For transverse wind direction
   A.1 Horizontal wind pressure on wall and roof projection

	Pressure (psf) (Table 4.4)			
Zone	Roof Angle $= 10°$	Roof Angle $= 15°$	Interpolated for 14°	$p_s = 1.15 p_{s30}$ (psf)
A. End zone wall	12.9	14.4	14.1	16.22
B. End zone roof	−5.4	−4.8	−4.92	−5.66
C. Interior wall	8.6	9.6	9.4	10.81
D. Interior roof	−3.1	−2.7	−2.78	−3.20

These pressures are shown in the section view in Figure 4.5a.

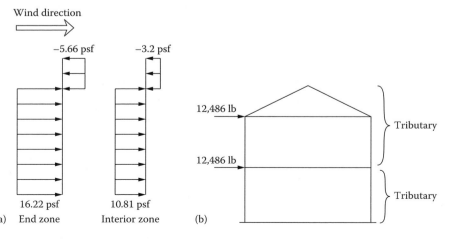

FIGURE 4.5 (a) Horizontal pressure distribution and (b) horizontal force on story—transverse wind.

A.2 Horizontal force at the roof level

Location	Zone	Tributary Height (ft)	Tributary Width (ft)	Tributary Area (ft²)	Pressure (psf)	Load (lb)
End	A	11[a]	$2a=10$ ft	110	16.22	1,784
	B	6.25	10	62.5	$-5.66 \rightarrow 0$	0
Interior	C	11	$L-2a=90$ ft	990	10.81	10,702
	D	6.25	90	562.5	$-3.2 \rightarrow 0$	0
Total						12,486

[a] It is also a practice to take 1/2 of the floor height for each level. In such case, the wind force on the 1/2 of the first floor height from the ground is not applied.
Taking pressures in zones B and D to be zero.

A.3 Horizontal force at the second floor level

Location	Zone	Tributary Height (ft)	Tributary Width (ft)	Tributary Area (ft²)	Pressure (psf)	Load (lb)
End	A	11	10	110	16.22	1,784
Interior	C	11	90	990	10.81	10,702
Total						12,486

The application of the forces is shown in Figure 4.5b.

B.1 Vertical wind pressure on the roof

Zone	Pressure (psf) (Table 4.4) Roof Angle =10°	Pressure (psf) (Table 4.4) Roof Angle =15°	Pressure (psf) (Table 4.4) Interpolated to 14°	$p_s=1.15p_{s30}$ (psf)
E: End, windward	−13.8	−13.8	−13.8	−15.87
F: End, leeward	−8.4	−9.0	−8.88	−10.21
G: Interior, windward	−9.6	−9.6	−9.6	−11.04
H: Interior, leeward	−6.5	−6.9	−6.82	−7.84

The pressures are shown in the sectional view in Figure 4.6a.

B.2 Vertical force on the roof

Zone		Tributary Length (ft)	Tributary Width (ft)	Tributary Area (ft²)	Pressure (psf)	Load (lb)
Windward	End	25	$2a=10$	250	−15.87	−3,968
	Interior	25	$L-2a=90$	2250	−11.04	−24,840
	Total					−28,808
Leeward	End	25	10	250	−10.21	2,552
	Interior	25	90	2250	−7.84	17,640
	Total					−20,192

The application of vertical forces is shown in Figure 4.6b.

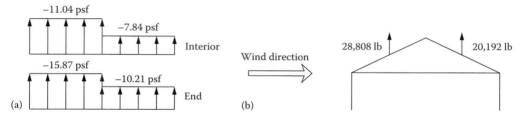

FIGURE 4.6 (a) Vertical pressure distribution on roof and (b) vertical force on roof—transverse wind.

  C. Minimum force on MWFRS by transverse wind
    The minimum pressure is 10 psf acting on the vertical projection the building. Thus,

$$\text{Minimum wind force} = 10 \times [(22 + 6.25) \times 100] = 28{,}250 \text{ lb}$$

  D. Applicable wind force
    The following two cases should be considered for maximum effect:
    1. The combined A.2, A.3, and B.2
    2. Minimum force C
III. For longitudinal wind direction
  A.1 Horizontal wind pressures on wall and roof projection
    Zones B and D do not exist, pressure on zone A = 16.22 psf and zone C = 10.81 psf from step A.1, p. 60.
  A.2 Horizontal force at the roof level
    From Figure 4.7

$$\text{Tributary area for end zone A} = \frac{1}{2}(11 + 13.5)(10) = 122.5 \text{ ft}^2$$

$$\text{Tributary area for interior zone C} = \frac{1}{2}(13.5 + 17.25)(15) + \frac{1}{2}(17.25 + 11)(25)$$

$$= 230.63 + 353.12 = 583.75 \text{ ft}^2$$

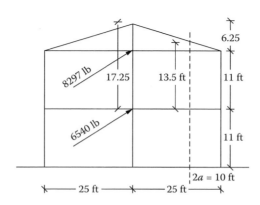

FIGURE 4.7 Horizontal wind force on wall and roof projection—longitudinal direction wind.

Zone	Tributary Area (ft²)	Pressure (psf)	Load[a] (lb)
A	122.5	16.22	1987
C	583.75	10.81	6310
Total			8297

[a] The centroids of area are different but the force is assumed to be acting at roof level.

A.3 Horizontal force at the second floor level

$$\text{Tributary area for end zone A} = 11 \times 10 = 110 \text{ ft}^2$$

$$\text{Tributary area for interior zone C} = 11 \times 40 = 440 \text{ ft}^2$$

Zone	Tributary Area (ft²)	Pressure (psf)	Load (lb)
A	110	16.22	1784
C	440	10.81	4756
Total			6540

The application of forces is shown in the sectional view in Figure 4.7.

B.1 Vertical wind on the roof (longitudinal case)

Zone	p_{s30} (psf)	$p_s = 1.15 p_{s30}$
End E	−13.8	−15.87
End F	−7.8	−8.97
Interior G	−9.6	−11.04
Interior H	−6.1	−7.02

Pressures with roof angle $\theta = 0$. The pressures are shown in Figure 4.8.

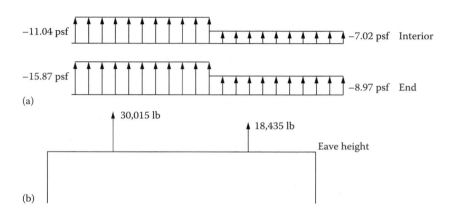

FIGURE 4.8 (a) Vertical distribution of pressures and (b) force on roof—longitudinal direction wind.

B.2 Vertical force on the roof

Zone		Tributary		Area (ft²)	Pressure (psf)	Load (lb)
		Length (ft)	Width (ft)			
Windward	E	$2a=10$	$L/2=50$	500	−15.87	−7,935
	G	$B-2a=40$	50	2000	−11.04	−22,080
	Total					−30,015
Leeward	F	10	50	500	−8.79	−4,395
	H	40	50	2000	−7.02	−14,040
	Total					−18,435

The application of forces is shown in Figure 4.8.

THE SIMPLIFIED PROCEDURES FOR COMPONENTS AND CLADDING

The C and C cover the individual structural elements that directly support a tributary area against the wind force. The conditions and the steps of the procedure are essentially similar to the MWFRS. The pressure, however, acts normal to each surface, i.e., horizontal on the wall and perpendicular to the roof. The following similar equation is used to determine the wind pressure. The adjustment factor, λ, the topographic factor, K_{zt}, and the importance factor, I, are determined from similar considerations:

$$p_{net} = \lambda K_{zt} I p_{net30} \tag{4.2}$$

where
   I is the importance factor (Table 4.1)
   λ is the adjustment factor for structure height and exposure (Tables 4.2 and 4.3)
   K_{zt} is the topographic factor
   p_{net30} is the simplified standard design wind pressure (Table 4.5)

However, the pressures p_{net30} are different from p_{s30}. Besides the basic wind speed, the pressures are a function of the roof angle, the effective wind area supported by the element, and the zone of the structure surface. The p_{net} represents the net pressures that are the algebraic summation of the internal and external pressures acting normal to the surface of the component and cladding.

The effective area is the tributary area of an element but need not be lesser than the span length multiplied by the width equal to one-third of the span length, i.e., $A = L^2/3$.

Table 4.5 reproduced from the ASCE7–05 lists p_{net30} values for effective wind areas of 10, 20, 50, and 100 ft² for roof and additionally 500 ft² for wall. A roof element having an effective area in excess of 100 ft² should use pressures corresponding to an area of 100 ft². Similarly, a wall element supporting an area in excess of 500 ft² should use pressures corresponding to 500 ft². A linear interpolation is permitted for intermediate areas. Table 4.6 lists p_{net30} values for roof overhang.

The following zones shown in Figure 4.9 have been identified for the C and C.

The dimension a is the smaller of the following two values:

1. 0.4 times the mean height to roof, h_{mean}
2. 0.1 times the smaller horizontal dimension

but, the value of a should not be less than the following:

1. 0.04 times the smaller horizontal dimension
2. 3 ft

TABLE 4.5
Net Design Wind Pressure, p_{net30} (psf)

Zone	Effective Wind Area (sf)	Basic Wind Speed V (mph)											
		85		90		100		105		110		120	
Roof 0°–7°													
1	10	5.3	−13.0	5.9	−14.6	7.3	−18.0	8.1	−19.8	8.9	−21.8	10.5	−25.9
1	20	5.0	−12.7	5.6	−14.2	6.9	−17.5	7.6	−19.3	8.3	−21.2	9.9	−25.2
1	50	4.5	−12.2	5.1	−13.7	6.3	−16.9	6.9	−18.7	7.6	−20.5	9.0	−24.4
1	100	4.2	−11.9	4.7	−13.3	5.8	−16.5	6.4	−18.2	7.0	−19.9	8.3	−23.7
2	10	5.3	−21.8	5.9	−24.4	7.3	−30.2	8.1	−33.3	8.9	−36.5	10.5	−43.5
2	20	5.0	−19.5	5.6	−21.8	6.9	−27.0	7.6	−29.7	8.3	−32.6	9.9	−38.8
2	50	4.5	−16.4	5.1	−18.4	6.3	−22.7	6.9	−25.1	7.6	−27.5	9.0	−32.7
2	100	4.2	−14.1	4.7	−15.8	5.8	−19.5	6.4	−21.5	7.0	−23.6	8.3	−28.1
3	10	5.3	−32.8	5.9	−36.8	7.3	−45.4	8.1	−50.1	8.9	−55.0	10.5	−65.4
3	20	5.0	−27.2	5.6	−30.5	6.9	−37.6	7.6	−41.5	8.3	−45.5	9.9	−54.2
3	50	4.5	−19.7	5.1	−22.1	6.3	−27.3	6.9	−30.1	7.6	−33.1	9.0	−39.3
3	100	4.2	−14.1	4.7	−15.8	5.8	−19.5	6.4	−21.5	7.0	−23.6	8.3	−28.1
Roof > 7°–27°													
1	10	7.5	−11.9	8.4	−13.3	10.4	−16.5	11.4	−18.2	12.5	−19.9	14.9	−23.7
1	20	6.8	−11.6	7.7	−13.0	9.4	−16.0	10.4	−17.6	11.4	−19.4	13.6	−23.0
1	50	6.0	−11.1	6.7	−12.5	8.2	−15.4	9.1	−17.0	10.0	−18.6	11.9	−22.2
1	100	5.3	−10.8	5.9	−12.1	7.3	−14.9	8.1	−16.5	8.9	−18.1	10.5	−21.5
2	10	7.5	−20.7	8.4	−23.2	10.4	−28.7	11.4	−31.6	12.5	−34.7	14.9	−41.3
2	20	6.8	−19.0	7.7	−21.4	9.4	−26.4	10.4	−29.1	11.4	−31.9	13.6	−38.0
2	50	6.0	−16.9	6.7	−18.9	8.2	−23.3	9.1	−25.7	10.0	−28.2	11.9	−33.6
2	100	5.3	−15.2	5.9	−17.0	7.3	−21.0	8.1	−23.2	8.9	−25.5	10.5	−30.3
3	10	7.5	−30.6	8.4	−34.3	10.4	−42.4	11.4	−46.7	12.5	−51.3	14.9	−61.0
3	20	6.8	−28.6	7.7	−32.1	9.4	−39.6	10.4	−43.7	11.4	−47.9	13.6	−57.1
3	50	6.0	−26.0	6.7	−29.1	8.2	−36.0	9.1	−39.7	10.0	−43.5	11.9	−51.8
3	100	5.3	−24.0	5.9	−26.9	7.3	−33.2	8.1	−36.6	8.9	−40.2	10.5	−47.9

Roof > 27°–45°

1	10	11.9	−13.0	13.3	−14.6	16.5	−18.0	18.2	−19.8	19.9	−21.8	23.7	−25.9
1	20	11.6	−12.3	13.0	−13.8	16.0	−17.1	17.6	−18.8	19.4	−20.7	23.0	−24.6
1	50	11.1	−11.5	12.5	−12.8	15.4	−15.9	17.0	−17.5	18.6	−19.2	22.2	−22.8
1	100	10.8	−10.8	12.1	−12.1	14.9	−14.9	16.5	−16.5	18.1	−18.1	21.5	−21.5
2	10	11.9	−15.2	13.3	−17.0	16.5	−21.0	18.2	−23.2	19.9	−25.5	23.7	−30.3
2	20	11.6	−14.5	13.0	−16.3	16.0	−20.1	17.6	−22.2	19.4	−24.3	23.0	−29.0
2	50	11.1	−13.7	12.5	−15.3	15.4	−18.9	17.0	−20.8	18.6	−22.9	22.2	−27.2
2	100	10.8	−13.0	12.1	−14.6	14.9	−18.0	16.5	−19.8	18.1	−21.8	21.5	−25.9
3	10	11.9	−15.2	13.3	−17.0	16.5	−21.0	18.2	−23.2	19.9	−25.5	23.7	−30.3
3	20	11.6	−14.5	13.0	−16.3	16.0	−20.1	17.6	−22.2	19.4	−24.3	23.0	−29.0
3	50	11.1	−13.7	12.5	−15.3	15.4	−18.9	17.0	−20.8	18.6	−22.9	22.2	−27.2
3	100	10.8	−13.0	12.1	−14.6	14.9	−18.0	16.5	−19.8	18.1	−21.8	21.5	−25.9

Wall

4	10	13.0	−14.1	14.6	−15.8	18.0	−19.5	19.8	−21.5	21.8	−23.6	25.9	−28.1
4	20	12.4	−13.5	13.9	−15.1	17.2	−18.7	18.9	−20.6	20.8	−22.6	24.7	−26.9
4	50	11.6	−12.7	13.0	−14.3	16.1	−17.6	17.8	−19.4	19.5	−21.3	23.2	−25.4
4	100	11.1	−12.2	12.4	−13.6	15.3	−16.8	16.9	−18.5	18.5	−20.4	22.0	−24.2
4	500	9.7	−10.8	10.9	−12.1	13.4	−14.9	14.8	−16.5	16.2	−18.1	19.3	−21.5
5	10	13.0	−17.4	14.6	−19.5	18.0	−24.1	19.8	−26.6	21.8	−29.1	25.9	−34.7
5	20	12.4	−16.2	13.9	−18.2	17.2	−22.5	18.9	−24.8	20.8	−27.2	24.7	−32.4
5	50	11.6	−14.7	13.0	−16.5	16.1	−20.3	17.8	−22.4	19.5	−24.6	23.2	−29.3
5	100	11.1	−13.5	12.4	−15.1	15.3	−18.7	16.9	−20.6	18.5	−22.6	22.0	−26.9
5	500	9.7	−10.8	10.9	−12.1	13.4	−14.9	14.8	−16.5	16.2	−18.1	19.3	−21.5

(continued)

TABLE 4.5 (continued)
Net Design Wind Pressure, p_{net30} (psf)

Zone	Effective Wind Area (sf)	Basic Wind Speed V (mph)											
		125		130		140		145		150		170	
Roof 0°–7°													
1	10	11.4	−28.1	12.4	−30.4	14.3	−35.3	15.4	−37.8	16.5	−40.5	21.1	−52.0
1	20	10.7	−27.4	11.6	−29.6	13.4	−34.4	14.4	−36.9	15.4	−39.4	19.8	−50.7
1	50	9.8	−26.4	10.6	−28.6	12.3	−33.2	13.1	−35.6	14.1	−38.1	18.1	−48.9
1	100	9.1	−25.7	9.8	−27.8	11.4	−32.3	12.2	−34.6	13.0	−37.0	16.7	−47.6
2	10	11.4	−47.2	12.4	−51.0	14.3	−59.2	15.4	−63.5	16.5	−67.9	21.1	−87.2
2	20	10.7	−42.1	11.6	−45.6	13.4	−52.9	14.4	−56.7	15.4	−60.7	19.8	−78.0
2	50	9.8	−35.5	10.6	−38.4	12.3	−44.5	13.1	−47.8	14.1	−51.1	18.1	−65.7
2	100	9.1	−30.5	9.8	−33.0	11.4	−38.2	12.2	−41.0	13.0	−43.9	16.7	−56.4
3	10	11.4	−71.0	12.4	−76.8	14.3	−89.0	15.4	−95.5	16.5	−102.2	21.1	−131.3
3	20	10.7	−58.5	11.6	−63.6	13.4	−73.8	14.4	−79.1	15.4	−84.7	19.8	−108.7
3	50	9.8	−42.7	10.6	−46.2	12.3	−53.5	13.1	−57.4	14.1	−61.5	18.1	−78.9
3	100	9.1	−30.5	9.8	−33.0	11.4	−38.2	12.2	−41.0	13.0	−43.9	16.7	−56.4
Roof >7°–27°													
1	10	16.2	−25.7	17.5	−27.8	20.3	−32.3	21.8	−34.6	23.3	−37.0	30.0	−47.6
1	20	14.8	−25.0	16.0	−27.0	18.5	−31.4	19.9	−33.7	21.3	−36.0	27.3	−46.3
1	50	12.9	−24.1	13.9	−26.0	16.1	−30.2	17.3	−32.4	18.5	−34.6	23.8	−44.5
1	100	11.4	−23.2	12.4	−25.2	14.3	−29.3	15.4	−31.4	16.5	−33.6	21.1	−43.2
2	10	16.2	−44.8	17.5	−48.4	20.3	−56.2	21.8	−60.3	23.3	−64.5	30.0	−82.8
2	20	14.8	−41.2	16.0	−44.6	18.5	−51.7	19.9	−55.4	21.3	−59.3	27.3	−76.2
2	50	12.9	−36.5	13.9	−39.4	16.1	−45.7	17.3	−49.1	18.5	−52.5	23.8	−67.4
2	100	11.4	−32.9	12.4	−35.6	14.3	−41.2	15.4	−44.2	16.5	−47.3	21.1	−60.8
3	10	16.2	−66.2	17.5	−71.6	20.3	−83.1	21.8	−89.1	23.3	−95.4	30.0	−122.5
3	20	14.8	−61.9	16.0	−67.0	18.5	−77.7	19.9	−83.3	21.3	−89.2	27.3	−114.5
3	50	12.9	−56.2	13.9	−60.8	16.1	−70.5	17.3	−75.7	18.5	−81.0	23.8	−104.0
3	100	11.4	−51.9	12.4	−56.2	14.3	−65.1	15.4	−69.9	16.5	−74.8	21.1	−96.0

Roof > 27°–45°

1	10	25.7	-28.1	27.8	-30.4	32.3	-35.3	34.6	-37.8	37.0	-40.5	47.6	-52.0
1	20	25.0	-26.7	27.0	-28.9	31.4	-33.5	33.7	-35.9	36.0	-38.4	46.3	-49.3
1	50	24.1	-24.8	26.0	-26.8	30.2	-31.1	32.4	-33.3	34.6	-35.7	44.5	-45.8
1	100	23.3	-23.3	25.2	-25.2	29.3	-29.3	31.4	-31.4	33.6	-33.6	43.2	-43.2
2	10	25.7	-32.9	27.8	-35.6	32.3	-41.2	34.6	-44.2	37.0	-47.3	47.6	-60.8
2	20	25.0	-31.4	27.0	-34.0	31.4	-39.4	33.7	-42.3	36.0	-45.3	46.3	-58.1
2	50	24.1	-29.5	26.0	-32.0	30.2	-37.1	32.4	-39.8	34.6	-42.5	44.5	-54.6
2	100	23.2	-28.1	25.2	-30.4	29.3	-35.3	31.4	-37.8	33.6	-40.5	43.2	-52.0
3	10	25.7	-32.9	27.8	-35.6	32.3	-41.2	34.6	-44.2	37.0	-47.3	47.6	-60.8
3	20	25.0	-31.4	27.0	-34.0	31.4	-39.4	33.7	-42.3	36.0	-45.3	46.3	-58.1
3	50	24.1	-29.5	26.0	-32.0	30.2	-37.1	32.4	-39.8	34.6	-42.5	44.5	-54.6
3	100	23.3	-28.1	25.2	-30.4	29.3	-35.3	31.4	-37.8	33.6	-40.5	43.2	-52.0
Wall 4	10	28.1	-30.5	30.4	-33.0	35.3	-38.2	37.8	-41.0	40.5	-43.9	52.0	-56.4
4	20	26.8	-29.2	29.0	-31.6	33.7	-36.7	36.1	-39.3	38.7	-42.1	49.6	-54.1
4	50	25.2	-27.5	27.2	-29.8	31.6	-34.6	33.9	-37.1	36.2	-39.7	46.6	-51.0
4	100	23.9	-26.3	25.9	-28.4	30.0	-33.0	32.2	-35.4	34.4	-37.8	44.2	-48.6
4	500	21.0	-23.3	22.7	-25.2	26.3	-29.3	28.2	-31.4	30.2	-33.6	38.8	-43.2
5	10	28.1	-37.6	30.4	-40.7	35.3	-47.2	37.8	-50.6	40.5	-54.2	52.0	-69.6
5	20	26.8	-35.1	29.0	-38.0	33.7	-44.0	36.1	-47.2	38.7	-50.5	49.6	-64.9
5	50	25.2	-31.8	27.2	-34.3	31.6	-39.8	33.9	-42.7	36.2	-45.7	46.6	-58.7
5	100	23.9	-29.2	25.9	-31.6	30.0	-36.7	32.2	-39.3	34.4	-42.1	44.2	-54.1
5	500	21.0	-23.2	22.7	-25.2	26.3	-29.3	28.2	-31.1	30.2	-33.6	38.8	-43.2

Source: Courtesy of American Society of Civil Engineers, Reston, VA.

Note: Exposure B at $h = 30$ ft with $l = 1.0$ and $K_{zt} = 1.0$.

TABLE 4.6
Roof Overhang Net Design Wind Pressure, p_{net30} (psf)

Zone	Effective Wind Area (sf)	Basic Wind Speed V (mph)							
		90	100	110	120	130	140	150	170
Roof 0°–7°									
2	10	−21.0	−25.9	−31.4	−37.3	−43.8	−50.8	−58.3	−74.9
2	20	−20.6	−25.5	−30.8	−36.7	−43.0	−49.9	−57.3	−73.6
2	50	−20.1	−24.9	−30.1	−35.8	−42.0	−48.7	−55.9	−71.8
2	100	−19.8	−24.4	−29.5	−35.1	−41.2	−47.8	−54.9	−70.5
3	10	−34.6	−42.7	−51.6	−61.5	−72.1	−83.7	−96.0	−123.4
3	20	−27.1	−33.5	−40.5	−48.3	−56.6	−65.7	−75.4	−96.8
3	50	−17.3	−21.4	−25.9	−30.8	−36.1	−41.9	−48.1	−61.8
3	100	−10.0	−12.2	−14.8	−17.6	−20.6	−23.9	−27.4	−35.2
Roof > 7°–27°									
2	10	−27.2	−33.5	−40.6	−48.3	−56.7	−65.7	−75.5	−96.9
2	20	−27.2	−33.5	−40.6	−48.3	−56.7	−65.7	−75.5	−96.9
2	50	−27.2	−33.5	−40.6	−48.3	−56.7	−65.7	−75.5	−96.9
2	100	−27.2	−33.5	−40.6	−48.3	−56.7	−65.7	−75.5	−96.9
3	10	−45.7	−56.4	−68.3	−81.2	−95.3	−110.6	−126.9	−163.0
3	20	−41.2	−50.9	−61.8	−73.3	−86.0	−99.8	−114.5	−147.1
3	50	−35.3	−43.6	−52.8	−62.8	−73.7	−85.5	−98.1	−126.1
3	100	−30.9	−38.1	−46.1	−54.9	−64.4	−74.7	−85.8	−110.1
Roof > 27°–45°									
2	10	−24.7	−30.5	−36.9	−43.9	−51.5	−59.8	−68.6	−88.1
2	20	−24.0	−29.6	−35.8	−42.6	−50.0	−58.0	−66.5	−85.5
2	50	−23.0	−28.4	−34.3	−40.8	−47.9	−55.6	−63.8	−82.0
2	100	−22.2	−27.4	−33.2	−39.5	−46.4	−53.8	−61.7	−79.3
3	10	−24.7	−30.5	−36.9	−43.9	−51.5	−59.8	−68.6	−88.1
3	20	−24.0	−29.6	−35.8	−42.6	−50.0	−58.0	−66.5	−85.5
3	50	−23.0	−28.4	−34.3	−40.8	−47.9	−55.6	−63.8	−82.0
3	100	−22.2	−27.4	−33.2	−39.5	−46.4	−53.8	−61.7	−79.3

Source: Courtesy of American Society of Civil Engineers, Reston, VA.
Note: Exposure B at $h = 30$ ft with $I = 1.0$.

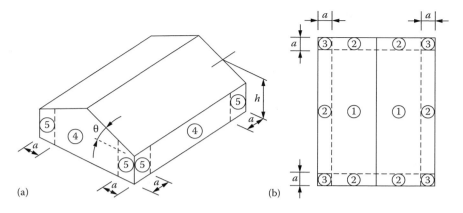

End zones		Interior zones		Corner zones	
Wall	5	Wall	4		
Roof	2	Roof	1	Roof	3

FIGURE 4.9 Zones for C and C. (a) Elevation and (b) plan.

There are two values of the net pressure that act on each element: a positive pressure acting inward (toward the surface) and a negative pressure acting outward (away from the surface). The two pressures must be considered separately for each element.

MINIMUM PRESSURES FOR COMPONENTS AND CLADDING

The positive pressure, p_{net}, should not be less than +10 psf and the negative pressure should not be less than −10 psf.

Example 4.2

Determine design wind pressures and forces for studs and rafters of Example 4.1.

Solution

A. Parameters
   1. $\theta = 14°$
   2. $a = 5$ ft (from Example 4.1) which is more than (1) 0.04 (50) = 2 ft and (2) 3 ft
   3. $p_{net} = 1.15 p_{net30}$ (from Example 4.1)
B. Wind pressures on studs (wall) at each floor level
   1. Effective area

$$A = L \times W = 11 \times \frac{16}{12} = 14.7 \, \text{ft}^2$$

$$A_{min} = \frac{L^2}{3} = \frac{(11)^2}{3} = 40.3 \, \text{ft}^2$$

   2. Net wall pressures for $V = 85$ mph

Zone	p_{net30} at Interpolated 40.3 ft² (psf)		$p_{net} = 1.15 p_{net30}$ (psf)	
End: 5	11.86	−15.19	13.64	−17.47
Interior: 4	11.86	−12.96	13.64	−14.90

C. Wind forces on studs
  C.1 On end studs with the higher pressures
    1. $W = p_{net}$ (tributary area)

$$= 13.64\,(14.7) = 200.5\ \text{lb (inward)}$$

    2. $W = p_{net}$ (tributary area)

$$= -17.47\,(14.7) = -256.8\ \text{lb (outward)}$$

These are shown in Figure 4.10

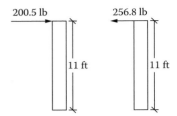

FIGURE 4.10 Wind force on end studs.

D. Wind pressures on rafters (roof)
  1. Length of rafter $= \dfrac{25}{\cos 14°} = 25.76\,\text{ft}$

  2. $A = (25.76)\left(\dfrac{16}{12}\right) = 34.35\,\text{ft}^2$

  3. $A_{min} = \dfrac{L^2}{3} = \dfrac{(25.76)^2}{3} = 221\,\text{ft}^2, \text{use}\,100\,\text{ft}^2$

  4. Net roof pressures

Zone	p_{net30} at 100 ft² (psf)		$p_{net} = 1.15 p_{net30}$ (psf)	
Corner 3	5.3	−24.0	6.1[a]	−27.6
End 2	5.3	−15.2	6.1[a]	−17.48
Interior 1	5.3	−10.8	6.1[a]	−12.42

[a] Use a minimum of 10 psf.

E. Wind forces on rafters
  E.1 On end rafters
    1. $W = p_{net}$ (tributary area)
      $= 10\,(34.35) = 343.5\ \text{lb (inward)}$

    2. $W = p_{net}$ (tributary area)
      $= -17.48\,(34.35) = -600\ \text{lb (outward)}$

    3. These are shown in Figure 4.11

PROBLEMS

4.1 Determine the horizontal wind pressures and the horizontal forces on the wall due to wind acting in the transverse direction on an MWFRS shown in Figure P4.1. It is a standard occupancy single-story building located in an urban area in Rhode Island where the basic wind speed is 100 mph. $K_{zt} = 1$.

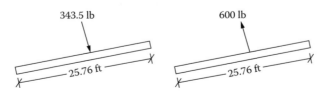

FIGURE 4.11 Wind force on end rafters.

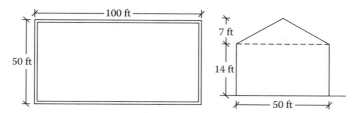

FIGURE P4.1 A single-story building in urban area.

4.2 An enclosed two-story heavily occupied building located in an open, flat terrain in Portland, Oregon is shown in Figure P4.2. Determine the wind pressures on the walls and roofs of the MWFRS in both principal directions. Also determine the design wind forces in both directions. $K_{zt}=1$.

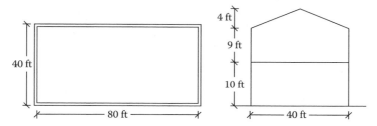

FIGURE P4.2 A double-story building in open terrain.

4.3 A three-story industrial steel building, located in open terrain in Honolulu, has a plan dimension of 200 ft × 90 ft. The structure consists of nine moment-resisting steel frames spanning 90 ft at 25 ft on center. It is roofed with steel deck, which is pitched at 1.25° on each side from the center. The building is 36 ft high with each floor height of 12 ft. Determine the MWFRS horizontal and vertical pressures and the forces due to wind in the transverse and longitudinal directions of the building. $K_{zt}=1$ (Figure P4.3).

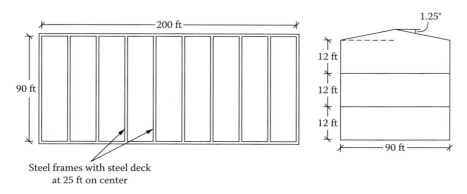

FIGURE P4.3 A three-story industrial building.

4.4 The building in Problem 4.1 has the wall studs and roof trusses spaced at 12 in. on center. Determine the elemental wind pressures and forces on the studs and roof trusses.

4.5 The building in Problem 4.2 has the wall studs and roof trusses spaced at 16 in. on center. Determine the elemental wind pressures and forces on the studs and roof trusses.

4.6 Determine the wind pressures and forces on the wall panel and roof decking of Problem 4.3. Decking is supported on joists that are 5 ft on center, spanning across the steel frames shown in Figure P4.3.

5 Earthquake Loads

SEISMIC FORCES

The earth's outer crust is composed of very big, hard plates as large or larger than a continent. These plates float on the molten rock beneath. When these plates encounter each other, appreciable horizontal and vertical ground motion of the surface occurs known as the *earthquake*. For example, in the western portion of the United States, earthquake is caused by the two plates comprising of the Northern American continent and the Pacific basin. The ground motion induces a very large inertia force known as the *seismic force* in a structure that often results in the destruction of the structure. The seismic force acts vertically like dead and live loads and laterally like wind load. But unlike the other forces that are proportional to the exposed area of the structure, the seismic force is proportional to the mass of the structure and is distributed in proportion to the structural mass at various levels.

In all other types of loads including the wind load, the structural response is static wherein the structure is subjected to a pressure applied by the load. However, in a seismic load, there is no such direct applied pressure.

If ground movement could take place slowly, the structure will ride it over smoothly moving along with it. But the quick movement of ground in an earthquake accelerates the mass of the structure. The product of the mass and acceleration is the internal force created within the structure. Thus, the seismic force is a dynamic entity.

There are three approaches to evaluate this dynamic seismic force:

1. Nonlinear dynamic analyses
2. Linear dynamic analyses
3. Static analyses

The first two consider the vibration modes of the structure and compute the induced forces for each mode. In the last approach, the seismic forces are represented by a set of supposedly equivalent static loads on the structure. It should be understood that no such simplified forces are truly equivalent to the complicated seismic forces but it is considered that a reasonable design of a structure can be produced by this approach.

The seismic analyses have been dealt in detail in ASCE 7-05 in eight chapters from Chapters 11 through 18. Six procedures of analysis have been presented, three of which pertain to the static analysis. The most commonly used method is the *equivalent lateral load procedure*, applicable to most light framed structures. This has been covered in the book.

SEISMIC PARAMETERS

FUNDAMENTAL PERIOD OF STRUCTURE

The basic dynamic property of a structure is its fundamental period of vibration. When a mass of body (in this case structure) is given a horizontal displacement (in this case due to earthquake), the mass oscillates back and forth. This is termed as the *free vibration*. The *fundamental period* is defined as the time (in seconds) it takes to go through one cycle of free vibration. The magnitude

(a)

FIGURE 5.1 (a) Maximum considered earthquake ground motion for the conterminous United States of 0.2 s spectral response acceleration (5% of critical damping), site class B.

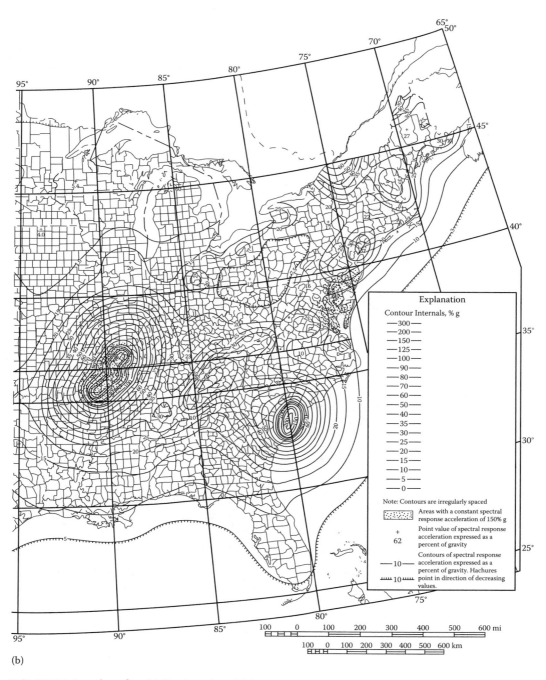

FIGURE 5.1 (continued) (b) Continuation of 0.2 s spectral response acceleration.

(*continued*)

(c)

FIGURE 5.1 (continued)   (c) Enlarged portion of Figure 5.1a. (Courtesy of American Society of Civil Engineers, Reston, VA.)

TABLE 5.1
Value of Parameters, C_t and x

Structure Type	C_t	x
Moment-resisting frame of steel	0.028	0.8
Moment-resisting frame of concrete	0.016	0.9
Braced steel frame	0.03	0.75
All other structures	0.02	0.75

depends upon the mass of structure and its stiffness. It can be determined by theory. The ASCE 7-05 provides the following formula to approximate the fundamental time t_a:

$$t_a = C_t h_n^x \qquad (5.1)$$

where

t_a is the approximate fundamental period in seconds
h_n is the height of the highest level of the structure in ft
C_t is the building period coefficient as given in Table 5.1
x is the exponential coefficient as given in Table 5.1

Example 5.1

Determine the approximate fundamental period for a five-story office building of moment-resisting steel, each floor having a height of 12 ft.

Solution

1. Height of building from ground $= 5 \times 12 = 60$ ft
2. $t_a = 0.028(60)^{0.8} = 0.74$ s

GROUND SPECTRAL RESPONSE MAPS

At the onset, the maximum considered earthquake (MCE)* spectral response accelerations for a place are read from the spectral maps of United States. There are two types of the mapped accelerations: (1) short-period (0.2 s) spectral acceleration, S_s which is used to study the acceleration-controlled portion of the spectra, and (2) 1 s spectral acceleration S_1 which is used to study the velocity-controlled portion of the spectra. These acceleration parameters represent 5% damped ground motions at 2% probability of exceedance in 50 years. The maps reproduced from Chapter 22 of the ASCE7-05 are given in Figures 5.1 and 5.2. A CD containing the spectral maps are distributed with the International Building Code (IBC). The maps are also available at the USGS site at http://eqhazmaps.usgs.gov. The values given in Figures 5.1 and 5.2 are percentages of the gravitational constant, g, i.e., 200 means 2.0g. The parameters, S_s and S_1 are used to apply the provisions of the ASCE7-05 with respect to earthquake analysis.

ADJUSTED SPECTRAL RESPONSE ACCELERATIONS

The mapped values of Figures 5.1 and 5.2 are for the site soil category B. The site soil classification is given in Table 5.2.

---

* For practical purposes, it represents the maximum earthquake that can reasonably occur at the site.

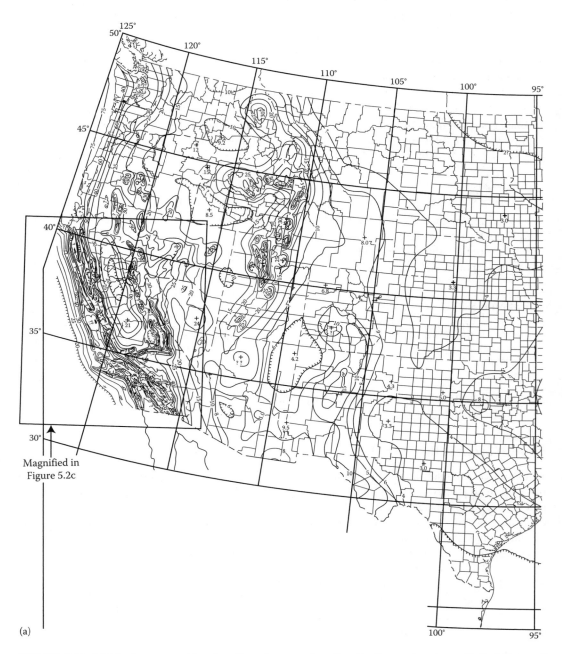

(a)

FIGURE 5.2 (a) Maximum considered earthquake ground motion for the conterminous United States of 1.0 s spectral response acceleration (5% of critical damping), site class B. (Courtesy of American Society of Civil Engineers, Reston, VA.)

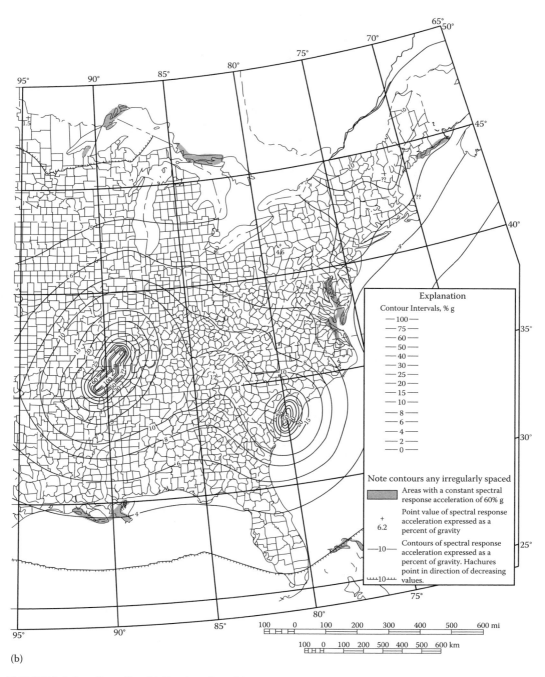

(b)

FIGURE 5.2 (continued) (b) Continuation of 1 s spectral response acceleration.

(*continued*)

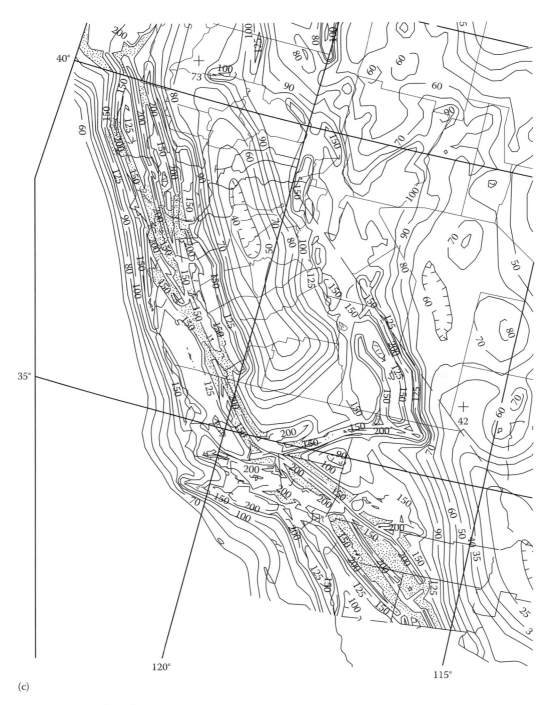

(c)

FIGURE 5.2 (continued)   (c) Enlarged portion of Figure 5.2a.

For a soil of the classification other than soil type B, the spectral response accelerations are adjusted as follows:

$$S_{MS} = F_a S_s \tag{5.2}$$

$$S_{M1} = F_v S_1 \tag{5.3}$$

The values of factors F_a and F_v reproduced from the ASCE 7-05 are given in Tables 5.3 and 5.4.

The factors are 0.8 for soil class A, 1 for soil class B, and higher than 1 for soils C onward; up to 3.5 for soil type E. The site class D should be used when the soil properties are not known in a sufficient detail.

TABLE 5.2
Soil Classification for Spectral Acceleration

Class	Type
A	Hard rock
B	Rock
C	Soft rock or very dense soil
D	Stiff soil
E	Soft soil
F	Requires the site-specific evaluation

DESIGN SPECTRAL ACCELERATION

These are the primary variables to prepare the design spectrum. The design spectral accelerations are 2/3 of the adjusted acceleration as follows:

$$S_{DS} = 2/3 S_{MS} \tag{5.4}$$

TABLE 5.3
Site Coefficient, F_a

Site Class	Mapped Maximum Considered Earthquake Spectral Response Acceleration Parameter at Short Period				
	$S_s \leq 0.25$	$S_s = 0.5$	$S_s = 0.75$	$S_s = 1.0$	$S_s \geq 1.25$
A	0.8	0.8	0.8	0.8	0.8
B	1.0	1.0	1.0	1.0	1.0
C	1.2	1.2	1.1	1.0	1.0
D	1.6	1.4	1.2	1.1	1.0
E	2.5	1.7	1.2	0.9	0.9
F	See Section 11.4.7 of ASCE 7-05				

Note: Use straight-line interpolation for intermediate values of S_s.

TABLE 5.4
Site Coefficient, F_v

Site Class	Mapped Maximum Considered Earthquake Spectral Response Acceleration Parameter at 1 s Period				
	$S_1 \leq 0.1$	$S_1 = 0.2$	$S_1 = 0.3$	$S_1 = 0.4$	$S_1 \geq 0.5$
A	0.8	0.8	0.8	0.8	0.8
B	1.0	1.0	1.0	1.0	1.0
C	1.7	1.6	1.5	1.4	1.3
D	2.4	2.0	1.8	1.6	1.5
E	3.5	3.2	2.8	2.4	2.4
F	See Section 11.4.7 of ASCE 7-05				

Note: Use straight-line interpolation for intermediate values of S_1.

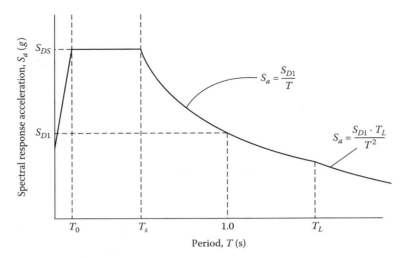

FIGURE 5.3   Design response spectrum. (Courtesy of American Society of Civil Engineers, Reston, VA.)

$$S_{D1} = 2/3S_{M1} \qquad (5.5)$$

DESIGN RESPONSE SPECTRUM

This is a graph that shows the design value of the spectral acceleration for a structure based on the fundamental period. A generic graph is shown in Figure 5.3 from which a site-specific graph is created based on the mapped values of accelerations and the site soil type.

The controlling time steps at which the shape of the design response spectrum graph changes are as follows:

1. Initial period

$$T_0 = 0.2 \frac{S_{D1}}{S_{DS}} \qquad (5.6)$$

2. Short-period transition for small structure

$$T_s = \frac{S_{D1}}{S_{DS}} \qquad (5.7)$$

3. Long-period transition for large structures

T_L = shown in Figure 5.4 which is reproduced from Figure 22-15 of the ASCE 7-05.
   The characteristics of the design response spectrum are as follows:

1. For the fundamental period, t_a having a value between 0 and T_0, the design spectral acceleration, S_a varies as a straight line from a value of $0.4S_{DS}$ and S_{DS}, as shown in Figure 5.3.
2. For the fundamental period, t_a having a value between T_0 and T_s, the design spectral acceleration, S_a is constant at S_{DS}.
3. For the fundamental period, t_a having a value between T_s and T_L, the design spectral acceleration, S_a is given by

$$S_a = \frac{S_{D1}}{T} \tag{5.8}$$

where T is the time period between T_s and T_L.

4. For the fundamental period, t_a having a value larger than T_L, the design spectral acceleration is given by

$$S_a = \frac{S_{D1}T_L}{T^2} \tag{5.9}$$

The complete graph is shown in Figure 5.3.

(a)

FIGURE 5.4   (a) Long-period transition period, T_L (s). (Courtesy of American Society of Civil Engineers, Reston, VA.)

(*continued*)

(b)

FIGURE 5.4 (continued)   (b) Continuation of long-period transition period, T_L (s). (Courtesy of American Society of Civil Engineers, Reston, VA.)

Example 5.2

At a location in California, the mapped values of the MCE accelerations S_s and S_1 are 1.5g and 0.75g, respectively. The site soil class is D. Prepare the design spectral response curve for this location.

Solution

1. Adjustment factors for soil class D are as follows:

$$F_a = 1.0$$

$$F_v = 1.5$$

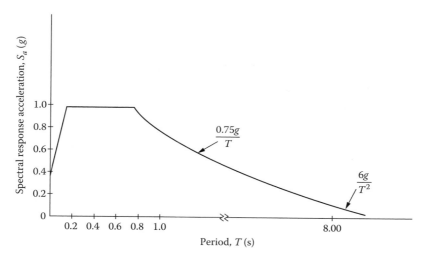

FIGURE 5.5 Design spectral acceleration Example 5.2.

2. $S_{MS} = F_a S_s$

$\quad = (1.0)(1.5g) = 1.5g$

$\quad S_{M1} = F_v S_1$

$\quad = (1.5)(0.75g) = 1.13g$

3. $S_{DS} = 2/3 S_{MS} = 2/3(1.5g) = 1g$

$\quad S_{D1} = 2/3 S_{M1} = 2/3(1.13g) = 0.75g$

4. $T_0 = 0.2 S_{D1}/S_{DS} = 0.2(0.75g)/(1g) = 0.15\,s$

$\quad T_s = 0.75g/1g = 0.75\,s$

$\quad T_L = 8\,s$ (from Figure 5.4)

5. The design spectral acceleration at time 0 is $0.4(1g)$ or $0.4g$. It linearly rises to $1g$ at time 0.15 s. It remains constant at $1g$ up to time 0.75 s. From time 0.75 to 8 s, it drops at a rate $0.75g/T$. At 0.75 s it is $0.75g/0.75 = 1g$ to a value of $0.75g/8 = 0.094g$ at time 8 s. Thereafter, the rate of drop is $S_{D1}T_L/T^2$ or $6g/T^2$. This is shown in Figure 5.5.

IMPORTANCE FACTOR, I

The importance factor, I for seismic coefficient which is based on the occupancy category of the structure is indicated in Table 5.5. The occupancy category is discussed in the "Classification of Buildings" section in Chapter 1).

SEISMIC DESIGN CATEGORIES

A structure is assigned a seismic design category (SDC) from A through F based on the occupancy category of the structure and the design spectral response acceleration parameters, S_{DS} and S_{D1} of the site. The seismic design categories are given in Tables 5.6 and 5.7.

TABLE 5.5
Importance Factor for Seismic Coefficient

Occupancy Category	Importance Factor
I and II	1.0
III	1.25
IV	1.5

TABLE 5.6
SDC Based on S_{DS}

	Occupancy Category		
S_{DS} Range	I or II (Low Risk and Standard Occupancy)	III (High Occupancy)	IV (Essential Occupancy)
0 to<0.167g	A	A	A
0.167g to<0.33g	B	B	C
0.33g to≤0.5g	C	C	D
>0.5g	D	D	D
When $S_1 \geq 0.75g$	E	E	F

TABLE 5.7
SDC Based on S_{D1}

	I or II (Low Risk and Standard Occupancy)	III (High Occupancy)	IV (Essential Occupancy)
Range S_{D1}			
0 to<0.067g	A	A	A
0.067g to<0.133g	B	B	C
0.133g to≤0.20g	C	C	D
>0.2g	D	D	D
When $S_1 \geq 0.75g$	E	E	F

A structure is assigned to the more severe of the category determined from the two tables except for the following cases:

1. When S_1 is 0.75g or more, a structure is assigned the category E for I, II, and III occupancies and it is assigned the category F for occupancy category IV.
2. When S_1 is less than 0.75g and the certain conditions of the small structure are met, as specified in 11.5.2 of the ASCE7-05, only Table 5.6 is applied.

EXEMPTIONS FROM SEISMIC DESIGNS

International Building Code exempts the following structures from seismic designs:

1. The structures belonging to SDC A
2. The detached one and two family dwellings in SDC A, SDC B, and SDC C
3. The conventional wood frame one and two family dwellings up to two stories in any SDC
4. The agriculture storage structures used only for incidental human occupancy
5. The structures located where $S_s \leq 0.15g$ and $S_1 \leq 0.04g$

EQUIVALENT LATERAL FORCE PROCEDURE TO DETERMINE SEISMIC FORCE

The design base shear, V, due to seismic force is expressed as

$$V = C_s W \tag{5.10}$$

where

W is the effective dead weight of structure discussed in the "Effective Weight of Structure, W" section below

C_s is the seismic response coefficient discussed in the "Seismic Response Coefficient" section below

EFFECTIVE WEIGHT OF STRUCTURE, W

Generally this is taken as the dead load of the structure. However, where a structure carries a large live load, a portion is included in W. For a storage warehouse, 25% of floor live load is included with the dead load in W. Where the location of partitions (nonbearing walls) are subject to relocation, a floor live load of 10 psf is added in W. When the flat roof snow load exceeds 30 psf, 20% of the snow load is included in W.

SEISMIC RESPONSE COEFFICIENT, C_s

The value of C_s for different time periods of the design spectrum is shown in Figure 5.6. Besides depending upon the fundamental period and design spectral accelerations, C_s is a function of the importance factor and the response modification factor. The importance factor, I is given in Table 5.5. The response modification factor, R is discussed in the "Response Modification Factor" section below.

RESPONSE MODIFICATION FACTOR, R

The response modification factor accounts for the following:

1. Ductility, which is the capacity to withstand stresses in the inelastic range.
2. Overstrength, which is the difference between the design load and the failure load.
3. Damping, which is the resistance to vibration by the structure.
4. Redundancy, which is an indicator that a component's failure does not lead to failure of the entire system.

A large value of the response modification factor reduces the seismic response coefficient and hence the design shear. The factor ranges from 1 to 8. The ductile structures have a higher value and the brittle ones have a lower value. The braced steel frames with moment-resisting connections

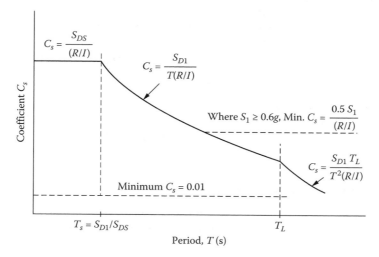

FIGURE 5.6 Seismic response coefficient for base shear.

have the highest value and the concrete and masonry shear walls have the smallest value. For wood-frame construction, the common R-factor is 6.5 for wood and light metal shear walls and 5 for special reinforced concrete shear walls. An exhaustive listing is provided in Table 12.2-1 of the ASCE 7-05.

Example 5.3

The five-story moment-resisting steel building of Example 5.1 is located in California where S_s and S_1 are 1.5g and 0.75g, respectively. The soil class is D. Determine (1) the SDC and (2) the seismic response coefficient, C_s.

Solution

1. From Example 5.1

   $t_a = 0.74\,\text{s}$

2. From Example 5.2

   $S_{DS} = 1g$ and $S_{D1} = 0.75g$

   $T_0 = 0.15\,\text{s}$ and $T_s = 0.75\,\text{s}$

3. To compute the SDC
   a. Occupancy category II
   b. From Table 5.6, for $S_1 \geq 0.75g$ and category II, SDC is E.
4. To compute the seismic coefficient:
   a. Importance factor from Table 5.5, $I = 1$
   b. Response modification factor, $R = 8$
   c. t_a (of 0.74 s) $< T_s$ (of 0.75 s)
   d. From Figure 5.6, for $t_a < T_s$, $C_s = S_{DS}/(R/I)$

   $C_s = 1g/(8/1) = 0.125g$

DISTRIBUTION OF SEISMIC FORCES

The seismic forces are distributed throughout the structure in a reverse order. The shear force at the base of the structure is computed from the base shear Equation 5.10. Then story forces are assigned at the roof and floor levels by distributing the base shear force over the height of the structure.

The primary lateral force resisting system (LFRS) consists of horizontal and vertical elements. In conventional buildings, the horizontal elements consist of roof and floors acting as horizontal diaphragms. The vertical elements consist of studs and end shear walls.

The seismic force distribution for vertical elements (e.g., walls), designated by F_x, is different from the horizontal distribution of forces designed by F_{px} that are applied to design the horizontal elements. It should be understood that both F_x and F_{px} are horizontal forces that are differently distributed at each story level. The forces acting on horizontal elements at different levels are not additive, whereas all of the story forces on vertical elements are considered to be acting concurrently and are additive from top to bottom.

Distribution of Seismic Forces on Vertical Wall Elements

The distribution of horizontal seismic forces acting on the vertical element (wall) is shown in Figure 5.7. The lateral seismic force induced at any level is determined from the following equations:

$$F_x = C_{vx}V \tag{5.11}$$

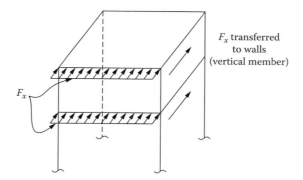

FIGURE 5.7   Distribution of horizontal seismic force to vertical elements.

and

$$C_{vx} = \frac{W_x h_x^k}{\sum W_i h_i^k}$$

(5.12)

Substituting Equation 5.12 into Equation 5.11

$$F_x = \frac{\left(V h_x^k\right) W_x}{\sum W_i h_i^k}$$

(5.13)

where
  i is the index for floor level, $i=1$, first level etc.
  F_x is the seismic force on vertical elements at floor level x
  C_{vx} is the vertical distribution factor
  V is the shear at the base of the structure from Equation 5.10
  W_i or W_x is the effective seismic weight of the structure at index level i or floor level x
  h_i or h_x is the height from base to index level i or floor x
  k is an exponent related to the fundamental period of structure, t_a, as follows: (a) for $t_a \le 0.5$, $k=1$
    and (b) for $t_a > 0.5$, $k=2$

The total shear force, V_x in any story is the sum of F_x from the top story up the x story. The shear force of an x story level, V_x is distributed among the various vertical elements in that story on the basis of the relative stiffness of the elements.

DISTRIBUTION OF SEISMIC FORCES ON HORIZONTAL ELEMENTS (DIAPHRAGMS)

The horizontal seismic forces transferred to the horizontal components (diaphragms) are shown in Figure 5.8. The floor and roof diaphragms are designed to resist the following minimum seismic force at each level:

$$F_{px} = \frac{\sum_{i=x}^{n} F_i W_{px}}{\sum_{i=x}^{n} W_i}$$

(5.14)

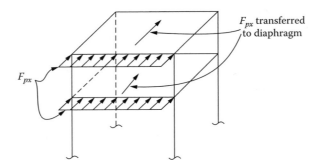

FIGURE 5.8   Distribution of horizontal seismic force to horizontal elements.

where
   F_{px} is the diaphragm design force
   F_i is the lateral force applied to level i which is the summation of F_x from level x (being evaluated) to the top level
   W_{px} is the effective weight of diaphragm at level x. The weight of walls parallel to the direction of F_{px} need not be included in W_{px}
   W_i is the effective weight at level i which is the summation of weight from level x (being evaluated) to the top

The force determined by Equation 5.14 is subject to the following two conditions:
The force should not be more than

$$F_{px}(\text{max}) = 0.4 S_{DS} I W_{px} \qquad (5.15)$$

The force should not be less than

$$F_{px}(\text{min}) = 0.2 S_{DS} I W_{px} \qquad (5.16)$$

DESIGN EARTHQUAKE LOAD

An earthquake causes the horizontal accelerations as well as the vertical accelerations. Accordingly, the earthquake load has two components. In load combinations, it appears in the following two forms:

$$E = E_{\text{horizontal}} + E_{\text{vertical}} \quad \text{(in Equation 1.25)} \qquad (5.17)$$

and

$$E = E_{\text{horizontal}} - E_{\text{vertical}} \quad \text{(in Equation 1.27)} \qquad (5.18)$$

when

$$E_{\text{horizontal}} = \rho Q_E \qquad (5.19)$$

and

$$E_{\text{vertical}} = 0.2S_{DS}D \qquad (5.20)$$

where
   Q_E is the horizontal seismic forces F_p or F_{px} as determined in the "Distribution of Seismic Forces" section
   D is the dead load W in the "Distribution of Seismic Forces" section
   ρ is the redundancy factor

The redundancy factor ρ is 1.00 for seismic design categories A, B, and C. It is 1.3 for SDC D, SDC E, and SDC F, except for special conditions. Redundancy factor is always 1.0 for F_{px} forces.
   $E_{\text{horizontal}}$ is combined with horizontal forces and E_{vertical} with vertical forces.
   The seismic forces are at the LRFD (strength) level as have a load factor of 1. To be combined for the allowable stress design (ASD), these should be multiplied by a factor of 0.7.

Example 5.4

A two-story wood-frame essential facility as shown in Figure 5.9 is located in Seattle, Washington. The structure is a bearing wall system with reinforced shear walls. The loads on the structures are as follows. Determine the earthquake loads acting on the structure.

   Roof dead load = 20 psf (in horizontal plane)
   Floor dead load = 15 psf
   Partition live load = 15 psf
   Exterior wall load = 60 psf

Solution

   A. Design parameters:
      1. Occupancy category = Essential, IV
      2. Importance factor from Table 5.5 for IV category = 1.5
      3. Mapped MCE response accelerations
         (ASCE 7-05 or USDS CD) $S_s = 1g$ or $S_1 = 0.4g$
      4. Site soil class (default) = D
      5. Seismic force resisting system
         Bearing wall with reinforced shear walls
      6. Response modification coefficient = 5

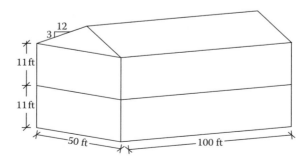

FIGURE 5.9   A two-story wood frame structure.

B. Seismic response parameters
1. Fundamental period $t_a = C_t h^x$. From Table 5.1, $C_t = 0.02$, $x = 0.75$.

$$t_a = 0.02(22)^{0.75} = 0.20 \text{ s}$$

2. From Table 5.3, $F_a = 1.1$

$$S_{MS} = F_a S_s = 1.1(1) = 1.1g$$

3. From Table 5.4, $F_v = 1.6$

$$S_{M1} = F_v S_1 = 1.6(0.4g) = 0.64\,g$$

4. $S_{DS} = 2/3\, S_{MS} = 2/3(1.1g) = 0.73\,g$

$$S_{D1} = 2/3\, S_{M1} = 2/3(0.64) = 0.43\,g$$

5. Based on occupancy and S_{DS}, SDC = D.
    Based on occupancy and S_{D1}, SDC = D.
6. $T_s = S_{D1}/S_{DS} = 0.43g/0.73\,g = 0.59$ s.
    Since $t_a < T_s$, $C_s = S_{DS}/(R/I) = 0.73\,g/(5/1.5) = 0.22\,g$*

C. Effective seismic weight at each level
1. W at roof level[†]
    (1) Area(roof DL) = $(50 \times 100)(20)/1000 = 100$ k
    (2) 2-longitudinal walls = 2(wall area)(wall DL)

    $$= 2(100 \times 11)(60)/1000 = 132\text{k}$$

    (3) 2-end walls = 2(wall area)(DL)

    $$= 2(50 \times 11)(60)/1000 = 66\text{k}$$

    Total = 298 k
2. W at second floor[‡]
    (1) Area(floor DL + partition load[§])
        = $(50 \times 100)(15 + 10)/1000 = 125$ k
    (2) 2-long walls = 132 k
    (3) 2-end walls = 323 k
        Total = 323 k
        Total effective building weight $W = 621$ k
D. Base shear

$$V = C_s W = 0.22(621) = 136.6\text{k}$$

E. Lateral seismic force distribution on the vertical shear walls
1. From Equation 5.13, since $t_a < 0.5$ s, $k = 1$

$$F_x = \frac{(Vh_x)W_x}{\sum W_i h_i}$$

2. The computations are arranged in Table 5.8.

---

* This is for the mass of structure. For weight the value is 0.22.
† It is also a practice to assign at the roof level one-half the second floor wall height.
‡ It is also a practice to assign at the second floor level, the wall load from one-half of the second floor wall and one-half of the first floor wall. This leaves the weight of one-half of the first floor wall not included in the effective weight.
§ ASCE-05 prescribes 15 psf for partition live load but it recommends that for seismic load computation the partition load should be taken as 10 psf.

F. Earthquake loads for the vertical members
   1. The redundancy factor ρ for SDC D is 1.3.
   2. The horizontal and vertical components of the earthquake loads for vertical members (walls) are given in Table 5.9.
   3. The earthquake forces are shown in Figure 5.10.

Example 5.5

For Example 5.4, determine the earthquake loads acting on the horizontal members (diaphragms)

A. Lateral seismic force distribution on the horizontal members
   1. From Equation 5.14

$$F_{px} = \frac{\left(\sum_{i=x}^{n} F_i\right) W_{px}}{\sum_{i=x}^{n} W_i}$$

   2. The computations are arranged in Table 5.10.

TABLE 5.8
Seismic Force Distribution on Vertical Members

Level, x	W_x, k	h_x, ft	$W_x h_x^a$, k	Vh_x or $136.6 h_x^b$, k ft	F_x^c, k	V_x^d (Shear at Story), k
(1)	(2)	(3)	(4)	(5)	(6)	(7)
Roof	298	22	6,556	3005.2	88.60	88.60
Second	323	11	3,553	1502.6	48.00	136.60
Σ	621		10,109			

a Column 2×column 3.
b 136.6×column 3.
c Column 2×column 5/summation of column 4.
d Cumulate column 6.

TABLE 5.9
Earthquake Loads on Vertical Elements

Level, x	W_x, k	F_x, k	$E_{horizontal} = \rho F_x$, k	$E_{vertical} = 0.2 S_{DS} W_x$, k
Roof	298	88.6	115.2	43.5
Second	323	48.0	62.4	47.2

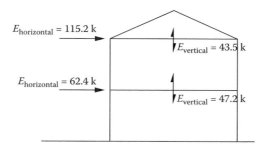

FIGURE 5.10 Earthquake loads on vertical elements—Example 5.4.

TABLE 5.10
Seismic Force Distribution on Horizontal Members

Level, x	W_x, k	W_{px}^a, k	F_x from Table 5.8, k	ΣF_i^b, k	ΣW_i^c, k	F_{px}^d, k	Max.[e] $0.4 S_{DS} I W_{px}$, k	Min. $0.2 S_{DS} I W_{px}$, k
(1)	(2)	(3)	(4)	(5)	(6)	(7)	(8)	(9)
Roof	298	232	88.60	88.60	298	89.09	101.62	58.8
Second	323	257	48.00	136.60	621	56.5[f]	112.57	56.3

[a] W_x – parallel exterior walls weight $= 298 - 66 = 232$ k.
[b] Summation of column 4.
[c] Summation of column 2.
[d] Column 3 × column 5/column 6.
[e] $0.4 S_{DS} I W_{px} = 0.4(0.73)(1.5)(232) = 101.62$ k.
[f] Since 56.5 k is less than 112.57 k and more than 56.3 k, it is OK.

TABLE 5.11
Earthquake Loads on Horizontal Elements

Level, x	W_{px}, k	F_{px}, k	$E_{horizontal} = \rho F_{px}$, k	$E_{vertical} = 0.2 S_{DS} W_{px}$, k
Roof	232	69.0	69.0	33.81
Second	257	56.5	56.5	37.52

B. Earthquake loads for vertical members
   1. The redundancy factor ρ for F_{px} is always 1.0
   2. The horizontal and vertical components of the earthquake loads for horizontal members (diaphragms) are given in Table 5.11.
   3. The earthquake forces on the horizontal members are shown in Figure 5.11.

PROBLEMS

5.1 Determine the approximate fundamental period for a five-story concrete office building with each floor having a height of 12 ft.

5.2 Determine the approximate fundamental period for a three-story wood framed structure having a total height of 25 ft.

5.3 At a location in California, the mapped values of MCE accelerations S_s and S_1 are 1.4g and 0.7g, respectively. The site soil class is C. The long-period transition period is 8 s. Prepare the design response acceleration curve for this location.

5.4 In Salt Lake City, UT, the mapped values of S_s and S are 1.8g and 0.75g. The site soil class is B. The long-period transition period is 6 s. Prepare the design response acceleration curve.

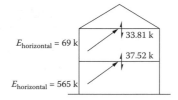

FIGURE 5.11 Earthquake loads on horizontal elements—Example 5.4.

$E_{horizontal} = 69$ k
33.81 k
37.52 k
$E_{horizontal} = 565$ k

5.5 For a five-story concrete office building with each floor of 12 ft height of Problem 5.1 located in California where S_s and S_1 are 1.4g and 0.7g respectively, and the site soil class is C, determine (1) the SDC, (2) the seismic response coefficient. Assume $R = 2.0$.

5.6 For the three-story wood framed building of total height 25 ft of Problem 5.2, located in Salt Lake City where S_s and S_1 are 1.8g and 0.75g, respectively, and the soil group is B, determine (1) the SDC and (2) the seismic response coefficient. Assume $R = 6.5$.

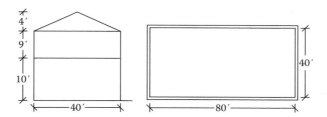

FIGURE P5.7 An office building in Portland, OR.

5.7 A two-story office building, as shown in Figure P5.7, is located in Portland, OR. The building has the plywood floor system and plywood sheathed shear walls ($R=6.5$). The soil in foundation is very dense. The loads on the building are as follows:

Roof dead load (on horizontal plane) = 20 psf
Floor load = 15 psf
Partition load = 15 psf
Exterior wall load = 50 psf

Determine the lateral and vertical earthquake loads that will act on the vertical elements and horizontal elements of the building.

5.8 A three-story industrial steel building (Figure P5.8) located where S_s and S_1 from figures in the ASCE7-05 are $0.61g$ and $0.18g$, respectively, has the plan dimension of $200\,\text{ft} \times 90\,\text{ft}$.

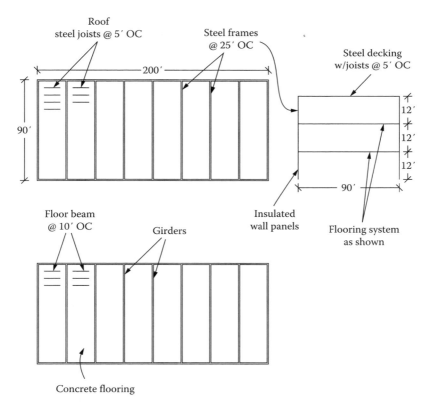

FIGURE P5.8 An industrial steel building.

The structure consists of nine gable moment-resisting steel frames spanning 90 ft at 25 ft on center; $R = 4.5$. The building is enclosed by insulated wall panels and is roofed with steel decking. The building is 36 ft high with each floor height of 12 ft. The building is supported on spread roofing on medium dense sand (soil class D).

The steel roof deck is supported by joists at 5 ft on center, between the main gable frames. The flooring consists of the concrete slab over steel decking, supported by floor beams at 10 ft on center. The floor beams rest on girders that are attached to the gable frames at each end.

The following loads have been determined in the building:

Roof dead load (horizontal plane) = 15 psf
Third floor storage live load = 120 psf
Slab and deck load on each floor = 40 psf
Weight of each framing = 10 K
Weight of non-shear-resisting wall panels = 10 psf
Include 25% of the storage live load for seismic force
Since the wall panels are non-shear resisting, these are not to be subtracted for F_{px}

Part II

Wood Structures

6 Wood Specifications

ENGINEERING PROPERTIES OF SAWN LUMBER

The National Design Specifications (NDS) for wood construction of the American Forest and Paper Association (AF&PA 2005 edition) provide the basic standards and specifications for the sawn lumber and the engineered wood (e.g., glued laminated timber) in the United States. The second part of the NDS, referred to as the NDS supplement, contains the numerical values for the strength of different varieties of wood grouped according to the species of trees. The pieces of wood sawn from the same species or even from the same source show a great variation in engineering properties. Accordingly, the lumber is graded to establish the strength values. The pieces of lumber having the similar mechanical properties are placed in the same class known as the *grade* of the wood. Most lumber is visually graded. However, a small percentage is graded mechanically. In each grade, the relative size of the wood section and the suitability of that size for a structural application are used as additional guides to establish the strength.

A lumber is referred to by the nominal size. However, the lumber used in construction is mostly a dressed lumber. In other words, the lumber is surfaced to a net size, which is taken to be 0.5 in. less than the nominal size. In the case of large sections, sometimes the lumber is rough sawed. The rough sawed dimensions are approximately 1/8 in. larger than the dressed size. The sectional properties of the standard dressed sawn lumber are given in Table B.1.

The sawed lumber is classified according to the size into (1) the dimension lumber and (2) the timber. The dimension lumber has smaller sizes. It has a nominal thickness of 2–4 in. and a width* of 2–16 in. Thus, the sizes of the dimension lumber range from 2 in. × 2 in. to 4 in. × 16 in. Timber has a minimum nominal thickness of 5 in.

The dimension lumber and the timber are further subdivided based on the suitability of the specific size for use as a structural member. The size and use categorization of the commercial lumber is given in Table 6.1.

REFERENCE DESIGN VALUES FOR SAWN LUMBER

The numerical values of the permissible level of stresses for design with respect to bending, tension, compression, shear, modulus of elasticity, and modulus of stability of a specific lumber are known as the *reference design values*. These values are arranged according to the species. Under each species, the size and use categories as listed in Table 6.1 are arranged. For each size and use category, the reference design values are listed for different grades of lumber. Thus, the design value may be different for the same grade name but in a different size category. For example, the select structural grade appears in SLP, SJ & P, B & S, and P & T categories and the design values for a given species are different for the select structural grade in all of these categories.

The following reference design values are provided in tables:

Table B.2: Reference design values for dimension lumber other than Southern Pine
Table B.3: Reference design values for Southern Pine dimension lumber
Table B.4: Reference design values for timber

---

* In the terminology of the lumber grading, the smaller cross-sectional dimension is thickness and the larger dimension is width. In the designation of engineering design, the dimension parallel to the neutral axis of a section as placed, is width and the dimension perpendicular to the neutral axis is depth. Thus, a member loaded about the strong axis (placed with the smaller dimension parallel to the neutral axis) has the "width" what is referred to as the "thickness" in the lumber terminology.

TABLE 6.1
Categories of Lumber and Timber

Name	Symbol	Nominal Dimension	
		Thickness (Smaller Dimension)	Width
A. Dimension lumber			
1. Light framing	LF	2–4 in.	2– in.
2. Structural light framing	SLF	2–4 in.	2–4 in.
3. Structural joist and plank	SJ & P	2–4 in.	5 in. or more
4. Stud		2–4 in.	2 in. or more
5. Decking		2–4 in.	4 in. or more
B. Timber			
1. Beam and stringer	B & S	5 in. or more	At least 2 in. more than thickness
2. Post and timber	P & T	5 in. or more	Not more than 2 in. than thickness

Although the reference design values are given according to the size and use combination, the values depend on the size of the member rather than its use. Thus, a section 6×8 listed under the post and timber (P & T) with its reference design values indicated therein can be used for beam and stringer (B & S) but its design values as indicated for P & T will apply.

ADJUSTMENTS TO THE REFERENCE DESIGN VALUES FOR SAWN LUMBER

The reference design values in the NDS tables are the basic values that are multiplied by many factors to obtain the adjusted design values. To distinguish an adjusted value from a reference value, a prime notation is added to the symbol of the reference value to indicate that the necessary adjustments have been made. Thus,

$$F'_{(\,)} = F_{(\,)} \times \left(\text{products of adjustment factors}\right) \tag{6.1}$$

The ( ) is replaced by a property like tensile, compression, and bending.

For wood structures, the allowable stress design (ASD) is a traditional basis of design. The LRFD provisions have been introduced in 2005. The reference design values given in the NDS are based on the ASD (i.e., these are permissible stresses). The reference design values for LRFD have to be converted from the ASD values.

To determine the nominal design stresses for LRFD, the reference design values of the NDS tables, as reproduced in the appendixes, are required to be multiplied by a format conversion factor, K_F. The format conversion factor serves a purpose reverse of the factor of safety to obtain the nominal strength values for LRFD application. In addition, the format conversion factor includes the effect of the load duration. It adjusts the reference design values of the normal (10 years) duration to the nominal strength values for a short duration (10 min) which have a better reliability.

In addition to the format conversion factor, a resistance factor, ϕ is applied to obtain the LRFD adjusted values. A subscript of n is added to recognize that it is a nominal (strength) value for the LRFD design. Thus, the adjusted nominal design stress is expressed as

$$F'_{(\,)n} = \phi F'_{(\,)} K_F \tag{6.2}$$

TABLE 6.2
Time Effect Factor

Load Combination	λ
$1.4D$	0.6
$1.2D+0.5$ (L_r or S)	0.6
$1.2D+1.6L+0.5$ (L_r or S)	0.7 when L is from storage
	0.8 when L is from occupancy
	1.25 when L is from impact
$1.2D+1.6$ (L_r or S) + (fL or $0.8W$)	0.8
$1.2D+1.6W+fL+0.5$ (L_r or S)	1.0
$1.2D+1.0E+fL+0.2S$	1.0
$0.9D+1.6W$	1.0
$0.9D+1.0E$	1.0

The adjustment factors are discussed below:

1. The wet-service factor is applied when the wood in a structure is not in a dry condition; its moisture content exceeds 19% (16% in the case of laminated lumber). Most structures use dry lumber for which $C_M=1$.
2. The reference design values are for the bending about the major axis, i.e., the load is applied on to the narrow face. The flat use factor refers to members that are loaded about the weak axis, i.e., the load is applied on the wider face. The reference value is increased by a factor, C_{fu} in such cases. Conservatively, $C_{fu}=1$ can be used.
3. The temperature factor is used if a prolonged exposure to higher than normal temperature is experienced by a structure. The normal condition covers the ordinary winter to summer temperature variations and the occasional heating up to 150°F. For normal conditions, $C_t=1$.
4. Some species of wood do not accept the pressure treatment easily and require incisions to make the treatment effective. An incision factor of 0.8 is applied for dimension lumber only and that too in a limited species that are subjected to the incision. The factor is applied to bending, tension, shear, and compression parallel to grains.
5. In addition, there are some special factors like the column stability factor, C_P and the beam stability factor, C_L that are discussed in the context of column and beam designs in Chapter 7.

The following are the other adjustment factors that are applied frequently.

TIME EFFECT FACTOR,* λ

Wood has a unique property that it can support a higher load when applied for a short duration. The nominal reference design values are representative of the short duration loading. For loading of long duration, the reference design value has to be reduced by a time effect factor. The different types of loads represent different load durations. Accordingly, the time effect factor depends on the combination of the loads. For various load combinations, the time effect factor is given in Table 6.2. It should be remembered that the factor is applied to the nominal reference (stress) value and not to the load.

---

* The time effect factor is relevant only to the LRFD. For ASD, this factor known as the *load duration factor*, C_D, has different values.

Size Factor, C_F

The size of a wood section has an effect on its strength. The factor for size is handled differently for the dimension lumber and for the timber.

Size Factor, C_F for Dimension Lumber

For visually graded dimension lumber, the size factors for species other than Southern Pine are given in Tables 4A and 4F of the NDS. These tables are presented together with the reference design values in Table B.2. For visually graded Southern Pine dimension lumber, the factors are generally built into the design values except for the bending values for 4 in. thick (breadth) dimension lumber. The factors for Southern Pine dimension lumber are given together with the reference design values in Table B.3. No size factor adjustment is required for the mechanically graded lumber.

Size Factor, C_F for Timber

For timber sections exceeding 12 in. depth, a reduction factor is applied only to bending as follows:

$$C_F = \left(\frac{12}{d}\right)^{1/9} \tag{6.3}$$

where d is the dressed depth of the section.
    This factor is not included in the reference design values in Table B.4.

Repetitive Member Factor, C_r

This factor is applied only to the dimension lumber and that only to the bending strength value. A repetitive member factor $C_r = 1.15$ is applied when all of the following three conditions are met:

1. The members are used as joists, truss chords, rafters, studs, planks, decking, or similar members that are joined by floor, roof, or other load distributing elements
2. The members are in contact or are spaced not more than 24 in. on center
3. The members are not less than 3 in number

Format Conversion Factor, K_F

The format conversion factors for the different types of stresses are reproduced in Table 6.3 from Table N1 of the NDS.

TABLE 6.3
Conversion Factors for Stresses

Application	Property	K_F
Member	Bending, tension, shear, compression parallel to grain	$2.16/\phi$
	Compression perpendicular to grain	$1.875/\phi$
	E_{min}	$1.5/\phi$
	E	1
Connection	All connections	$2.16/\phi$

Source: Courtesy of American Forest & Paper Association, Washington, DC.

Resistance Factor, ϕ

The resistance factor, also referred to as the *strength reduction factor*, is used to account for all uncertainties whether related to the materials manufacturing, structural construction, or design computations that may cause actual values to be less than the theoretical values. The resistance factor, given in Table 6.4, is a function of the mode of failure.

The applicable factors for different loading and types of lumber are summarized in Table 6.5.

TABLE 6.4
Resistance Factor

Property	ϕ
Bending	0.85
Tension	0.80
Shear	0.75
Compression (parallel, perpendicular)	0.90
Stability, E_{min}	0.85

LRFD DESIGN WITH WOOD

As discussed in the "Working Stress, Strength Design, and Unified Design of Structures" section in Chapter 1, the LRFD designs are performed at the strength level in terms of the force and moment. Accordingly, the adjusted nominal design stress values from the "Reference Design Values for Sawn Lumber" section are changed to the strength values by multiplying by the cross-section area or the section modulus. Thus, the basis of design in LRFD is as follows:

$$\text{Bending: } M_u = \phi M_n = F'_{bn}S = \phi F_b \lambda C_M C_t C_F C_r C_{fu} C_i (C_L) K_F S \tag{6.4}$$

$$\text{Tension: } T_u = \phi T_n = \phi F_t \lambda C_M C_t C_F C_i K_F A \tag{6.5}$$

$$\text{Compression: } P_u = \phi P_n = \phi F_c \lambda C_M C_t C_F C_i (C_P) K_F A \tag{6.6}$$

$$P_{u\perp} = \phi P_n = \phi F_{c\perp} \lambda C_M C_t C_i K_F A \tag{6.7}$$

$$\text{Shear: } V_u = \phi V_n = \phi F_v \lambda C_M C_t C_i K_F (2/3A^*) \tag{6.8}$$

$$\text{Stability: } E_{min(n)} := \phi E_{min} C_M C_t C_i K_F \tag{6.9}$$

$$\text{Modulus of elasticity: } E_{(n)} = E C_M C_t C_i \tag{6.10}$$

The LHS of the above equations are the factored loads combination and the factored moments combination. In Table 6.3, the format conversion factors contain ϕ in the denominator, which is cancelled with the multiplication by factor ϕ in Equations 6.4 through 6.10. Hence, the factor ϕ does not appear in wood design since it is a built in component of the format conversion factor.[*]

---

[*] $\tau_{max} = \dfrac{3V}{2A}$ or $V = \tau_{max}(2/3A)$.

TABLE 6.5
Applicability of Adjustment Factors for Sawn Lumber

Loading Condition	Type of Lumber	Time Effect	Size Effect	Repetitive	Wet Service	Temperature	Incision	Flat Use	Beam Stability	Column Stability
					Factor				**Special Factor**	
Bending	Dimension lumber: visually graded	λ	C_F	C_r	C_M	C_t	C_i	C_{fu}	C_L	
	Dimension lumber: mechanically graded	λ		C_r	C_M	C_t	C_i		C_L	
	Timber	λ	C_F		C_M	C_t			C_L	
	Decking	λ			C_M	C_t			C_L	
Tension	Dimension lumber	λ	C_F		C_M	C_t	C_i			
Compression—parallel to grain	Dimension lumber	λ	C_F		C_M	C_t	C_i			C_P
	Timber	λ			C_M	C_t	C_i			C_P
Compression—normal to grain	Both	λ			C_M	C_t	C_i			
Shear parallel to grain	Both	λ			C_M	C_t				
Modulus of elasticity	Both				C_M	C_t				
Modulus of elasticity for stability	Both				C_M	C_t				

Example 6.1

Determine the adjusted nominal reference design values and the nominal strength capacities of the Douglas Fir-Larch #1 2 in. × 8 in. roof rafters at 18 in. on center that support the dead and roof live loads. Consider the dry-service conditions, normal temperature range, and no-incision application.

Solution

1. The reference design values of Douglas Fir-Larch #1 2 in. × 8 in. section from Appendix B.2
2. The adjustment factors are given below and the adjusted nominal reference design values are computed in the following table:

Property	Reference Design Value (psi)	Adjustment Factors				$F'_{()n}$ (psi)
		λ for $D+L_r$	C_F	C_r	$\dfrac{(K_F)}{\phi} \times \phi$	
Bending	1000	0.6	1.2	1.15	2.16	1788.48
Tension	675	0.6	1.2		2.16	1049.76
Shear	180	0.6			2.16	233.28
Compression	1500	0.6	1.05		2.16	2041.2
Compression$^\perp$	625	0.6			1.875	703.13
E	1.7×10^6					1.7×10^6
E_{min}	0.62×10^6				1.5	0.93×10^6

3. Strength capacities
   For section 2 in. × 8 in. section, $S = 13.14$ in.^{3}, $A = 10.88$ in.^{2}
   $M_u = F'_{bn}S = (1788.48)(13.14) = 23500.6$ lb
   $T_u = F'_{tn}A = (1049.76)(10.88) = 11{,}421$ lb
   $V_u = F'_{vn}(2/3A) = (233.28)(2/3 \times 10.88) = 1692.1$ lb
   $P_u = F'_{cn}A = (2041.2)(10.88) = 22{,}208$ lb

Example 6.2

Determine the adjusted nominal reference design values and the nominal strength capacities of a Douglas Fir-Larch #1 6 in. × 16 in. floor beam supporting a combination of load comprising of dead, live, and snow loads. Consider the dry-service conditions, normal temperature range, and no-incision application.

Solution

1. The reference design values of Douglas Fir-Larch #1 6 in. × 16 in. beams and stringers from Table B.4
2. The adjustment factors and the adjusted nominal reference design values are given in the table below:

		Adjustment Factors				
	Reference Design Value (psi)	λ for D, L, S	$C_F{}^a$	$\dfrac{(K_F)}{\phi} \times \phi$	F'_{0n} (psi)	
Property						
Bending	1350	0.8	0.97	2.16	2262.8	
Tension	675	0.8		2.16	1166.4	
Shear	170	0.8		2.16	293.8	
Compression	925	0.8		2.16	1598.4	
E	1.6×10^6				1.6×10^6	
E_{min}	0.58×10^6			1.5	0.87×10^6	

a $\quad C_F = \left(\dfrac{12}{d}\right)^{1/9} = \left(\dfrac{12}{15.5}\right)^{1/9} = 0.97.$

3. Strength capacities
   For section 6 in. × 16 in., $S = 220.2$ in.^{3}, $A = 85.25$ in.^{2}
   $M_u = F'_{bn}S = (2262.8)(220.2) = 498,270$ lb
   $T_u = F'_{tn}A = (1166.4)(85.25) = 99,436$ lb
   $V_u = F'_{vn}(2/3A) = (293.8)(2/3 \times 85.25) = 16,700$ lb
   $P_u = F'_{cn}A = (1598.4)(85.25) = 136,264$ lb

Example 6.3

Determine the unit load (per square foot load) that can be imposed on a floor system consisting of 2 in. × 6 in. Southern Pine select structural joists spaced at 24 in. on center spanning 12 ft. Assume that the dead load is one-half of the live load. Ignore the beam stability factor.

Solution

1. For Southern Pine 2 in. × 6 in. structural dressed lumber, the reference design value

   $F_b = 2550$ psi

2. Size factor is included in the tabular value
3. Time effect factor for dead and live loads $= 0.8$
4. Repetitive factor $= 1.15$
5. Format conversion factor $= 2.16/\phi$
6. Nominal reference design value

   $F'_{bn} = \phi F_b \lambda C_M C_t C_r C_F C_{fu} C_i K_F$

   $= \phi(2550)(0.8)(1.15)(2.16/\phi) = 5067.36$ psi

7. For 2 in. × 6 in. $S = 7.56$ in.^{3}
8. $M_u = F'_{bn}S = (5067.36)(7.56) = 38309.24$ in-lb or 3192.44 ft-lb

9. $M_u = \dfrac{w_u l^2}{8}$

or

$$w_u = \frac{8M_u}{L^2} = \frac{8(3192.44)}{(12)^2} = 177.36 \text{ lb/ft}$$

10. Tributary area/ft of joists $= \dfrac{24}{12} \times 1 = 2\,\text{ft}^2/\text{ft}$

11. $w_u =$ (design load/ft^2)(tributary area/ft)

$$177.36 = \left(1.2D + 1.6L\right)(2)$$

or

$$177.36 = \left[1.2D + 1.6\left(2D\right)\right](2)$$

or

$$D = 20.15 \text{ lb/ft}^2$$

and

$$L = 40.3 \text{ lb/ft}^2$$

Example 6.4

For a Southern Pine No. 1 floor system, determine the size of the joists at 18 in. on center, spanning 12 ft and the column receiving loads form an area of 100 ft^2 acted upon by a dead load of 30 psf and a live load of 40 psf. Assume that the beam and column stability factors are not a concern.

Solution

A. Joist design
   1. Factored unit combined load $= 1.2(30) + 1.6(40) = 100\,\text{psf}$
   2. Tributary area/ft $= (18/12) \times 1 = 1.5\,\text{ft}^2/\text{ft}$
   3. Design load/ft $w_u = 100(1.5) = 150\,\text{lb/ft}$
   4. $M_u = \dfrac{w_u L^2}{8} = \dfrac{(150)(12)^2}{8} = 2{,}700$ ft-lb or 32,400 in.-lb
   5. For a trial section, select the reference design value of 2–4 in. wide section and assume the nominal reference design value to be one-and-half times of the table value.
      From Table B.3, for Southern Pine No. 1, $F_b = 1850\,\text{psi}$
      Nominal reference design value $= 1.5(1850) = 2775\,\text{psi}$
   6. Trial size

$$S = \frac{M_u}{F'_{bn}} = \frac{32{,}400}{2{,}775} = 11.68 \text{ in.}^3$$

   Use 2 in. × 8 in. $S = 13.14\,\text{in.}^3$
   7. From Table B.3, $F_b = 1500\,\text{psi}$

8. Adjustment factors

   $\lambda = 0.8$

   $C_r = 1.15$

   $K_F = 2.16/\phi$

9. Adjusted nominal reference design value

   $$F'_{bn} = \phi(1500)(0.8)(1.15)(2.16/\phi) = 2980.8 \text{ psi}$$

10. $M_u = F'_{bn}S$

    or

    $$S_{reqd} = \frac{M_u}{F'_{bn}} = \frac{32,400}{2980.8} = 10.87 < 13.14 \text{ in.}^3$$

    Selected size 2 in. × 8 in. is **OK**.

B. Column design
   1. Factored unit load (Step A.1 above) = 100 psf
   2. Design load = (unit load)(tributary area)
      $$= (100)(100) = 10,000 \text{ lb}$$
   3. For a trial section, select the reference design value of 2–4 in. wide section and assume the nominal reference design value to be one-and-half times of the table value
      From Table B.3, for Southern Pine No. 1, $F_c = 1850$ psi
      Nominal reference design value = 1.5(1850) = 2775 psi
   4. Trial size

      $$A = \frac{P_u}{F'_{cn}} = \frac{10,000}{2,775} = 3.6 \text{ in.}^2$$

      Use 2 in. × 4 in. $A = 5.25$ in.^{2}
   5. $F_b = 1850$ psi
      $\lambda = 0.8$
      $K_F = 2.16/\phi$
   6. Adjusted nominal reference design value

      $$F'_{cn} = \phi(1850)(0.8)(2.16/\phi) = 3196.8 \text{ psi}$$

      $$A_{reqd} = \frac{P_u}{F'_{cn}} = \frac{10,000}{3196.8} = 3.13 < 5.25 \text{ in.}^2$$

      Selected size 2 in. × 4 in. is **OK**.

STRUCTURAL GLUED LAMINATED TIMBER

Glued laminated timber (GLULAM) members are composed of individual pieces of dimension lumber that are bonded together by an adhesive to create the required size. For Western species, the common width* (breadth) are 3-⅛, 5-⅛, 6-¾, 8-¾, 10-¾, and 12-½ in (there are other interim sections as well). The laminations are typically in 1½ in. incremental depth. For Southern Pine, the common widths are 3, 5, 6-¾, 8-½ and 10-½ in. and the depth of each lamination is 1-⅜ in. Usually the lamination of GLULAM is horizontal (the wide faces are horizontally oriented). A typical cross section is shown in Figure 6.1.

The sectional properties of Western species structural glued laminated timber are given in Appendix B.5 and for Southern Pine structural glued laminated timber in Appendix B.6.

Because of their composition, large GLULAM members can be manufactured from smaller trees from a variety of species such as Douglas Fir, Hem Fir, and Southern Pine. GLULAM has much greater strength and stiffness than sawn lumber.

REFERENCE DESIGN VALUES FOR GLULAM

The reference design values for GLULAM is given in Appendix B.7 for members stressed primarily in bending (beams) and that in Appendix B.8 for members stressed primarily in axial tension or compression.

Appendix B.7 relating to bending members is a summary table based on the stress class. The first part of the stress class symbol refers to the bending stress value for the grade in hundreds of psi followed by the letter F. For example, 24F indicates a bending stress of 2400 psi for normal duration loaded in the normal manner, i.e., loads are applied perpendicular to the wide face of lamination. The second part of the symbol is the modulus of elasticity in millions of psi. Thus, 24F-1.8E indicates a class with the bending stress in 2400 psi and the modulus elasticity of 1.8×10^6 psi. For each class, the NDS provides the expanded tables that are according to the combination symbol and the types of species making up the GLULAM. The first part of the combination symbol is the bending stress level, i.e., 24F referring to 2400 psi bending stress. The second part of the symbol refers to the lamination stock; V standing for visually graded and E for mechanically graded or E-rated. Thus, the combination symbol 24F-V5 refers to the grade of 2400 psi bending stress of visually graded

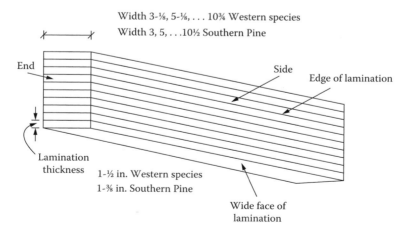

FIGURE 6.1   Illustration of a structural glued laminated timber section.

---

* Not in terms of the lumber grading terminology.

lumber stock. Under this, the species are indicated by abbreviations, i.e., DF for Douglas Fir, SP for Southern Pine, and HF for Hem Fir.

The values listed in Appendix B.7 are more complex than sawn lumber. The first six columns are the values for bending about strong ($X–X$) axis when the loads are perpendicular to wide face of lamination. These are followed up values for bending about $Y–Y$ axis. The axially loaded values are also listed in case the member is picked up for the axial load conditions.

For F_{bx}, two values have been listed in column 1 and 2 of Appendix B.7 (for bending) as $F_{bx}+$ and $F_{bx}-$. In a rectangular section, the compression and tension stresses are equal in extreme fibers. However, it has been noticed that the outer tension laminations are in a critical state and therefore high-grade laminations are placed at the bottom of the beam which is recognized as the tensile zone of the beam. The other side is marked as "top" of the beam in the lamination plant. Placed in this manner, the top marked portion is subjected to compression and the bottom to tension. This is considered as the condition in which "tension zone is stressed in tension" and the $F_{bx}+$ value of first column is used for bending stress. This is a common condition.

However, if the beam is installed upside down or in the case of a continuous beam when the negative bending moment condition develops, i.e., the top fibers are subject to tension, the reference values in the second column known as the "compression zone stressed in tension" $F_{bx}-$ should be used.

Appendix B.8 lists the reference design values for principally axially load-carrying members. Here members are identified by the numbers such as 1, 2, 3 followed by the species such as DF, HF, SP, and by the grade. The values are not complex like Appendix B.7 (the bending case).

It is expected that the members with bending combination in Appendix B.7 will be used as beam as they make efficient beams. However, it does not mean that they cannot be used for axial loading. Similarly, an axial combination member can be used for a beam. The values with respect to all types of loading modes are covered in both tables.

ADJUSTMENT FACTORS FOR GLULAM

The reference design values of Appendices B.7 and B.8 are applied by the same format conversion factors and the time effect factors, as discussed in the "Adjustments to the Reference Design Values Sawn Lumber" section.

Additionally, the other adjustment factors listed in Table 6.6 are applied to the structural GLULAM.

For GLULAM, when moisture content is more than 16% (as against 19% for sawn lumber), the wet-service factor, has been specified in a table in the NDS. The values are different for sawn lumber and GLULAM. The temperature factor is the same for GLULAM as for sawn lumber.

The other factors that are typical to GLULAM are described below:

FLAT USE FACTOR FOR GLULAM, C_{FU}

The flat use factor is applied to the reference design value only for the case bending that is loaded parallel to the laminations. The factor is

$$C_{fu} = \left(\frac{12}{d}\right)^{1/9} \tag{6.11}$$

where d is the depth of the section.

Equation 6.11 is similar to the size factor (Equation 6.3) of sawn timber lumber that is applied for depths exceeding 12 in. But the GLULAM (Equation 6.11) is applicable to any depth. Thus, the flat use factor is more than 1 for less than 12 in. depth and less than 1 for depths exceeding 12 in.

TABLE 6.6
Applicability of Adjustment Factors for GLULAM

Loading Condition	Loading Case	Factor						Special Factor	
		Time Effect	Wet Service	Temperature	Size (Volume)	Flat Use	Curvature	Beam Stability	Column Stability
Bending	Load perpendicular to lamination	λ	C_M	C_t	C_v		C_c	C_L (or C_V)	
	Load parallel to lamination	λ	C_M	C_t		C_{fu}	C_c	C_L	
Tension		λ	C_M	C_t					
Compression	Load perpendicular to lamination	λ	C_M	C_t					
	Load parallel to lamination	λ	C_M	C_t					C_P
Shear		λ	C_M	C_t					
Modulus of elasticity			C_M	C_t					
Modulus of elasticity for stability			C_M	C_t					

Volume Factor for GLULAM, C_v

The volume factor is applied to bending only for horizontally laminated timber (for loading applied perpendicular to laminations). The beam stability factor, C_L and the volume factor, C_v are not used together; only the smaller of the two is applied to adjust F'_{bn}. The concept of the volume factor for GLULAM is similar to the size factor of sawn lumber because the test data have indicated that the size effect extends to volume in the case of GLULAM. The volume factor is

$$C_v = \left(\frac{5.125}{b}\right)^{1/x} \left(\frac{12}{d}\right)^{1/x} \left(\frac{21\text{ ft}}{L\text{ ft}}\right)^{1/x} \tag{6.12}$$

where
   b is the width (in.)
   d is the depth (in.)
   L is the length of member between points of zero moments (ft)
   $x=20$ for Southern Pine and 10 for other species

Curvature Factor for GLULAM, C_c

The curvature factor is applied to bending stress only to account for the stresses that are introduced in laminations when they are bent into curved shapes during manufacturing. The curvature factor is

$$C_c = 1 - 2000\left(\frac{t}{R}\right)^2 \tag{6.13}$$

where
   t is the thickness of lamination, 1-½ in. or 1-⅜ in.
   R is the radius of curvature of the inside face of lamination

The ratio t/R may not exceed 1/100 for Southern Pine and 1/125 for other species. The curvature factor is not applied to straight portion of a member regardless of curvature in the other portion.

Example 6.5

Determine the adjusted nominal reference design stresses and the strength capacities of a 6-¾ in. × 18 in. GLULAM from Doulas Fir-Larch of stress class 24F-1.7E, used primarily for bending. The span is 30 ft. The loading consists of the dead and live loads combination along the major axis.

Solution

1. The adjusted reference design values are computed in the table below:

Property	Reference Design Value (psi)	Adjustment Factors			$F'_{()n}$ (psi)
		λ	C_v	ϕK_F	
Bending	2400	0.8	0.90[a]	2.16	3732.48
Tension	775	0.8		2.16	1339.2
Shear	210	0.8		2.16	362.88
Compression	1000	0.8		2.16	1728.0
E	1.7×10^6				1.7×10^6
E_{min}	0.88×10^6			1.5	1.32×10^6

[a] $C_v = \left(\frac{5.125}{6.75}\right)^{1/10} \left(\frac{12}{18}\right)^{1/10} \left(\frac{21}{30}\right)^{1/10} = 0.90$.

2. Strength capacities
   For section 6-¾ in. × 18 in. section, $S_x = 364.5$ in.^{3}, $A = 121.5$ in.^{2}
   Bending: $\phi M_n = F'_{bn}S = (3732.48)(364.5) = 1.36 \times 10^6$ in.-lb
   Tension: $\phi T_n = F'_{tn}A = (1339.2)(121.5) = 162.71 \times 10^3$ lb
   Shear: $\phi V_n = F'_{vn}(2/3A) = (362.88)(2/3 \times 121.5) = 29.39 \times 10^3$ lb
   Compression: $\phi P_n = F'_{cn}A = (1728)(121.5) = 210 \times 10^3$ lb

Example 6.6

The beam in Example 6.5 is installed upside own. Determine the design strengths.

Solution

1. Bending reference design value for compression zone stressed in tension = 1450 psi from Appendix B.7
2. Adjustment factors from Example 6.5

   $\lambda = 0.80$

   $C_v = 0.90$

   $\phi K_F = 2.16$

3. Adjusted nominal design value

   $F'_{bn} = \phi 1450(0.80)(0.9)(2.16/\phi) = 2255$ psi

4. Strength capacity

   $F'_{bn}S = (2255)(364.5) = 0.822 \times 10^6$ in.-lb.

5. The other values are the same as in Example 6.5.

Example 6.7

The beam used in Example 6.5 is flat with loading along minor axis; determine the design strengths.

Solution

1. The adjusted reference design values are computed in the table below:

Property	Reference Design Value (psi)	Adjustment Factors			
		λ	C_{fu}	ϕK_F	$F'_{()n}$ (psi)
Bending	1050	0.8	1.066[a]	2.16	1934.15
Tension	775	0.8		2.16	1339.2
Shear	185	0.8		2.16	319.68
Compression	1000	0.8		2.16	1728.0
E	1.3×10^6				1.3×10^6
E_{min}	0.67×10^6			1.5	1.01×10^6

[a] $C_{fu} = \left(\dfrac{12}{6.75}\right)^{1/9} = 1.066.$

2. Strength capacities
   For section 6-¾ in. × 18 in. section, $S_y = 136.7$ in.^{3}, $A = 121.5$ in.^{2}
   Bending: $\phi M_n = F'_{bn}S = (1934.15)(136.7^*) = 0.26 \times 10^6$ in.-lb
   Tension: $\phi T_n = F'_{tn}A = (1339.2)(121.5) = 162713$ lb
   Shear: $\phi V_n = F'_{vn}(2/3A) = (319.68)(2/3 \times 121.5) = 25.89 \times 10^3$ lb
   Compression: $\phi P_n = F'_{cn}A = (1728)(121.5) = 209 \times 10^3$ lb

Example 6.8

What are the unit dead and live loads (per ft^2) that are resisted by the beam in Example 6.5 that is spaced 10 ft on center. Assume that the unit dead load is one-half of the live load?

Solution

1. From Example 6.5

$$M_u = \phi M_n = F'_{bn}S = (3732.48)(364.5) = 1.36 \times 10^6 \text{ in.-lb. or } 113333.33 \text{ ft-lb}$$

2. $M_u = 113333.33 = \dfrac{w_u L^2}{8}$

   or

   $$w_u = \frac{113333.33(8)}{(30)^2} = 1007.41 \text{ lb/ft}$$

3. Tributary area/ft of beam $= 10 \times 1 = 10$ ft^2/ft
   11. $w_u = $ (design load/ft^2)(tributary area, ft^2/ft)

   $$1007.41 = (1.2D + 1.6L)(10)$$

   or

   $$1007.41 = \left[1.2D + 1.6(2D)\right](10)$$

   or

   $$D = 22.9 \text{ lb/ft}^2$$

   and

   $$L = 45.8 \text{ lb/ft}^2$$

STRUCTURAL COMPOSITE LUMBER

The structural composite lumber (SCL) commonly includes the *laminated veneer lumber* (LVL) and the *parallel strand lumber* (PSL). Like the GLULAM, the SCL also have laminations. However, the LVL is manufactured from thin veneers that are rotary peeled from a log, and then dried and laminated together by an adhesive under heat and pressure. The grains on each ply are oriented in the same direction as the length of the member. The PSL is manufactured from wood strands (veneers that have been further sized up into long and narrow strips) glued into parallel lamination.

---

* S_y value.

The lamination of LVL and PSL is vertical (wide faces of lams are oriented vertically) as compared to the horizontal lamination of GLULAM (wide faces oriented horizontally). The strength and stiffness of LVL and PSL are higher than GLULAM.

The typical reference design values for LVL are listed in Appendix B.9. A wide brand of LVL and PSL are available. The reference values and technical specifications for the specific brand might be obtained from the manufacturer's literature.

The same time effect factors and the format conversion factors are applied to the reference design values as for the sawn lumber and the GLULAM, as discussed in the "Adjustments to the Reference Design Values Sawn Lumber" section.

In addition, the adjustment factors listed in Table 6.7 are applied to the SCL. The wet-service factors, C_M, and the temperature factors, C_t are the same GLULAM and SCL. To the members used in repetitive assembly, as defined in the "Repetitive Member Factor, C_r," section, a repetitive factor C_r of 1.04 is applied. The value of the size (volume) factor is obtained from the SCL manufacturer's literature.

PROBLEMS

6.1–6.5 Determine the adjusted reference design values and the strength capacities for the following members. In all cases, consider the dry-service conditions, the normal temperature range, and the no-incision application. In practice, all loading combinations must be checked. However, in these problems only a single load condition should be considered for each member, as indicated in the problem.

6.1 Floor joists are 2 in. × 6 in. at 18 in. on center (OC) of Douglas Fir-Larch #2. They support the dead and live loads.

6.2 Roof rafters are 2 in. × 8 in. at 24 in. OC of Southern Pine #2. The loads are dead load and roof live load.

6.3 Five floor beams are of 4 in. × 8 in. dimension lumber Hem Fir #1, spaced 5 ft apart. The loads are dead and live loads.

6.4 Studs are 2 in. × 8 in. at 20 in. OC of Hem Fir #2. The loads are dead load, live load, and wind load.

6.5 Interior column is 5 in. × 5 in. of Douglas Fir-Larch #2 to support the dead and live loads.

6.6 Determine the unit dead and live loads (on per sq ft area) that can be resisted by a floor system consisting of 2 in. × 4 in. joists at 18 in. OC of Douglas Fir-Larch #1. The span is 12 ft. The dead and live loads are equal.

6.7 Determine the unit dead load on the roof. The roof beams are 4 in. × 10 in. of Hem Fir #1. The beams are located at 5 ft on center and the span is 20 ft apart. They support the dead load and a snow load of 20 psf.

6.8 A 6 in. × 6 in. column of Douglas Fir-Larch #1 supports the dead load and live load on an area of 100 ft². Determine the per square ft load if the unit dead load is one-half of the unit live load.

6.9 A floor system is acted upon by a dead load of 20 psf and a live load of 40 psf. Determine the size of the floor joists of Douglas Fir-Larch Structural. They are located 18 in. OC and span 12 ft. Assume that the beam stability factor is not a concern.

6.10 Determine the size of a column of Southern Pine #2 that receives loads from an area of 20 ft × 25 ft. The unit service loads are 20 psf dead load and 30 psf live load. Assume that the column stability factor is not a concern.

6.11 A GLULAM beam section is 6-¾ in. × 37.5 in. from Douglas Fir 24F-1.7E Class. The loads combination comprises the dead load, snow load, and wind load. The bending is about the X-axis. Determine the adjusted nominal reference design stresses and the strength capacities for bending, tension, shear, compression, modulus of elasticity, and modulus of stability (E_{min}). The span is 30 ft.

TABLE 6.7
Applicability of Adjustment Factors for SCL

Loading Condition	Loading Case	Factor					Special Factor	
		Time Effect	Wet Service	Temperature	Size (Volume)	Repetitive	Beam Stability	Column Stability
Bending	Load perpendicular to lamination	λ	C_M	C_t	C_v	C_r	C_L (or C_v)	
	Load parallel to lamination	λ	C_M	C_t		C_r	C_L	
Tension		λ	C_M	C_t				
Compression	Load perpendicular to lamination	λ	C_M	C_t				
	Load parallel to lamination	λ	C_M	C_t				C_P
Shear		λ	C_M	C_t				
Modulus of elasticity			C_M	C_t				
Modulus of elasticity for stability			C_M	C_t				

6.12 Determine the wind load for Problem 6.11 if the unit dead load is 50 psf and the unit snow and wind loads are equal. The beams are 10 ft apart.

6.13 The beam in Problem 6.11 is installed upside down. Determine the strength capacities.

6.14 The beam in Problem 6.11 is used flat with bending about minor axis. Determine the design capacities.

6.15 A 5-⅛ in. × 28.5 in., 26F-1.9E Southern Pine GLULAM is used to span 35 ft. The beam has a radius of curvature of 10 ft. The load combination is the dead load and the snow load. Determine the adjusted nominal reference design stresses and the strength capacities for loading perpendicular to the laminations for the beam installed according to the specifications.

6.16 The beam in Problem 6.15 is installed upside down. Determine the percent reduction in the strength capacities.

6.17 The beam in Problem 6.15 is loaded along the laminations, about the minor axis. Determine the percent change in the strength capacities.

6.18 A 1-¾ in. × 7-¼ in. size LVL of 1.9E. Class is used for the roof rafters spanning 20 ft, located 24 in. on center. Determine the strength design capacities for the dead and snow loads combinations. The size factor is given by $(12/d)^{1/7.5}$.

6.19 2–1-¾ in. × 16 in. (two sections side by side) of LVL of 2.0E class are used for a floor beam spanning 32 ft, spaced 8 ft on center. The loading consists of the dead and live loads. Determine the strength capacities for bending, tension, composition, and shear. The size factor is given by $(12/d)^{1/7.5}$.

6.20 Determine the unit loads (per ft²) on the beam in Problem 6.19 if the live load is one-and-half times of the dead load.

7 Flexure and Axially Loaded Wood Structures

INTRODUCTION

The conceptual design of wood members was presented in Chapter 6. The underlying assumption of design in that chapter was that an axial member was subjected to axial tensile stress or axial compression stress only and a flexure member to normal bending stress only. However, the compression force acting on a member tends to buckle a member out of the plane of loading, as shown in Figure 7.1. This buckling occurs in the columns and in the compression flange of the beams unless the compression flange is adequately braced. The beam and column stability factors C_L and C_P, mentioned in the "Reference Design Values for Sawn Lumber" section of Chapter 6, are applied to account for the effect of this lateral buckling.

This chapter presents the detailed designs of flexure members, axially loaded tensile and compression members, and the members subjected to the combined flexure and axial force made of sawn lumber, GLULAM, and LVL.

DESIGN OF BEAMS

In most cases, for the design of a flexure member or beam, the bending capacity of the material is a critical factor. Accordingly, the basic criterion for the design of a wood beam is developed from a bending consideration.

In a member subjected to flexure, compression develops on one side of the section; under compression, lateral stability is an important factor. It could induce a buckling effect that will undermine the moment capacity of the member. An adjustment factor is applied in wood design when the buckling effect could prevail, as discussed subsequently.

A beam is initially designed for the bending capacity. It is checked for the shear capacity. It is also checked from the serviceability consideration of the limiting state of deflection. If the size is not found adequate for the shear capacity or the deflection limits, the design is revised.

The bearing strength of a wood member is considered at the beam supports or where loads from other members frame onto the beam. The bearing length (width) is designed on this basis.

BENDING CRITERIA OF DESIGN

For the bending capacity of a member as discussed before

$$M_u = F'_{bn}S \tag{7.1}$$

M_u represents the design moment due to the factored combination of loads. The design moment for a uniformly distributed load, w_u, is given by $M_u = w_u L^2/8$ and for a concentrated load, P_u centered at mid-span, $M_u = P_u L/4$. For other cases, M_u is ascertained from the analysis of structure.

The span length, L is taken as the distance from the center of one support to the center of the other support. However, when the provided (furnished) width of a support is more than what is required from the bearing consideration, it is permitted to take the span length to be the clear distance between the supports plus one-half of the required bearing width at each end.

F'_{bn} is the adjusted LRFD reference design value for bending. To start with, the reference bending design value, F_b for the appropriate species and grade is obtained. These values are listed in Appendices B.2 through B.4 for sawn lumber and Appendices B.7 through B.9 for GLULAM and LVL. Then the value is adjusted by multiplying the reference value by a string of factors. The applicable adjustment factors for sawn lumber were given in Table 6.5, and for GLULAM in Table 6.6 and for LVL in Table 6.7.

For sawn lumber, the adjusted reference bending design value is restated as

$$F'_{bn} = \phi F_b \lambda C_M C_t C_F C_r C_{fu} C_i C_L K_F \tag{7.2}$$

FIGURE 7.1 Buckling due to compression.

For GLULAM and SCL, the adjusted reference bending design value is restated as

$$F'_{bn} = \phi F_b \lambda C_M C_t C_c C_{fu}(C_v \text{ or } C_L)K_F \tag{7.3}$$

where
F_b is the tabular reference bending design value
ϕ is the resistance factor for bending $= 0.85$
λ is the time factor (Table 6.2)
C_M is the wet-service factor
C_t is the temperature factor
C_F is the size factor
C_r is the repetitive member factor
C_{fu} is the flat use factor
C_i is the incision factor
C_L is the beam stability factor
C_c is the curvature factor
C_v is the volume factor
K_F is the format conversion factor $= 2.16/\phi$

Using the assessed value of F'_{bn}, from Equations 7.2 or 7.3, based on the adjustment factors known initially, the required section modulus, S is determined from Equation 7.1 and a trial section is selected having the section modulus S higher than the computed value. In the beginning, some section-dependent factors such as C_F, C_v, and C_L will not be known while the others such as λ, K_F, and ϕ will be known. The design is performed considering all possible load combinations along with the relevant time factor. If loads are of one type only, i.e., all vertical or all horizontal, the highest value of the combined load divided by the relevant time factor determines which combination is critical for design.

Based on the trial section, all adjustment factors including C_L are then computed and the magnitude of F'_{bn} is reassessed. A revised S is obtained from Equation 7.1 and the trial section is modified, if necessary.

BEAM STABILITY FACTOR, C_L

As stated earlier, the compression stress, besides causing an axial deformation, can cause a lateral deformation if the compression zone of the beam is not braced against the lateral movement. In the presence of the stable one-half tensile portion, the buckling in the plane of loading is prevented. However, the movement could take place sideways (laterally), as shown in Figure 7.2.

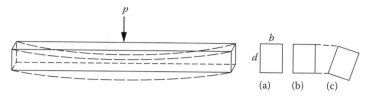

FIGURE 7.2 Buckling of a bending member. (a) Original position of the beam. (b) Deflected position without lateral instability. (c) Compression edge buckled laterally.

The bending design described in Chapter 6 had assumed that no buckling was present and adjustments were made for other factors only. The condition of no buckling is satisfied when the bracing requirements, as listed in Table 7.1, are met. In general, when the depth-to-breadth ratio is 2 or less, no lateral bracings are required. When the depth-to-breadth ratio is more than 2 but does not exceed 4, the ends of the beam should be held in position by one of these methods: full-depth solid blocking, bridging, hangers, nailing, or bolting to other framing members. The stricter requirements are stipulated to hold the compression edge in line for a depth-to-breadth ratio of higher than 4.

When the requirements of Table 7.1 are not met, the following beam stability factor has to be applied to account for the buckling effect.

$$C_L = \left(\frac{1+\alpha}{1.9}\right) - \sqrt{\left(\frac{1+\alpha}{1.9}\right)^2 - \left(\frac{\alpha}{0.95}\right)} \tag{7.4}$$

TABLE 7.1
Bracing Requirements for Lateral Stability

Depth/Breadth Ratio[a]	Bracing Requirements
	Sawn lumber
≤2	No lateral bracing required
>2 but ≤4	The ends are to be held in position, as by full-depth solid blocking, bridging, hangers, nailing, or bolting to other framing members, or other acceptable means
>4 but ≤5	The compression edge is to be held in line for its entire length to prevent lateral displacement, as by sheathing or subflooring, and the ends at points of bearing are to be held in position to prevent rotation and/or lateral displacement
>5 but ≤6	Bridging, full-depth solid blocking, or diagonal cross bracing is to be installed at intervals not exceeding 8 ft, the compression edge is to be held in line for its entire length to prevent lateral displacement, as by sheathing or subflooring, and the ends at points of bearing are to be held in position to prevent rotation and/or lateral displacement
>6 but ≤7	Both edges of a member are to be held in line for their entire length, and the ends at points of bearing are to be held in position to prevent rotation and/or lateral displacement
Combined bending and compression	The depth/breadth ratio may be as much as 5 if one edge is held firmly in line. If under all load conditions, the unbraced edge is in tension, the depth/breadth ratio may be as much as 6
	Glued laminated timber
≤1	No lateral bracing required
>1	The compression edge is supported throughout its length to prevent lateral displacement, and the ends at point of bearing are laterally supported to prevent rotation

[a] Nominal dimensions.

where

$$\alpha = \frac{F_{bEn}}{F_{bn}^{'*}}$$ (7.5)

where

$F_{bn}^{'*}$ is the reference bending design value adjusted for all factors except C_v, C_{fu}, and C_L

F_{bEn} is the Euler-based LRFD critical buckling stress for bending

$$F_{bEn} = \frac{1.2E'_{y\,min(n)}*}{R_B^2}$$ (7.6)

where

$E'_{min(n)}$ is the adjusted nominal stability modulus of elasticity

R_B is the slenderness ratio for bending

$$R_B = \sqrt{\frac{L_e d}{b^2}} \leq 50$$ (7.7)

where L_e is the effective unbraced length, discussed in the "Effective Unbraced Length" section below.

When R_B exceeds 50 in Equation 7.7, the beam dimensions should be revised to limit the slenderness ratio to 50.

EFFECTIVE UNBRACED LENGTH

The effective unbraced length is a function of several factors such as the type of span (simple, cantilever, continuous), the type of loading (uniform, variable, concentrated loads), the unbraced length, L_u which is the distance between the points of lateral supports, and the size of the beam.

For a simple one span or cantilever beam, the following values can be conservatively used for the effective length.

$$\text{For} \quad \frac{L_u}{d} < 7, \quad L_e = 2.06L_u$$ (7.8)

$$\text{For} \quad 7 \leq \frac{L_u}{d} \leq 14.3, \quad L_e = 1.63L_u + 3d$$ (7.9)

$$\text{For} \quad \frac{L_u}{d} > 14.3, \quad L_e = 1.84L_u$$ (7.10)

Example 7.1

A 5-½ in. × 24 in. glued laminated timber beam is used for a roof system having a span of 32 ft, which is braced only at the ends. GLULAM consists of the Douglas Fir 24F-1.8E. Determine the beam stability factor. Use the dead and live conditions only.

---

* Use Y-axis.

Solution

1. Reference design values

   $F_b = 2400\,\text{psi}$

   $E = 1.8 \times 10^6\,\text{psi}$

   $E_{y\,(min)} = 0.83 \times 10^6\,\text{psi}$

2. Adjusted design values

$$F_{bn}^* = \lambda F_b \phi K_F$$

$$= (0.8)(2400)(\phi)\frac{(2.16)}{\phi} = 4147\,\text{psi or }4.15\,\text{ksi}$$

$$E'_{min(n)} = E_{y(min)}K_F$$

$$= (0.83 \times 10^6)(\phi)\left(\frac{1.5}{\phi}\right) = 1.25 \times 10^6\,\text{psi or }1.25 \times 10^3\,\text{ksi}$$

3. Effective unbraced length

$$\frac{L_u}{d} = \frac{32 \times 12}{24} = 16 > 14.3$$

   From Equation 7.10

$$L_e = 1.84 L_u = 1.84(32) = 58.88\,\text{ft}$$

4. From Equation 7.7

$$R_B = \sqrt{\frac{L_e d}{b^2}}$$

$$= \sqrt{\frac{(58.88 \times 12)(14)}{(5.5)^2}}$$

$$= 23.68 < 50 \quad \textbf{OK}$$

5. $F_{bEn} = \dfrac{1.2 E'_{min(n)}}{R_B^2}$

$$= \frac{1.2(1.25 \times 10^3)}{(23.68)^2} = 2.675\,\text{ksi}$$

6. $\alpha = \dfrac{F_{bEn}}{F_{bn}^*} = \dfrac{2.675}{4.15} = 0.645$

7. $C_l = \left(\dfrac{1+\alpha}{1.9}\right) - \sqrt{\left(\dfrac{1+\alpha}{1.9}\right)^2 - \left(\dfrac{\alpha}{0.95}\right)}$

$$= \frac{1.645}{1.9} - \sqrt{\left(\frac{1.645}{1.9}\right)^2 - \left(\frac{0.645}{0.95}\right)}$$

$$= 0.6$$

SHEAR CRITERIA

A transverse loading applied to a beam results in vertical shear stresses in any transverse (vertical) section of a beam. Due to the complimentary property of shear, an associated longitudinal shear stress acts along the longitudinal plane (horizontal face) of a beam element. In any mechanics of materials text, it can be seen that the longitudinal shear stress distribution across the cross section is given by

$$f_v = \frac{VQ}{I} \tag{7.11}$$

where
f_v is the shear stress at any plane across the cross section
V is the shear force along the beam at the location of the cross section
Q is the moment of the area above the plane where stress is desired to the top or bottom edge of the section. Moment is taken at neutral axis
I is the moment of inertia along the neutral axis

Equation 7.11 also applies for the transverse shear stress at any plane of the cross section as well because the transverse and the longitudinal shear stresses are complimentary, numerically equal and opposite in sign.

For a rectangular cross section that is usually the case with wood beams, the shear stress distribution by above relation is parabolic with the following maximum value at the center.

$$f_{v\,max} = F'_{vn} = \frac{3}{2}\frac{V_u}{A} \tag{7.12}$$

In terms of V_u, the basic equation for shear design of beam is

$$V_u = 2/3 F'_{vn} A \tag{7.13}$$

where
V_u is the maximum shear force due to factored load on beam
F'_{vn} is the adjusted reference shear design value
A is the area of the beam

The NDS permits that the maximum shear force, V_u, might be taken to be the shear force at a distance equal to the depth of the beam from the support. However, usually V_u is usually taken to be the maximum shear force from the diagram, which is at the support for a simple span.

For sawn lumber, the adjusted reference shear design value is

$$F'_{vn} = \phi F_v \lambda C_M C_t C_i K_F \tag{7.14}$$

For GLULAM and SCL, the adjusted reference shear design value is

$$F'_{vn} = \phi F_v \lambda C_M C_t K_F \tag{7.15}$$

where
F_v is the tabular reference shear design value
ϕ is the resistance factor for shear $= 0.75$

λ is the time factor (see the "Time Effect Factor, λ" section)
C_M is the wet-service factor
C_t is the temperature factor
C_i is the incision factor
K_F is the format conversion factor $= 2.16/\phi$

DEFLECTION CRITERIA

The actual maximum deflection for the beam is calculated using the service load (not the factored load). The maximum deflection is a function of the type of loading, type of beam span, moment of inertia of the section, and modulus of elasticity. For a uniformly loaded simple span member, the maximum deflection at mid-span is

$$\delta = \frac{5wL^4}{384E'I} \tag{7.16}$$

where
w is the uniform combined service load per unit length
L is the span of beam
E' is the adjusted modulus of elasticity

$$E' = EC_MC_tC_i$$

E is the reference modulus of elasticity
I is the moment of inertia along neutral axis

The actual maximum deflection should be less than or equal to the allowable deflections, Δ. Often a check is made for live load alone as well as for the total load. Thus,

$$\text{Max. } \delta_L \leq \text{allow. } \Delta_L \tag{7.17}$$

$$\text{Max. } \delta_{TL} \leq \text{allow. } \Delta_{TL} \tag{7.18}$$

The allowable deflections are given in Table 7.2.

TABLE 7.2
Recommended Deflection Criteria

Classification	Live or Applied Load Only	Dead Load Plus Applied Load
Roof beams		
No ceiling	Span/180	Span/120
Without plaster ceiling	Span/240	Span/180
With plaster ceiling	Span/360	Span/240
Floor beams[a]	Span/360	Span/240
Highway bridge stringers	Span/300	
Railway bridge stringers	Span/300–Span/400	

Source: American Institute of Timber Association, *Timber Construction Manual*, 5th edn., John Wiley & Sons, New York, 2005.

[a] Additional limitations are used where increased floor stiffness or reduction of vibrations is desired.

When the above criteria are not satisfied, a new beam size is determined using the allowable deflection as a guide and computing the desired moment of inertia on that basis.

Example 7.2

Design the roof rafters spanning 16 ft and spaced 16 in. on center (OC). The plywood roof sheathing prevents the local buckling. The dead load is 12 psf and the roof live load is 20 psf. Use Douglas Fir-Larch #1 wood.

Solution

A. Loads

1. Tributary area/ft $= \dfrac{16}{12} \times 1 = 1.333 \text{ ft}^2/\text{ft}$
2. Loads per feet
   $w_D = 12 \times 1.333 = 16 \text{ lb/ft}$
   $w_L = 20 \times 1.333 = 26.66 \text{ lb/ft}$
3. Loads combination

$$w_u = 1.2 w_D + 1.6 w_L$$

$$= 1.2(16) + 1.6(26.66) = 61.86 \text{ lb/ft}$$

4. Maximum BM

$$M_u = \frac{w_u L^2}{8} = \frac{(61.86)(16)^2}{8} = 1974.52 \text{ ft lb or } 23.75 \text{ in.-k}$$

5. Maximum shear

$$V_u = \frac{w_u L}{2} = \frac{(61.86)(16)}{2} = 494.9 \text{ lb}$$

B. Reference design values (Douglas Fir-Larch #1, 2 in. and wider)

1. $F_b = 1000 \text{ psi}$
2. $F_v = 180 \text{ psi}$
3. $E = 1.7 \times 10^6 \text{ psi}$
4. $E_{min} = 0.62 \times 10^6 \text{ psi}$

C. Preliminary design

1. Initially adjusted bending design value

$$F'_{bn} \text{ (estimated)} = \phi \lambda F_b K_F C_r$$

$$= \phi(0.8)(1000)(2.16/\phi)(1.15) = 1987 \text{ psi}$$

2. $S_{reqd} = \dfrac{M_u}{F'_{bn} \text{ (estimated)}} = \dfrac{(23.75 \times 1000)}{1987} = 11.95 \text{ in.}^3$

3. Try 2 in. $\times$ 8 in. $S = 13.14$ in.^{3}

$$A = 10.88 \text{ in.}^2$$
$$I = 47.63 \text{ in.}^4$$

D. Revised design

1. Adjusted reference design values

	Reference Design Values (psi)	λ	C_F	C_r	K_F	F'_{On} (psi)
F'_{bn} [a]	1000	0.8	1.2	1.15	2.16/ϕ	2384.6
F'_{vn}	180	0.8			2.16/ϕ	311
E'	1.7×10^6	—			—	1.7×10^6
$E'_{min(n)}$	0.62×10^6	—			1.5/ϕ	0.93×10^6

[a] Without the C_L factor.

2. Beam stability factor $C_L = 1.0$

E. Check for bending strength

Bending capacity $= F'_{bn}S$ in.-k

$$= \frac{(2384.6)(13.14)}{1000} = 31.33 \text{ in.-k} > 23.75 \text{ in.k} \quad \textbf{OK}$$

F. Check for shear strength

Shear capacity $= F'_{vn}\left(\frac{2A}{3}\right) = 311\left(\frac{2}{3} \times 10.88\right) = 2255 \text{ lb} > 494.5 \text{ lb} \quad \textbf{OK}$

G. Check for deflection

1. Deflection is checked for service load, $w = 16 + 26.66 = 42.66$ lb/ft

2. $\delta = \dfrac{5}{384}\dfrac{wL^4}{E'I} = \dfrac{5}{384}\dfrac{(42.66)(16)^4(12)^3}{(1.7 \times 10^6)(47.63)} = 0.78$ in.

3. Allowable deflection

$$\Delta = \frac{L}{180} = \frac{16 \times 12}{180} = 1.07 \text{ in.} > 0.78 \text{ in.} \quad \textbf{OK.}$$

Example 7.3

A structural glued laminated timber is used as a beam to support a roof system. The tributary width of the beam is 16 ft. The beam span is 32 ft. The floor dead load is 15 psf and the live load is 40 psf. Use Douglas Fir GLULAM 24F-1.8E. The beam is braced only at the supports.

Solution

A. Loads

1. Tributary area/ft $= 16 \times 1 = 16\,\text{ft}^2/\text{ft}$
2. Loads per ft
   $w_D = 15 \times 16 = 240\,\text{lb/ft}$
   $w_L = 40 \times 16 = 640\,\text{lb/ft}$
3. Design load, $w_u = 1.2 w_D + 1.6 w_L$

$$= 1.2(240) + 1.6(640) = 1312\,\text{lb/ft or } 1.31\,\text{k/ft}$$

4. Design bending moment

$$M_u = \frac{w_u L^2}{8} = \frac{(1.31)(32)^2}{8} = 167.88\,\text{ft-k or } 2012.16\,\text{in.-k}$$

5. Design shear

$$V_u = \frac{w_u L}{2} = \frac{1.31(32)}{2} = 20.96\,\text{k}$$

B. Reference design values
   $F_b = 2400\,\text{psi}$
   $F_v = 265\,\text{psi}$
   $E = 1.8 \times 10^6\,\text{psi}$
   $E_{y(min)} = 0.83 \times 10^6\,\text{psi}$
C. Preliminary design

1. Initially adjusted bending reference design value
   $= (0.8)(2400)\phi(2.16/\phi) = 4147\,\text{psi or } 4.15\,\text{ksi}$
2. $S_{reqd} = \dfrac{2012.16}{4.15} = 484.86\,\text{in.}^3$

   Try 5 1/2 in. × 24 in. $S = 528\,\text{in.}^3$
   $A = 132\,\text{in.}^2$
   $I = 6336\,\text{in.}^4$

D. Revised adjusted design values

Type	Reference Design Values (psi)	λ	K_F	$F'_{()n}$ (psi)
$F'_{bn\ a}$	2400	0.8	2.16/ϕ	4147.2 ($F_b{}^a$)
F'_{vn}	265	0.8	2.16/ϕ	457.9
E'	1.8×10^6	—	—	1.8×10^6
$E'_{min(n)}$	0.83×10^6	—	1.5/ϕ	1.25×10^6

E. Volume factor, C_v

$$C_v = \left(\frac{5.125}{b}\right)^{1/10} \left(\frac{12}{d}\right)^{1/10} \left(\frac{21}{L}\right)^{1/10}$$

$$= \left(\frac{5.125}{5.5}\right)^{1/10} \left(\frac{12}{24}\right)^{1/10} \left(\frac{21}{32}\right)^{1/10} = 0.89$$

F. Beam stability factor, C_L
   From Example 7.1, $C_L = 0.60$
   Since $C_L < C_v$, use the C_L factor
G. Bending capacity

   1. $F'_{bn} = (4147.2)\,(0.6) = 2488.32$ psi or 2.488 ksi
   2. Moment capacity $= F'_{bn}S$
         $= 2.488(528)$
         $= 1313.7$ in.-k $< 2012.16(M_u)$ **NG**

A revised section should be selected and steps E, F, and G should be repeated.
H. Check for shear strength*

$$\text{Shear capacity} = F'_{vn}\left(\frac{2A}{3}\right) = 457.9\left(\frac{2}{3}\times132\right) = 40295\,\text{lb or } 40.3\,\text{k} > 20.29\,\text{k}\quad\textbf{OK}$$

I. Check for deflection

   1. Deflection checked for service load $w = 240 + 640 = 880$ lb/ft
   2. $\delta = \dfrac{5}{384}\dfrac{wL^4}{EI} = \dfrac{5}{384}\dfrac{(880)(32)^4(12)^3}{(1.8\times10^6)(6336)} = 1.82$ in.

   3. Permissible deflection

$$\Delta = \frac{L}{180} = \frac{32\times12}{180} = 2.13\,\text{in.} > 1.82\,\text{in.}\quad\textbf{OK}$$

BEARING AT SUPPORTS

The bearing perpendicular to the grains occurs at the supports or wherever a load bearing member rests onto the beam, as shown in Figure 7.3. The relation for bearing design is

$$P_u = F'_{C^\perp n}A \tag{7.19}$$

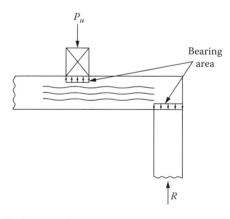

FIGURE 7.3   Bearing perpendicular to grain.

---

* Based on the original section.

The adjusted compressive design value perpendicular to grain is obtained by multiplying the reference design value by the adjustment factors. Including these factors, Equation 7.19 becomes
    For sawn lumber

$$P_u = \phi F_{c\perp} \lambda C_M C_t C_i C_b K_F A \tag{7.20}$$

For GLULAM

$$P_u = \phi F_{c\perp} \lambda C_M C_t C_b K_F A \tag{7.21}$$

where
    P_u is the reaction at the bearing surface due to factored load on the beam
    $F_{c\perp}$ is the reference compressive design value perpendicular to grain
    $F'_{c\perp_n}$ is the adjusted compressive design value perpendicular to grain
    ϕ is the resistance factor for compression = 0.9
    λ is the time effect factor (see the "Time Effect Factor, λ" section in Chapter 6)
    C_M is the wet-service factor
    C_t is the temperature factor
    C_i is the incision factor
    C_b is the bearing area factor as discussed below
    K_F is the format conversion factor for bearing = 1.875/ϕ
    A is the area of bearing surface

BEARING AREA FACTOR, C_b

The bearing area factor is applied only to a specific case when the bearing length l_b is less than 6 in. and also the distance from the end of the beam to the start of the contact area is larger than 3 in., as shown in Figure 7.4. The factor is not applied to the bearing surface at the end of a beam that may be of any length or where the bearing length is 6 in. or more at any other location than the end. This factor accounts for the additional wood fibers that could resist the bearing load. It increases the bearing length by 3/8 in. Thus,

$$C_b = \frac{l_b + 3/8}{l_b} \tag{7.22}$$

where l_b is the bearing length is the contact length parallel to the grain.

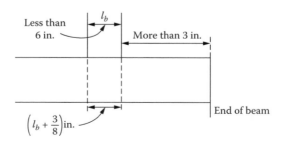

FIGURE 7.4   Bearing area factor.

Since C_b is only for a specific case and is always greater than 1, conservatively it could be taken as unity and disregarded.

Example 7.4

For Example 7.3, determine the bearing surface area at the beam supports.

Solution

1. Reaction at the supports

$$R_u = \frac{w_u L}{2} = \frac{1.31(32)}{2} = 20.96 \text{ k}$$

2. Reference design value for compression perpendicular to grains $F_{c\perp n} = 650 \text{ psi}$
3. Adjusted perpendicular compression reference design value

$$F'_{c\perp n} = \phi F_{c\perp} \lambda C_M C_t C_i C_b K_F$$

$$= (\phi)(650)(0.8)(1)(1)(1)(1.875/\phi) = 975 \text{ psi or } 0.975 \text{ ksi}$$

4. $A_{reqd} = \dfrac{R_u}{F'_{c\perp n}} = \dfrac{20.96}{0.975} = 21.50 \text{ in.}^2$

5. Bearing length

$$l_b = \frac{A}{b} = \frac{21.50}{5.5} = 3.9 \text{ in.}$$

DESIGN OF AXIAL TENSION MEMBERS

Axially loaded wood members generally comprise of studs, ties, diaphragms, shear walls, and trusses where loads directly frame into joints to pass through the member's longitudinal axis or with a very low eccentricity. These loads exert either tension or compression without any appreciable bending in members. For example, a truss has some members in compression and some in tension. The treatment of a tensile member is relatively straightforward because only the direct axial stress is exerted on the section. However, the design is typically governed by the net section at the connection since in a stretched condition, an opening separates out from the fastener.

The tensile capacity of a member is given by

$$T_u = F'_{tn} A_n \tag{7.23}$$

Axial tension members in wood generally involve relatively small force for which a dimensional lumber section is used which requires inclusion of a size factor.

Including the adjustment factors, the tensile capacity is represented as follows:

For sawn lumber

$$T_u = \phi F_t \lambda C_M C_t C_F C_i K_F A_n \tag{7.24}$$

For GLULAM

$$T_u = \phi F_t \lambda C_M C_t K_F A_n \tag{7.25}$$

where

    T_u is the factored tensile load on member
    F_t is the reference tension design value parallel to grain
    F'_{tn} is the adjusted tension design value parallel to grain
    ϕ is the resistance factor for tension = 0.8
    λ is the time effect factor (see the "Time Effect Factor, λ" section in Chapter 6)
    C_M is the wet-service factor
    C_t is the temperature factor
    C_i is the incision factor
    C_F is the size factor for sawn dimension lumber only
    K_F is the format conversion factor for tension = 2.16/ϕ
    A_n is the net cross-sectional area as follows

$$A_n = A_g - \Sigma A_h \tag{7.26}$$

where

    A_g is the gross cross-sectional area
    ΣA_h is the sum of projected area of holes

In determining the net area of a nail or a screw connection, the projected area of nail or screw is neglected. For bolted connection, the projected area consists of rectangles given by

$$\Sigma A_h = nbh \tag{7.27}$$

where

    n is the number of bolts in a row
    b is the width (thickness) of the section
    h is the diameter of hole, usually $d + 1/16$, in.
    d is the diameter of bolt

Example 7.5

Determine the size of the bottom (tension) chord of the truss shown in Figure 7.5. The service loads acting on the horizontal projection of the roof are dead load = 20 psf and snow load = 30 psf.

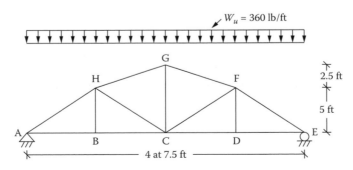

FIGURE 7.5   Roof truss of Example 7.5.

The trusses are 5 ft on center. The connection is made by one bolt of 3/4 in. diameter in each row. Lumber is Douglas-Fir Larch #1.

Solution

A. Design loads

1. Factored unit loads $= 1.2D + 1.6S = 1.2(20) + 1.6(30) = 72$ psf
2. Tributary area, ft²/ft $= 5 \times 1 = 5$ ft²/ft
3. Load/ft, $w_u = 72(5) = 360$ lb/ft
4. Load at joints

$$\text{Exterior} = 360\left(\frac{7.5}{2}\right) = 1350 \text{ lb or } 1.5 \text{ k}$$

$$\text{Interior} = 360(7.5) = 2700 \text{ lb or } 2.7 \text{ k}$$

B. Analysis of truss

1. Reaction at A: $A_y = 1.35 + 2.7 + \dfrac{2.7}{2} = 5.4$ k

2. For members at joint A, taking moment at H,
   $(5.4 - 1.35)7.5 - F_{AB}(5) = 0$
   $F_{AB} = 6.075$ k
   $F_{BC} = F_{AB} = 6.075$ k

C. Reference design value and the adjustment factors

1. $F_t = 675$ psi
2. $\lambda = 0.8$
3. $\phi = 0.8$
4. Assume a size factor $C_F = 1.5$, which will be checked later
5. $K_F = 2.16/\phi$
6. $F'_{tn} = \phi\, 675(0.8)(1.5)(2.16/\phi) = 1750$ psi or 1.75 ksi

D. Design

1. $A_{n\,reqd} = \dfrac{P_u}{F'_{tn}} = \dfrac{6.075}{1.75} = 3.47$ in.^{2}

2. For one bolt in a row and an assumed 2 in. wide section

$$h = \frac{3}{4} + \frac{1}{16} = 0.813 \text{ in.}$$

$$\Sigma nbh = (1)(1.5)(0.813) = 1.22 \text{ in.}^2$$

3. $A_g = A_n + A_h = 3.47 + 1.22 = 4.69$ in.²
   Select a 2 in. × 4 in. section, $A = 5.25$ in.²
4. Verify the size factor and revise the adjusted value if required
   For 2 in. × 4 in., $C_F = 1.5$ as assumed

DESIGN OF COLUMNS

The axial compression capacity of a member in terms of the nominal strength is

$$P_u = F'_{cn} A \tag{7.28}$$

In Equation 7.28, F'_{cn} is the adjusted LRFD reference design value for compression. To start with, the reference design compression value, F_c for the appropriate species and grade is ascertained. These values are listed in Appendices B.2 through B.4 for sawn lumber and Appendices B.7 through B.9 for GLULAM and LVL. Then the adjusted value is obtained multiplying the reference value by a string of factors. The applicable adjustment factors for sawn lumber are given in Table 6.5, for GLULAM in Table 6.6, and for LVL in Table 6.7.

For sawn lumber, the adjusted reference compression design value is

$$F'_{cn} = \phi F_c \lambda C_M C_t C_F C_i C_P K_F \tag{7.29}$$

For GLULAM and SCL, the adjusted reference compression design value is

$$F'_{cn} = \phi F_b \lambda C_M C_t C_P K_F \tag{7.30}$$

where
  F_c is the tabular reference compression design value parallel to grain
  ϕ is the resistance factor for compression $=0.90$
  λ is the time factor (see the "Time Effect Factor, λ" section in Chapter 6)
  C_M is the wet-service factor
  C_t is the temperature factor
  C_F is the size factor for dimension lumber only
  C_i is the incision factor
  C_P is the column stability factor discussed below
  K_F is the format conversion factor $=2.16/\phi$

Depending upon the relative size of a column, it might act as a *short column* when only the direct axial stress will be borne by the section or it might behave as a *long column* with a possibility of buckling and a corresponding reduction of the strength. This latter effect is considered by a column stability factor, C_P. Since this factor can be ascertained only when the column size is known, the column design is a trial procedure.

The initial size of column is decided using an estimated value of F'_{cn} by adjusting the reference design value, F_c for whatever factors are initially known in Equation 7.29 or 7.30.

Based on the trial section, the adjusted F'_{cn} from Equations 2.29 or 7.30 is assessed again and the revised section is determined from Equation 7.28.

COLUMN STABILITY FACTOR, C_P

As stated, the column stability factor accounts for buckling. The slenderness ratio expressed as KL/r is a limiting criteria of buckling. For wood, the slenderness ratio is adopted in a simplified form as KL/d, where d is the least dimension of the column section. The factor, K known as the *effective length factor*, depends upon the end support conditions of the column. The column end conditions are identified in Figure 7.6 and the values of the effective length factors for these conditions, adopted from Appendix G of the NDS, are also indicated therein.

When a column is supported differently along the two axes, the slenderness ratio K is determined with respect to each axis and the larger of the ratios is used in design.

Condition	a	b	c	d	e	f
Buckling modes	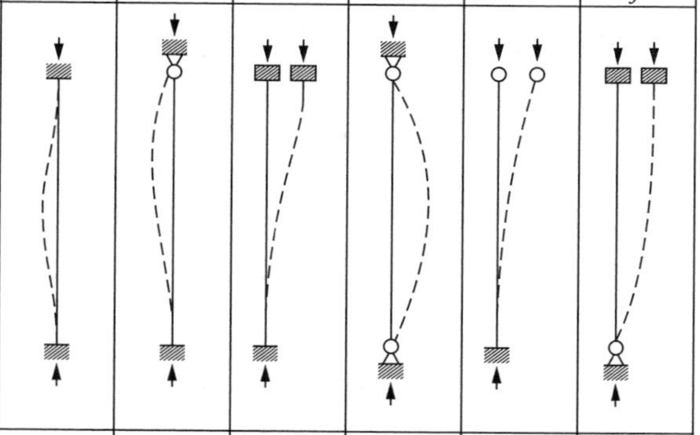					
Theoretical K_e value	0.5	0.7	1.0	1.0	2.0	2.0
Recommended design K_e when ideal conditions approximated	0.65	0.80	1.2	1.0	2.10	2.0
End condition code		Rotation fixed, translation fixed				
		Rotation free, translation fixed				
		Rotation fixed, translation free				
		Rotation free, translation free				

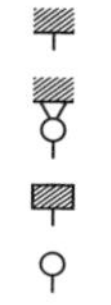

FIGURE 7.6 Buckling length coefficients, K_e. (Courtesy of American Forest & Paper Association, Washington, DC.)

The slenderness ratio should not larger than 50.

The expression for a column stability factor is similar to that of the beam stability factor, as follows:

$$C_P = \left(\frac{1+\beta}{2c}\right) - \sqrt{\left(\frac{1+\beta}{2c}\right)^2 - \left(\frac{\beta}{c}\right)} \tag{7.31}$$

where

$$\beta = \frac{F_{cEn}}{F_{cn}'^*} \tag{7.32}$$

where

$F_{cn}'^*$ is the reference design value for compression parallel to grain adjusted by all factors except C_P.
F_{cEn} is the Euler critical buckling stress

$$F_{cEn} = \frac{0.822 E_{min(n)}'}{(KL/d)^2} \tag{7.33}$$

Use $E_{min(n)}'$ value corresponding to d dimension in the equation. Determine F_{cEn} for both axes and use the smaller value.

$$\frac{KL}{d} \leq 50 \tag{7.34}$$

where

$E_{min(n)}' =$ adjusted modulus of elasticity for buckling
$\quad\quad\quad = \phi E_{min} C_M C_t C_i K_F$

where

c = buckling–crushing interaction factor

= 0.8 for sawn lumber

= 0.85 for round timber poles

= 0.9 for GLULAM or SCL

ϕ is the resistance factor for stability modulus of elasticity = 0.85

K_F is the format conversion factor for stability modulus of elasticity = $1.5/\phi$

The column behavior is dictated by the interaction of the crushing and buckling modes of failure. When C_P is 1, the strength of a column is $F_{cn}^{\prime *}$ (the adjusted reference compressive design value without C_P), and the mode of failure is by crushing. As the C_P reduces, i.e., the slenderness ratio is effective, the column fails by the buckling mode.

Example 7.6

Design a 12 ft long simply supported column. The axial loads are dead load = 1500 lb, live load = 1700 lb, and snow load = 2200 lb. Use Southern Pine #1.

Solution

A. Loads

The controlling combination is the highest ratio of the factored loads to the time effect factor.

1. $\dfrac{1.4D}{\lambda} = \dfrac{1.4(1500)}{0.6} = 3500\ \text{lb}$

2. $\dfrac{1.2D + 1.6L + 0.5S}{\lambda} = \dfrac{1.2(1500) + 1.6(1700) + 0.5(2200)}{0.8} = 7025\ \text{lb}$

3. $\dfrac{1.2D + 1.6S + 0.5L}{\lambda} = \dfrac{1.2(1500) + 1.6(2200) + 0.5(1700)}{0.8} = 7713\ \text{lb} \leftarrow \text{Controls}$

So $P_u = 1.2D + 1.6S + 0.5L = 6170\ \text{lb}$

B. Reference design values:

For 2–4 in. wide section

$F_c = 1850\ \text{psi}$

$E = 1.7 \times 10^6\ \text{psi}$

$E_{y\,min} = 0.62 \times 10^6\ \text{psi}$

C. Preliminary design

$$F_{cn}^{\prime} = \phi F_c \lambda K_F = (\phi)1850(0.8)(2.16/\phi) = 3196.8\ \text{psi}$$

$$A_{reqd} = \frac{6170}{3196.8} = 1.93\ \text{in.}^2$$

Try 2 in. × 4 in. section, $A = 5.25\ \text{in.}^2$

D. Adjusted design values

Type	Reference Design Values (psi)	ϕ	λ	C_F	K_F	$F_{()n}^{\prime}$ (psi)
Compression	1850	0.9	0.8	1	$2.16/\phi$	3196.8 ($F_{cn}^{\prime *}$)
E	1.7×10^6	—			—	1.7×10^6
E_{min}	0.62×10^6	0.85	—		$1.5/\phi$	0.93×10^6

E. Column stability factor

1. Both ends hinged, $K = 1.0$

2. $\dfrac{KL}{d} = \dfrac{1(12 \times 12)}{1.5} = 96 > 50$ **NG**

3. Revise the section to 4 in. × 4 in., $A = 12.25$ in.^{2}

4. $\dfrac{KL}{d} = \dfrac{1(12 \times 12)}{3.5} = 41.14 < 50$ **OK**

5. $F_{cEn} = \dfrac{0.822\left(0.93 \times 10^6\right)}{(41.14)^2} = 451.68$ psi

6. $\beta = \dfrac{F_{cEn}}{F_{cn}^{\prime *}} = \dfrac{451.68}{3196.8} = 0.14$

7. $C_P = \left(\dfrac{1+\beta}{2c}\right) - \sqrt{\left(\dfrac{1+\beta}{2c}\right)^2 - \left(\dfrac{\beta}{c}\right)}$

$\qquad = \dfrac{1.14}{1.6} - \sqrt{\left(\dfrac{1.14}{1.6}\right)^2 - \left(\dfrac{0.14}{0.8}\right)}$

$\qquad = 0.713 - \sqrt{(0.508) - (0.175)} = 0.136$

F. Compression capacity

1. $P_u = F_{cn}^{\prime *} C_p A$

$\qquad = (3196.8)(0.136)(12.25) = 5325 \text{ lb} < 6170 \text{ lb}$ **NG**

Use section 4 in. × 6 in., $A = 19.25$ in.^{2}

2. $KL/d = 41.14$
3. $F_{cEn} = 451.68$ psi
4. $\beta = 0.14$
5. $C_P = 0.136$
6. Capacity $= (3196.8)(0.136)(19.25) = 8369 > 6170$ lb **OK**

DESIGN FOR COMBINED BENDING AND COMPRESSION

The members stressed simultaneously in bending and compression are known as *beam-columns*. The effect of combined stresses is considered through an interaction equation. When bending occurs simultaneously with axial compression, a *second order effect* known as the P–Δ moment takes place. This can be explained this way. First consider only the transverse loading that causes a deflection, Δ. Now, when an axial load P is applied, it causes an additional bending moment equal to $P \cdot \Delta$. In a simplified approach, this additional bending stress is not computed directly. Instead, it is accounted for indirectly by amplifying the bending stress component in the interaction equation. This approach is similar to the design of the steel structures.

The amplification is defined as follows:

$$\text{Amplification factor} = \frac{1}{\left(1 - \dfrac{P_u}{F_{cEx(n)}A}\right)} \qquad (7.35)$$

where $F_{cEx(n)}$ is the Euler-based stress with respect to the x-axis slenderness as follows

$$F_{cEx(n)} = \frac{0.822 E'_{x\,min(n)}}{(KL/d)^2_x} \tag{7.36}$$

$E'_{x\,min(n)}$ = adjusted stability modulus of elasticity along x-axis

$\qquad = \phi E_{x\,min} C_M C_t C_i K_F$

$E_{x\,min}$ is the stability modulus of elasticity along the x-axis

$(KL/d)_x$ is the slenderness ratio along the x-axis

As P–Δ increases, the amplification factor or the secondary bending stresses increases.

From Equation 7.35, the amplification factor increases with the larger value of P_u. The increase of Δ is built into the reduction of the term $F_{cEx(n)}$.

In terms of the load and bending moment, the interaction formula is expressed as follows:

$$\left(\frac{P_u}{F'_{cn} A}\right)^2 + \frac{1}{\left(1 - \dfrac{P_u}{F_{cEx(n)} A}\right)} \left(\frac{M_u}{F'_{bn} S}\right) \leq 1 \tag{7.37}$$

where

F'_{cn} is the reference design value for compression parallel to grain adjusted for all factors. See Equations 7.29 and 7.30

$F_{cEx(n)}$ = see Equation 7.36

F'_{bn} is the reference bending design value adjusted for all factors. See Equations 7.2 and 7.3

P_u is the factored axial load

M_u is the factored bending moment

A is the area of cross section

S is the section modulus along the major axis

It should be noted that while determining the column adjustment factor C_P, for F'_{cn} in Equation 7.33, the maximum slenderness ratio (generally with respect to the y-axis) is used, whereas the $F_{cEx(n)}$ (in Equation 7.36) is based on the x-axis slenderness ratio.

Equation 7.37 should be evaluated for all the load combinations.

The design proceeds with a trial section which in the first iteration is checked by the interaction formula with the initial adjusted design values (without the column and beam stability factors) and without the amplification factor. The resultant value should be only a fraction of 1, preferably not exceeding 0.5.

Then the final check is made with the fully adjusted design values including the column and beam stability factors together with the amplification factor.

Example 7.7

A 16 ft long column in a building is subjected to a total tributary dead load of 4 k, and the roof live load of 5 k, in addition to the lateral wind force of 125 lb/ft. Design the column of 2DF GLULAM.

Solution

A. Load combinations

a. Vertical loads

1. $1.4D = 1.4(4) = 5.6\,k$
2. $1.2D + 1.6L + 0.5L_r = 1.2(4) + 1.6(0) + 0.5(5) = 7.3\,k$
3. $1.2D + 1.6L_r + 0.5L = 1.2(4) + 1.6(5) + 0.5(0) = 12.8\,k$

b. Vertical and lateral loads

4. $1.2D + 1.6L_r + 0.8W$ broken down into (4a) and (4b) as follows:
4a. $1.2D + 1.6L_r = 1.2(4) + 1.6(5) = 12.8\,k$ (vertical)
4b. $0.8W = 0.8(125) = 100\,lb/ft$ (lateral)
5. $1.2D + 1.6W + 0.5L + 0.5L_r$ broken down into (5a) and (5b) as follows:
5a. $1.2D + 0.5L_r = 1.2(4) + 0.5(5) = 7.3\,k$ (vertical)
5b. $1.6W = 1.6(125) = 200\,lb/ft$ (lateral)

Either 4 (4a + 4b) or 5 (5a + 5b) could be critical. Both will be evaluated.

B. Initially adjusted reference design values

Property	Reference Design Values (psi)	λ	K_F	$F'_{()n}$ (psi)	$F'_{()n}$ (ksi)
Bending	1,700	0.8	2.16/ϕ	2937.6	2.94
Compression	1,950	0.8	2.16/ϕ	3369.6	3.37
E	1.6×10^6	—	—	1.6×10^6	1.6×10^3
$E_{x\,min}$	830,000	—	1.5/ϕ	12,450,000	1.245×10^3
$E_{y\,min}$	830,000	—	1.5/ϕ	12,450,000	1.245×10^3

I. Load case 4:

C. Design loads

$P_u = 12.8\,k$

$$M_u = \frac{w_u L^2}{8} = \frac{100(16)^2}{8} = 3200\ \text{ft-lb}\ 38.4\ \text{in.-k}$$

D. Preliminary design

1. Try a section 5-⅛ in. × 7-½ in. section, $S_x = 48.05\ \text{in.}^3\ A = 38.44\ \text{in.}^2$
2. Equation 7.37 with the initial design values but without the amplification factor

$$\left[\frac{12.8}{3.37(38.44)}\right]^2 + \left[\frac{38.4}{2.94(48.05)}\right] = 0.27\ \text{a small fraction of 1}\quad\textbf{OK}$$

E. Column stability factor, C_p

1. Hinged ends, $K = 1$
2. $(KL/d)_y = \dfrac{(1)(16 \times 12)}{5.125} = 37.46 < 50$ **OK**
3. $F_{cEn} = \dfrac{0.822(1.245 \times 10^3)}{(37.46)^2} = 0.729\ \text{ksi}$

4. $\beta = \dfrac{F_{cEn}}{F_{cn}^{\prime *}} = \dfrac{0.729}{3.37} = 0.216$

5. $c = 0.9$ for GLULAM

6. $C_P = \left[\dfrac{1+0.216}{(2)(0.9)}\right] - \sqrt{\left(\dfrac{1+.216}{2(0.9)}\right)^2 - \left(\dfrac{0.216}{0.9}\right)} = 0.21$

7. $F_{cn}' = 3.37(0.21) = 0.71$ ksi

F. Volume factor, C_v

$$C_v = \left(\dfrac{5.125}{b}\right)^{1/10}\left(\dfrac{12}{d}\right)^{1/10}\left(\dfrac{21}{L}\right)^{1/10} = \left(\dfrac{5.125}{5.125}\right)^{1/10}\left(\dfrac{12}{7.5}\right)^{1/10}\left(\dfrac{21}{16}\right)^{1/10} = 1.07,\quad \text{use } 1.0.$$

G. Beam stability factor:

1. $\dfrac{L_u}{d} = \dfrac{16(12)}{7.5} = 25.6 > 14.3$

$L_e = 1.84 L_u = 1.84(16 \times 12) = 353.28$ in.

2. $R_B = \sqrt{\dfrac{L_e d}{b^2}} = \sqrt{\dfrac{(353.28)(7.5)}{(5.125)^2}} = 10.04$

3. $F_{bEn} = \dfrac{1.2\left(1.245 \times 10^3\right)}{(10.04)^2} = 14.82$ ksi

4. $\alpha = \dfrac{F_{bEn}}{F_{bn}^{\prime *}} = \dfrac{14.82}{2.94} = 5.04$

5. $C_L = \left(\dfrac{1+5.04}{1.9}\right) - \sqrt{\left(\dfrac{1+5.04}{1.9}\right)^2 - \left(\dfrac{5.04}{0.95}\right)} = 0.99$

6. $F_{bn}' = (2.94)(0.99) = 2.91$ ksi

H. Amplification factor

1. Based on the x-axis $\left(KL/d\right)_x = \dfrac{1(16 \times 12)}{7.5} = 25.6$

2. $F_{cEx(n)} = \dfrac{0.822 E_{x\,min(n)}}{\left(KL/d\right)_x^2}$

$$= \dfrac{0.822\left(1.245 \times 10^3\right)}{(25.6)^2} = 1.56 \text{ ksi}$$

3. Amplification factor $= \dfrac{1}{\left(1 - \dfrac{P_u}{F_{cEx(n)}A}\right)}$

$$= \dfrac{1}{\left[1 - \dfrac{12.8}{(1.56)(38.44)}\right]} = \dfrac{1}{0.787} = 1.27$$

I. Interaction equation, Equation 7.36

$$\left[\frac{12.8}{(0.71)(38.44)}\right]^2 + \left[\frac{1.27(38.4)}{(2.91)(48.05)}\right] = 0.22 + 0.35 = 0.57 < 1 \quad \textbf{OK}$$

II. Design load case 5:

J. Design loads:

$$P_u = 7.3 \text{ k}$$

$$M_u = \frac{w_u l^2}{8} = \frac{200(16)^2}{8} = 6400 \text{ ft-lb or } 76.8 \text{ in.-k}$$

K. Column stability factor $C_p = 0.21$ and $F'_{cn} = 0.71$ ksi from step E

L. Beam stability factor $C_L = 0.99$ and $F'_{bn} = 2.91$ ksi from step G

M. Amplification factor

$$= \frac{1}{\left(1 - \dfrac{P_u}{F_{cEx(n)}A}\right)}$$

$$= \frac{1}{\left[1 - \dfrac{7.3}{(1.56)(38.44)}\right]} = \frac{1}{0.878} = 1.14$$

L. Interaction equation, Equation 7.36

$$\left[\frac{7.3}{(0.71)(38.44)}\right]^2 + \left[\frac{1.14(76.8)}{(2.91)(48.05)}\right] = 0.07 + 0.626 = 0.7 < 1 \quad \textbf{OK}$$

PROBLEMS

7.1 Design the roof rafters with the following information:

1. Span: 10 ft
2. Spacing: 16 in. on center (OC)
3. Species: Southern Pine #1
4. Dead load = 15 psf
5. Roof live load = 20 psf
6. Roof sheathing provides the full lateral support

7.2 Design beam of Problem 7.1 except that the beam is supported only at the ends.

7.3 Design the roof rafters in Figure P7.3 with the following information:

1. Spacing 24 in. on center
2. Species: Douglas Fir Larch #1

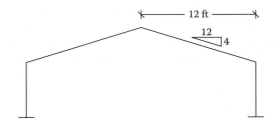

FIGURE P7.3   Roof rafters for Problem 7.3.

    3. Dead load: 15 psf
    4. Snow Load: 40 psf
    5. Wind load (vertical): 18 psf
    6. Unbraced length: support at ends only.

7.4   Design the floor beam in Figure P7.4 for the following conditions:

    1. Span, $L = 12$ ft
    2. $P_D = 500$ lb (service)
    3. $P_L = 1000$ lb (service)
    4. Unbraced length: one-half of the span
    5. Species: Hem Fir #1

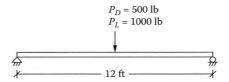

FIGURE P7.4   Floor beam for Problem 7.4.

7.5   Design the beam of Problem 7.4 for the unbraced length equal to the span.
7.6   Design the floor beam in Figure P7.6 with the following information:

    1. $w_D = 100$ lb/ft (service)
    2. $P_L = 400$ lb (service)
    3. Species: Douglas Fir Larch Select Structural
    4. Unbraced length: at the supports
    5. The beam section should not be more than 10 in. deep

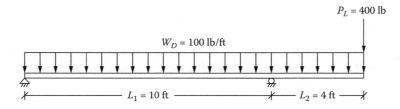

FIGURE P7.6   Floor beam for Problem 7.6.

7.7 The floor framing plan of a building is shown in Figure P7.7. Dead loads are as follows:

Floor = 12 psf
Joists = 7 psf
Beams = 9 psf
Girders = 10 psf
Live load = 40 psf

Design the beams of Southern Pine select structural timber. The beam is supported only at the ends. The beam should not have more than 12 in. depth.

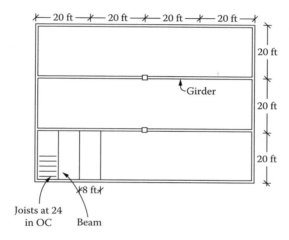

FIGURE P7.7   Floor framing plan for Problem 7.7.

7.8 Design girders for Problem 7.7 of 24F-1.8E Southern Pine GLULAM of 6-¾ in. width having a lateral bracing at the supports only.

7.9 A Douglas Fir structural glued laminated timber of 24F-1.8E is used to support a floor system. The tributary width of the beam is 12 ft and the span is 40 ft. The dead and live loads are 15 psf and 40 psf, respectively. Design the beam of 10-¾ width which is braced only at the supports.

7.10 To the beam shown in Figure P7.10 the loads are applied by purlins spaced at 10 ft on center. The beam has the lateral supports at the ends and at the locations where the purlins frame onto the beam. Design the beam of 24F-1.8E. Douglas Fir GLULAM. Use 8-¾ wide section.

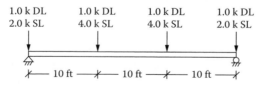

FIGURE P7.10   Load on beam by parlins for Problem 7.10.

7.11 Design Problem 7.10. The beam is used flat with bending along the minor axis. Use 10-¾ wide section.

7.12 Design the bearing plate for the supports of Problem 7.4.

7.13 Design the bearing plate for the support of Problem 7.9.

7.14 Determine the length of bearing plate placed under the interior loads of the beam of Problem 7.10.

7.15 Roof trusses, spanning 24 ft at 4 ft on center, support a deal load of 16 psf and a snow load of 50 psf only. The lumber is Hem Fir #1. The truss members are connected by a single row of ¾ in. bolts. Design the bottom chord. By truss analysis, the tensile force due to the service loads in the bottom chord members is 5.8 k. Assume the dry wood and normal temperature conditions.

[*Hint*: Divide the force in the chord between dead and snow loads in the above ratio of unit loads for factored load determination]

7.16 A warren type truss supports only dead load. The lumber is Douglas Fir-Larch #2. The end connection consists of two rows of ½ in. bolts. Determine the size of the tensile member. By truss analysis, the maximum force due to service load in the bottom chord is 5.56 k tension. Assume the dry wood and normal temperature conditions.

7.17 Design a simply supported 10 ft long column using Douglas Fir-Larch #1. The loads comprise of 10 k of dead load and 10 k of roof live load.

7.18 Design a 12 ft long simply supported column of Southern Pine #2. The axial loads are dead load = 1000 lb, live load = 2000 lb, and snow load = 2200 lb.

7.19 Design the column of Problem 7.18. A full support is provided by the sheeting about the smaller dimension.

7.20 What is the largest axial load that can be applied to a 4 in. × 6 in. #1 Hem Fir Column? The column is 15 ft long fixed at the both ends.

7.21 A 6 in. × 8 in. column carries the dead and snow loads of equal magnitude. Lumber is Douglas Fir-Larch #1. If the unbraced length of the column, which is fixed at the one end and hinged at the other end, is 9 ft, what is the load capacity of the column?

7.22 Determine the axial compression capacity of a 20 ft long GLULAM 6-¾ in. × 11 in. column of 24F-1.7E Southern Pine.

7.23 Determine the capacity column of Problem 7.22. It is braced at the center in the weaker direction.

7.24 A GLULAM column of 24F-1.8E Douglas Fir carries a dead load of 20 k and a roof live load of 40 k. The column has a simply supported length of 20 ft. Design a 8-¾ in. wide column.

7.25 The column in Problem 7.24 is braced along the weaker axis at 8 ft from the top. Design a 6-¾ in. wide column.

7.26 A 2 in. × 6 in. exterior stud wall is 12 ft tall. The studs are 16 in. on center. The studs carries the following vertical loads per foot horizontal distance of the wall:

Dead = 400 lb/ft
Live = 1000 lb/ft
Snow = 1500 lb/ft

The sheathing provides the lateral support in the weaker direction. The lumber is No. 1 Douglas Fir-Larch. Check the studs. Assume a simple end support condition and the loads on studs act axially.

7.27 The first floor (10 ft high) bearing wall of a building consists of 2 in. × 6 in. studs at 16 in. on center. The following roof loads are applied: roof dead load = 10 psf, roof live load = 20 psf, wall dead load = 5 psf, floor dead load = 7 psf, live load = 40 psf, lateral wind load = 25 psf. The tributary width of the roof framing to the bearing wall is 8 ft. The sheathing provides a lateral support to studs in the weaker direction. Check whether the wall studs made of Douglas Fir Larch #2, are adequate?

7.28 A beam column is subjected to an axial dead load of 1 k, a snow load of 0.8 k, and the lateral wind load of 100 lb/ft. The column height is 10 ft. Design the beam-column of Southern Pine #1.

7.29 A tall 20 ft long building column supports a dead load of 4 k and a live load of 5 k along with the lateral load of 150 lb/ft. Design the beam-column of 5 1/8 in. × ____ section made of 2-DF GLULAM.

7.30 A vertical 4 in. × 12 in. Southern Pine dense #1, 12 ft long member is embedded at the base to provide the fixidity. The other end is free to sway without rotation along the strong axis and is hinged along the weaker axis. The bracing about the weak axis is provided at every 4 ft by wall girts and only at the ends about the strong axis. The dead load of 1000 lb and the roof live load of 4000 lb act axially. A uniform wind load of 150 lb/ft acts along the strong axis. The sheathing provides a continuous lateral support to the compression side. Check the member for adequacy.

[*Hint*: Consider that the member is fixed at one end and has a spring support at the other end. Take the spring supported bending moment to be 70% of the maximum bending moment on the column acting like a cantilever.]

7.31 Solve Problem 7.30 when no lateral support to the compression side is provided. If 4 in. × 12 in. section in not adequate, select a new section of a maximum 12 in. depth.

7.32 Choose a Southern Pine E GLULAM column supporting two beams, as shown in Figure P7.32. The beam reactions cause bending about the major axis only. The bottom is fixed and the top is hinged.

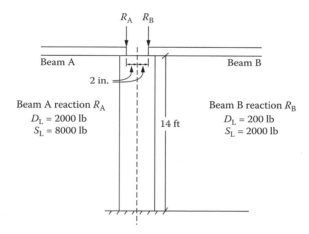

FIGURE P7.32   A column supporting two beams—Problem 7.32.

8 Wood Connections

TYPES OF CONNECTIONS AND FASTENERS

Broadly there are two types of wood connections: (1) the mechanical connections that attach members with some kind of fasteners and (2) the adhesive connections that bind members chemically together under the controlled environmental conditions such as the GLULAM. The mechanical connections, with the exception of moment splices, are not expected to transfer any moment from one element to another. The mechanical connections are classified according to the direction of load on the connector. Shear connections or lateral load connections have the load or the load component applied perpendicular to the length of the fastener. The withdrawal connections have the tensile load applied along (parallel to) the length of the fastener. When the load along the fastener length is in compression, a washer or a plate of sufficient size is provided so that the compressive strength of the wood perpendicular to the grain is not exceeded.

The mechanical type of connectors can be grouped as

1. Dowel-type connectors
2. Split ring and shear plate connectors
3. Timber rivets
4. Pre-engineered metal connectors

Dowel-type connectors comprising of nails, staples and spikes, bolts, lag bolts, and lag screws are the common type of fasteners; these are discussed in this book. The split ring and shear plate connectors fit into precut grooves and are used in shear-type connections to provide additional bearing area for added load capacity. Timber rivets or GLULAM rivets are nail-like fasteners of hardened steel (minimum strength of 145 ksi) with a countersunk head and rectangular-shaped cross section; they have no similarity to the steel rivets. These are primarily used in GLULAM members for large loads.

Pre-engineered metal connectors comprise of joist hangers, straps, ties, and anchors. These are used as accessories along with the dowel-type fasteners. They make connections simpler and easier to design and in certain cases, like earthquakes and high winds, as an essential requirement. The design strength values for specific connectors are available from the manufacturers.

DOWEL-TYPE FASTENERS (NAILS, SCREWS, BOLTS, PINS)

The basic design equation for dowel-type fasteners is

$$R_Z \text{ or } R_W \leq N Z'_n \tag{8.1}$$

where
R_Z is the factored lateral design force on a shear-type connector
R_W is the factored axial design force on a withdrawal-type connector
N is the number of fasteners
Z'_n is the adjusted reference design value of fastener given as

$$Z'_n = \text{reference design value } (Z) \times \text{adjustment factors} \tag{8.2}$$

The term reference design value, Z, refers to the basic load capacity of a fastener. The shear-type connections rely on the bearing strength of wood against the metal fastener or the bending yield strength of the fastener (not the shear rupture of the fastener as in steel design). The withdrawal-type connections rely on the frictional or interfacial resistance to the transfer of loads. Until the 1980s, the capacities of fasteners were obtained from the empirical formulas based on the field and laboratory tests. However, in the subsequent approach, the yield mechanism is considered from the principles of engineering mechanics. The yield-related approach is limited to the shear-type or laterally loaded connections. The withdrawal-type connections are still designed from the empirical formulas.

YIELD LIMIT THEORY FOR LATERALLY LOADED FASTENERS

The yield limit theory considers the various modes by which a connection can yield under a lateral load. The capacity is computed for each mode of yielding. Then the reference value is taken as the smallest of these capacities.

In yield limit theory, the primary factors that contribute to the reference design value comprise the following:

1. Fastener diameter, D
2. Bearing length, l
3. Dowel-bearing strength of wood, F_{ew}, controlled by the (1) specific gravity of wood; (2) angle of application of load to the wood grain, θ; and (3) relative size of the fastener
4. Bearing strength of metal side plates, F_{ep}
5. Bending yield strength, F_{yb}

A subscript m or s is added to the above factors to indicate whether they apply to the main member or the side member. For example, l_m and l_s refer to bearing lengths of main member and side member, respectively. For bolted connections, the bearing length l and member thickness are identical, as shown in Figure 8.1.

For nail, screw, or lag bolt connections, the bearing length of main member, l_m, is less than the main member thickness, as shown in Figure 8.2.

Depending upon the mode of yielding, one of the strength terms corresponding to items 3, 4, or 5 above or their combinations are the controlling factor(s) for the capacity of the fastener. For example, in the bearing-dominated yield of the wood fibers in contact with the fastener, the term F_{ew} for wood will be a controlling factor; for a metal side member used in a connection, the bearing strength of metal plate F_{ep} will control.

For a fastener yielding in bending with the localized crushing of the wood fibers, both F_{yb} and F_{ew} will be the relevant factors. The various yield modes are described in the next section.

1. The *dowel-bearing strength of wood*, also known as the *embedded strength*, F_{ew} (item 3 above), is the crushing strength of the wood member. Its value depends on the specific gravity of wood. For large-diameter fasteners ($\geq\frac{1}{4}$ in.), the bearing strength also depends

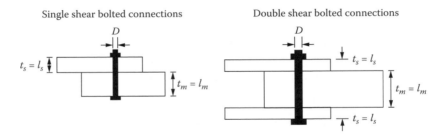

FIGURE 8.1 Bolted connection bearing length.

on the angle of load to grains of wood. The NDS provide the values of specific gravity, G, for various species and their combinations and also include the formulas and tables for the dowel-bearing strength, F_{ew}, for the two cases of loading—the load acting parallel to the grains and the load applied perpendicular to the grains.

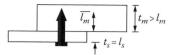

FIGURE 8.2 Nail or screw connection bearing length.

2. The *bearing strength of steel members* (item 4 above) is based on the ultimate tensile strength of steel. For hot-rolled steel members (usually of thickness $\geq\frac{1}{4}$ in.), $F_{ep} = 1.5\ F_u$, and for cold-formed steel members (usually $<\frac{1}{4}$ in.) $F_{ep} = 1.375 F_u$.

3. The *fastener bending yield strength*, F_{yb} (item 5 above), has been listed by the NDS for various types and diameters of fasteners. These values can be used in the absence of the manufacturer's data.

YIELD MECHANISMS AND YIELD LIMIT EQUATIONS

The dowel-type fasteners have the following four possible modes of yielding.

MODE I: Bearing yield of wood fibers when stress distribution is uniform over the entire thickness of the member.

In this case, due to the high lateral loading, the dowel-bearing stress of a wood member uniformly exceeds the strength of wood. This mode is classified as I_m if the bearing strength is exceeded in the main member and as I_s if the side member is overstressed, as shown in Figure 8.3.

MODE II: Bearing yield of wood by crushing due to maximum stress near the outer fibers.

The bearing strength of wood is exceeded in this case also. However, the bearing stress is not uniform. In this mode, the fastener remains straight but undergoes a twist that causes the flexure-like nonuniform distribution of stress with the maximum stress at the outer fibers. The wood fibers are accordingly crushed at the outside face of both members, as shown in Figure 8.4.

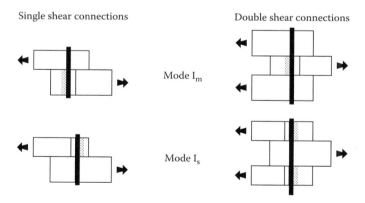

FIGURE 8.3   Mode I yielding. (Courtesy of American Forest & Paper Association, Washington, DC.)

FIGURE 8.4   Mode II yielding. (Courtesy of American Forest & Paper Association, Washington, DC.)

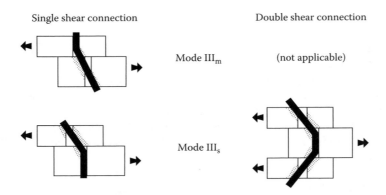

FIGURE 8.5 Mode III yielding. (Courtesy of American Forest & Paper Association, Washington, DC.)

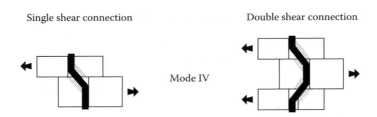

FIGURE 8.6 Mode IV yielding. (Courtesy of American Forest & Paper Association, Washington, DC.)

Mode II yield occurs simultaneously in the main and side members. It is not applicable to a double-shear connection because of symmetry by the two side plates.

MODE III: Fastener bends at one point within a member and wood fibers in contact with the fastener yield in bearing.

This is classified as III_m when the fastener bending occurs in the main member and the wood bearing strength is exceeded in the main member. Likewise, III_s indicates the bending and crushing of wood fibers in the side member, as shown in Figure 8.5.

Mode III_m is not applicable to a double-shear connection because of symmetry by the two side plates.

MODE IV: Fastener bends at two points in each shear plane and wood fibers yield in bearing near the shear plane(s).

Mode IV occurs simultaneously in the main and side members in a single shear, as shown in Figure 8.6. However, in a double shear, this can occur in each plane, hence yielding can occur separately in the main member and the side member.

To summarize, in a single-shear connection, there are six modes of failures comprising of I_m, I_s, II, III_m, III_s, and IV. Correspondingly, there are six yield limit equations derived for the single-shear connections. For a double-shear connection, there are four modes of failures comprising of I_m, I_s, IV_m, and IV_s. There are four corresponding yield limit equations for the double-shear connections.

REFERENCE DESIGN VALUES FOR LATERAL LOADS (SHEAR CONNECTIONS)

For a given joint configuration, depending upon the single or the double-shear connection, six- or four-yield limit equations are evaluated and the smallest value obtained from these equations is used as a reference design value, Z.

Instead of using the yield limit equations, the NDS provide the tables for the reference design values that evaluate all relevant equations and adopt the smallest values for various fastener properties and specific gravity of species. The selected reference design values for the lateral loading are included in Appendices B.10 through B.16 for different types of fasteners.

As stated previously, for fasteners of ¼ in. or larger, the angle of loading with respect to the wood grains also affects the reference design values. The NDS tables include two cases: one for the loads parallel to the grains and one for the perpendicular loads. The loads that act at other angles involve the application of Hankinson formula, which has not been considered in this book.

A reference design value, Z, obtained by the yield limit equations or from the NDS tables, is then subjected to the adjustment factors to get the adjusted reference design value, Z'_n, to be used in Equation 8.1. The adjustment factors are discussed in the "Adjustments of the Reference Design Values" section.

REFERENCE DESIGN VALUES FOR WITHDRAWAL LOADS

Dowel-type fasteners are much less stronger in withdrawal capacity. The reference design values for different types of fasteners in lb/in. of penetration is given by the empirical formulas that are the functions of specific gravity of species and diameter of the fasteners. The NDS provide the tables based on these formulas. The selected reference design values for withdrawal loading are included in the tables for different types of fasteners.

ADJUSTMENTS OF THE REFERENCE DESIGN VALUES

Table 8.1 specifies the adjustment factors that apply to the lateral loads and withdrawal loads for dowel-type fasteners.

The last three factors, K_F, ϕ_z, and λ are relevant to the LRFD only. For connections, their values are

$K_F = 2.16/\phi_z$

$\phi_z = 0.65$

$\lambda =$ as given in the "Time Effect Factor, λ" section in Chapter 6

The other factors are discussed below.

WET SERVICE FACTOR, C_M

For connections, the listed reference design values are for seasoned wood having a moisture content of 19% or less. For wet woods or those exposed to wet conditions, the multiplying factors of less than 1 are specified in the NDS Table 10.3.3 of the *National Design Specifications for Wood Construction* cited in References and Bibliography.

TEMPERATURE FACTOR, C_t

For connections that will experience sustained exposure to higher than 100°F temperature, a factor of less than 1 shall be applied, as specified in the NDS Table 10.3.4 of the *National Design Specifications for Wood Construction* cited in References and Bibliography.

GROUP ACTION FACTOR, C_g

The load carried by a row of fasteners is not equally divided among the fasteners; the end fasteners in a row carry a larger portion load as compared to the interior fasteners. A row consists of a number of fasteners in a line parallel to the direction of loading. The unequal sharing of loads is accounted for by the group action factor, C_g.

TABLE 8.1
Adjustment Factors for Dowel-Type Fasteners

	LRFD Only					Factors				
Loads	Format Conversion	Resistance Factor	Time Effect	Wet Service	Temperature	Group Action	Geometry	End Grain	Diaphragm	Toenail
Lateral loads, $Z'_n = Z \times$ (nails and spikes only)	K_F	ϕ_z	λ	C_M	C_t	C_g	C_Δ	C_{eg}	C_{di}	C_{tn}
Withdrawal, $Z'_n = Z \times$ (nails and spikes only)	K_F	ϕ_z	λ	C_M	C_t	—	—	C_{eg}	—	C_{tn}

For dowel-type fasteners of diameter less than ¼ in. (i.e., nails and wood screws), $C_g=1$. For ¼ in. or larger diameter fasteners, C_g is given by a formula, which is quite involved. The NDS provide tabulated values for simplified connections. The number of fasteners in a single row is the primary consideration. For bolts and lag screws, conservatively, C_g has the following values (nails and screws have $C_g=1$) (Table 8.2).

TABLE 8.2
Conservative Value of the Group Action Factor

Number of Fasteners in One Row	C_g
2	0.97
3	0.90
4	0.80

GEOMETRY FACTOR, C_Δ

When the diameter of a fastener is less than ¼ in. (nails and screws), $C_\Delta=1$. For larger diameter fasteners, the geometry factor accounts for the end distance, edge distance, and spacing of fasteners, as defined in Figure 8.7.

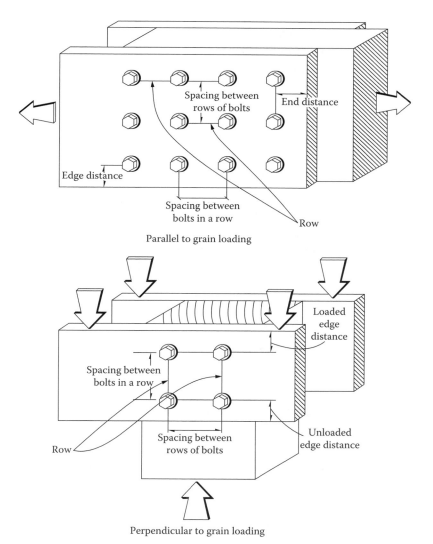

FIGURE 8.7 Connection geometry. (Courtesy of American Forest & Paper Association, Washington, DC.)

1. The edge distance requirements, according to the NDS, are given in Table 8.3, where l/D is lesser of the following two:

   a. $\dfrac{l_m}{D} = \dfrac{\text{bearing length of bolt in main member}}{\text{bolt diameter}}$

   b. $\dfrac{l_s}{D} = \dfrac{\text{combined bearing length of bolt in all side members}}{\text{bolt diameter}}$

2. The spacing requirements between rows, according to the NDS, are given in Table 8.4, where l/D are defined above
3. The end distance requirements, according to the NDS, are given in Table 8.5
4. The spacing requirements for fasteners along a row, according to the NDS, are given in Table 8.6

TABLE 8.3
Minimum Edge Distance

Direction of Loading	Minimum Edge Distance
1. Parallel to grains	
When $l/D \leq 6$	$1.5D$
When $l/D > 6$	$1.5D$ or ½ spacing between rows, whichever is greater
2. Perpendicular to grains	
Loaded edge	$4D$
Unloaded edge	$1.5D$

TABLE 8.4
Minimum Spacing Between Rows

Direction of Loading	Minimum Spacing
1. Parallel to grains	$1.5D$
2. Perpendicular to grains	
When $l/D \leq 2$	$2.5D$
When $l/D > 2$ but <6	$(5l + 10D)/8$
When $l/D \geq 6$	$5D$

TABLE 8.5
Minimum End Distance

Direction of Loading	End Distance for $C_\Delta = 1$	Minimum End Distance for $C_\Delta = 0.5$
1. Parallel to grains		
Compression	$4D$	$2D$
Tension—softwood	$7D$	$3.5D$
Tension—hardwood	$5D$	$2.5D$
2. Perpendicular to grains	$4D$	$2D$

TABLE 8.6
Minimum Spacing in a Row

Direction of Loading	Spacing for $C_\Delta = 1$	Minimum Spacing
1. Parallel to grains	$4D$	$3D$
2. Perpendicular to grains	On side plates (attached member) spacing should be $= 4D$	$3D$

The provisions for C_Δ are based on the assumption that the edge distance and the spacing between rows are met in accordance with Tables 8.3 and 8.4, respectively.

The requirements of the end distance and the spacing along a row for $C_\Delta = 1$ are given in the second column of Tables 8.5 and 8.6. The tables also indicate the (absolute) minimum requirements. When the actual end distance and the actual spacing along a row are less than those indicated for $C_\Delta = 1$, the value of C_Δ should be computed by the ratio as follows:

$$C_\Delta = \frac{\text{actual end distance or actual spacing along a row}}{\text{end distance for } C_\Delta = 1 \text{ from Table 8.5 or spacing } C_\Delta = 1 \text{ from Table 8.6}}$$

END GRAIN FACTOR, C_{eg}

In a shear connection, load is perpendicular to the length (axis) of the fastener, and in a withdrawal connection, load is parallel to the length of the fastener. But in both cases, the length (axis) of the fastener is perpendicular to the wood fibers (fastener is installed in the side grains). However, when a fastener penetrates into an end grain so that the fastener axis is parallel to the wood fibers, as shown in Figure 8.8, it is a weaker connection.

For a withdrawal-type loading, $C_{eg} = 0.75$. For a lateral (shear)-type loading, $C_{eg} = 0.67$.

DIAPHRAGM FACTOR, C_{di}

This applies to nails and spikes only. When nails or spikes are used in a diaphragm construction, $C_{di} = 1.1$.

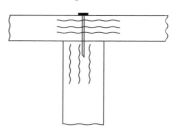

FIGURE 8.8 End grain factor.

TOENAIL FACTOR, C_{tn}

This applies to nails and spikes only. In many situations, it is not possible to directly nail a side member to a holding member. Toenails are used in the side member at an angle of about 30° and starts at about 1/3 of the nail length from the intersection of the two members, as shown in Figure 8.9.

For lateral loads, $C_{tn} = 0.83$. For withdrawal loads, $C_{tn} = 0.67$. For withdrawal loads, the wet service factor is not applied together with C_{tn}.

Example 8.1

The reference lateral design value for the parallel-to-grain loaded lag screw connection shown in Figure 8.10 is 1110 lb. Determine the adjusted reference design value. Diameter of screws is

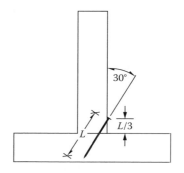

FIGURE 8.9 Toenail factor.

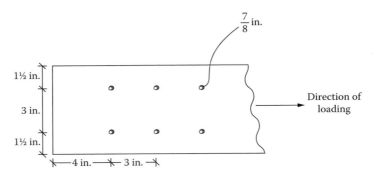

FIGURE 8.10   Parallel-to-grain loaded connection.

7/8 in. The connection is subjected to dead and live tensile loads in a dry softwood at normal temperatures.

Solution

1. Adjusted reference design value, $Z'_n = Z \times (\phi_z \lambda K_F C_g C_\Delta)$; since C_M and $C_t = 1$
2. $\phi_z = 0.65$
3. $\lambda = 0.8$
4. $K_F = 2.16/\phi_z$
5. Group action factor, C_g
   For three fasteners in a row, $C_g = 0.90$ (from Table 8.2)
6. Geometry factor, C_Δ
   a. End distance = 4 in.
   b. End distance for $C_\Delta = 1$, $7D = 7\left(\dfrac{7}{8}\right) = 6.125$ in.
   c. End factor $= \dfrac{4.0}{6.125} = 0.65$
   d. Spacing along a row = 3 in.
   e. Spacing for $C_\Delta = 1$, $4D = 3.5$ in.
   f. Spacing factor $= \dfrac{3.0}{3.5} = 0.857$
   g. $C_\Delta = (0.65)(0.857) = 0.557$
7. $Z'_n = 1110(\phi_z)(0.8)(2.16/\phi_z)\,(0.9)(0.557) = 961.5$ lb

Example 8.2

The reference lateral design value for the perpendicular-to-grain loaded bolted connection shown in Figure 8.11 is 740 lb. Determine the adjusted reference design value. Bolt diameter is 7/8 in. Use soft dry wood and normal temperature condition. The connection is subjected to dead and live loads.

Solution

1. Adjusted reference design value, $Z'_n = Z \times (\phi_z \lambda K_F C_g C_\Delta)$; since C_M and $C_t = 1$
2. $\phi_z = 0.65$
3. $\lambda = 0.8$
4. $K_F = 2.16/\phi_z$
5. Group action factor, C_g
   For two fasteners in a row, $C_g = 0.97$ (from Table 8.2)
6. Geometry factor, C_Δ

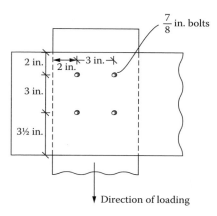

FIGURE 8.11   Perpendicular-to-grain loaded connection.

    a. End distance $= 2$ in.

    b. End distance for $C_\Delta = 1$, $4D = 4\left(\dfrac{7}{8}\right) = 3.5$ in.

    c. End factor $= \dfrac{2.0}{3.5} = 0.57$

    d. Spacing along a row $= 3$ in.

    e. Spacing for $C_\Delta = 1$, $4D = 4\left(\dfrac{7}{8}\right) = 3.5$ in.

    f. Spacing factor $= \dfrac{3.0}{3.5} = 0.857$

    g. $C_\Delta = (0.57)(0.857) = 0.489$

  7. $Z'_n = 740(\phi_z)(0.8)(2.16/\phi_z)\,(0.97)(0.489) = 606.5$ lb

Example 8.3

The connection of Example 8.1 when loaded in withdrawal mode has a reference design value of 500 lb. Determine the adjusted reference withdrawal design value.

Solution

  1. Adjusted reference design value, $Z'_n = Z \times (\phi_z \lambda K_F)$;

  2. $\phi_z = 0.65$

  3. $\lambda = 0.8$

  4. $K_F = 2.16/\phi_z$

  5. $Z'_n = 500(\phi_z)(0.8)\,(2.16/\phi_z) = 864$ lb

NAIL AND SCREW CONNECTIONS

Once the adjusted reference design value is determined, Equation 8.1 can be used with the factored load to design a connection for any dowel-type fasteners. The nails and wood screws generally fall into small-size fasteners having a diameter of less than ¼ in. For small-size fasteners, the angle of load with respect to grains of wood is not important. Moreover, the group action factor, C_g, and the geometry factor, C_Δ, are not applicable. The end grain factor, C_{eg}, the diaphragm factor, C_{di}, and the toenail factor, C_{tn} apply to specific cases. Thus, for a common type of dry wood under normal temperature conditions, no adjustment factors are required to be used except for the special LRFD factors of ϕ_z, λ, and K_F.

The basic properties of nails and wood screws are described below.

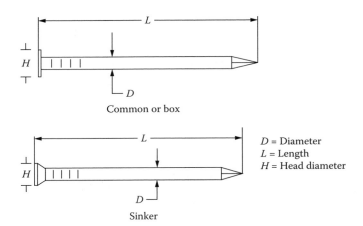

Common or box

Sinker

D = Diameter
L = Length
H = Head diameter

FIGURE 8.12 Typical specifications of nails.

NAILS

Nails are specified by the pennyweight, abbreviated as d. A nail of a specific pennyweight has a fixed length, L, the shank diameter, D, and head size, H. There are three kinds of nails: common, box, and sinker. The common and box nails have a flat head and the sinker nails have a countersunk head, as shown in Figure 8.12.

For the same pennyweight, the box and sinker nails have a smaller diameter and, hence, a lower capacity as compared to common nails.

The reference lateral design values for the simple nail connector is given in Appendix B.10. The values for the other cases are included in the NDS specifications. The reference withdrawal design values for nails of different sizes for various wood species are given in Appendix B.11.

WOOD SCREWS

Wood screws are identified by a number. A screw of a specific number has a fixed diameter (outside to outside of threads) and a fixed root diameter, as shown in Figure 8.13. Screws of each specific number are available in different lengths. There are two types of screws known as *cut thread screws* and *rolled thread screws*. The thread length, T, of a cut thread screw is approximately 2/3 of the screw length, L. In rolled thread screw, the thread length, T, is at least four times the screw diameter, D, or 2/3 of the screw length, L, whichever is greater. The screws that are too short to accommodate the minimum thread length have threads extended as close to the underside of the head as practical.

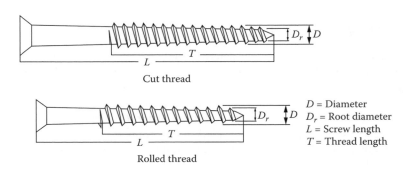

Cut thread

Rolled thread

D = Diameter
D_r = Root diameter
L = Screw length
T = Thread length

FIGURE 8.13 Typical specifications of wood screws.

The screws are inserted in its lead hole by turning with a screw driver and not driven by a hammer. The minimum penetration of the wood screw into the main member for single shear or into the side member for double shear should be six times the diameter of the screw. Wood screws are not permitted to be used in withdrawal-type connection in end grain.

The reference lateral design values for simple wood screw connection are given in Appendix B.12. The values for other cases are included in the NDS specifications. The reference withdrawal design values for wood screws are given in Appendix B.13.

Example 8.4

A 2 in. × 6 in. diagonal member of No. 1 Southern pine is connected to a 4 in. × 6 in. column. It is acted upon by a service wind load component of 1.25 k, as shown in Figure 8.14. Design the nailed connection. Neglect the dead load.

Solution

1. Factored design load, $R_z = 1.6\,(1.25) = 2\,k$ or 2000 lb
2. Use 30d nails, 3 in a row
3. Reference design value for a side thickness of 1.5 in.
   From Appendix B.10, $Z = 203$ lb
4. For nails, the adjusted reference design value

$$Z_n' = Z \times (\phi_z \lambda K_F)$$

where
   $\phi_z = 0.65$
   $\lambda = 1.00$
   $K_F = 2.16/\phi_z$

$$Z_n' = 203\,(\phi_z)(1)(2.16/\phi_z) = 438.5\ lb$$

5. From Equation 8.1

$$N = \frac{R_z}{Z_n'} = \frac{2000}{438.5} = 4.56\ nails$$

6. For number of nails per row, $n = 3$

$$\text{Number of rows} = \frac{4.56}{3} = 1.52\ (\text{use 2})$$

Provide 2 rows of 3 nails each of 30d size.

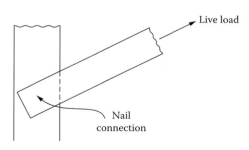

FIGURE 8.14 Diagonal member nail connection.

BOLT AND LAG SCREW CONNECTIONS

Bolts and lag screws are used for larger loads. The angle of load to grains is an important consideration in large diameter (≥1/4 in.) connections comprising of bolts and lag screws. However, this book makes use of the reference design tables, in lieu of the yield limit equations, that include only the two cases of parallel-to-grain and perpendicular-to-grain conditions. The group action factor, C_g, and the geometry factor, C_Δ, apply to bolts and lag screws. Although the end grain factor, C_{eg}, is applicable, it is typical to a nail connection. The other two factors, the diaphragm factor, C_{di}, and the toenail factor, C_{tn}, also apply to nails.

The larger diameter fasteners often involve the use of prefabricated steel accessories or hardware. The NDS provide details of the typical connections involving various kinds of hardware.

Bolts

In steel structures, the trend is to use high strength bolts. However, this is not the case in wood structures where low strength A307 bolts are commonly used. Bolt sizes used in wood construction range from ½ in. through 1 in. diameter, in increments of 1/8 in. The NDS restrict the use of bolts to a largest size of 1 in. The bolts are installed in the predrilled holes. The NDS specify that the hole size should be a minimum of 1/32 in. to a maximum of 1/16 in. larger than the bolt diameter for uniform development of the bearing stress.

Most bolts are used in the lateral-type connections. They are distinguished by the single-shear (two members) and the double-shear (three members) connections. For more than double shear, the single-shear capacity at each shear plane is determined and the value of the weakest shear plane is multiplied by the number of shear planes.

The connections are further recognized by the types of main and side members, such as wood-to-wood, wood-to-metal, wood-to-concrete, and wood-to-masonry connections. The last two are simply termed as *anchored* connections.

Washers of adequate size are provided between the wood member and the bolt head, and between the wood member and the nut. The size of washer is not of a significance in shear. For bolts in tension and compression, the size should be adequate so that the bearing stress is within the compression strength perpendicular to the wood grain.

The reference lateral design values for simple bolted connection are given in Appendix B.14.

Lag Screws

Lag screws are relatively larger than wood screws. They have wood screw threads and a square or hexagonal bolt head. The dimensions for lag screws include the nominal length, L; diameter, D; root diameter, D_r; unthreaded shank length, S; minimum thread length, T; length of tapered tip, E; number of threads per in., N; height of head, H; and width of head across flats, F, as shown in Figure 8.15.

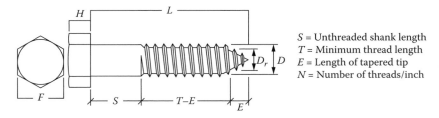

S = Unthreaded shank length
T = Minimum thread length
E = Length of tapered tip
N = Number of threads/inch

FIGURE 8.15 Typical specifications of lag screws.

Lag screws are used when an excessive length of bolt will be required to access the other side or when the other side of a through bolted connection is not accessible. Lag screws are used in shear as well as withdrawal applications.

Lag screws are installed with a wrench as against wood screws by screwdrivers. Lag screws involved prebored holes with two different diameter bits. The larger diameter hole has the same diameter and length as the unthreaded shank of the lag screw and the lead hole for the threaded portion is similar to that for wood screw, the size of which depends on the specific gravity of wood. The minimum penetration (excluding the length of the tapered tip) into the main member for single shear and into the side member for double shear should be four times the lag screw diameter, D.

The reference lateral design values for simple lag screw connection are given in Appendix B.15. The other cases are included in the NDS specifications. The reference withdrawal design values for lag screws are given in Appendix B.16.

Example 8.5

The diagonal member of Example 8.4 is subjected to a wind load component of 2.5 k. Design the bolted connection. Use 5/8 in. bolts.

Solution

1. Factored design load, $R_Z = 1.6(2.5) = 4$ k or 4000 lb
2. Use 5/8 in. bolts, three in a row
3. Reference design value
   a. For a side thickness of 1.5 in.
   b. Main member thickness of 3.5 in.
   c. From Appendix B.14, $Z = 940$ lb
4. Adjusted reference design value, $Z'_n = Z \times (\phi_z \lambda K_F C_g C_\Delta)$
5. $\phi_z = 0.65$
   $\lambda = 1.0$
   $K_F = 2.16/\phi_z$
6. Group action factor, C_g
   For three fasteners in a row, $C_g = 0.90$ (from Table 8.2)
7. Geometry factor, C_Δ
   a. End distance for $C_\Delta = 1$, $7D = 4.375$ in.
   b. Spacing for $C_\Delta = 1$, $4D = 2.5$ in.
   c. $C_\Delta = 1$ (assuming above conditions are satisfied)
8. $Z'_n = 940(\phi_z)(1)(2.16/\phi_z)(0.90) = 1827.4$ lb
9. From Equation 8.1

$$N = \frac{R_Z}{Z'_n} = \frac{4000}{1827.4} = 2.2 \text{ (use 3)}$$

10. For number of bolts per row, $n = 3$

$$\text{Number of rows} = \frac{3}{3} = 1$$

Provide 1 row of three 5/8 in. bolts

PROBLEMS

8.1 The reference lateral design value of a parallel-to-grain loaded lag screw connection shown in Figure P8.1 is 740 lb. The screw diameter is 5/8 in. The loads comprise of dead and live loads. Determine the adjusted reference design value for soft dry wood at normal temperature.

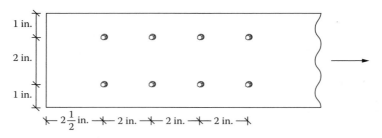

FIGURE P8.1   A parallel-to-grain lag screw connection.

8.2 The reference lateral design value of a perpendicular-to-grain loaded lag screw connection shown in Figure P8.2 is 500 lb. The screw diameter is 5/8 in. The loads comprise of dead and live loads. Determine the adjusted reference design value for soft dry wood at normal temperature.

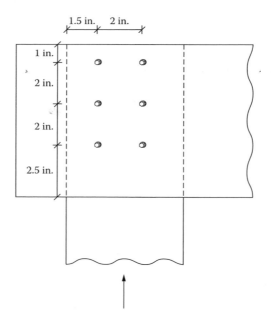

FIGURE P8.2   A perpendicular-to-grain lag screw connection.

8.3 The connection of Problem 8.1 has a reference withdrawal design value of 400 lb. Determine the adjusted reference design value.

8.4 Problem 8.2 has the nailed connection of 0.225 in. diameter nails. The holding member has fibers parallel to the nail axis. The reference design value is 230 lb. Determine the adjusted reference design value.

8.5 A parallel-to-grain loaded connection uses two rows of 7/8 in. lag screws with three fasteners in each row, as shown in Figure P8.5. The load carried is 1.2D + 1.6L. The reference design value is 1500 lb. The connection is in hard dry wood at normal temperature. Determine the adjusted reference design value.

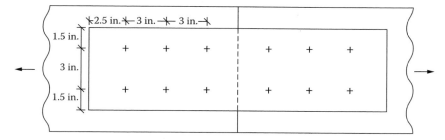

FIGURE P8.5   Spliced parallel-to-grain connection.

8.6 The connection of Problem 8.5 is subjected to a perpendicular-to-grain load from top only. The reference design value is 1000 lb. Determine the adjusted reference design value.

8.7 The connection of Problem 8.5 is subjected to withdrawal loading. The reference design value is 500 lb. Determine the adjusted reference design value.

8.8 The connection shown in Figure P8.8 uses ¾ in. diameter bolts in a single shear. There are two bolts in each row. The reference design value is 2000 lb. It is subjected to lateral wind load only (no live load). Determine the adjusted reference design value for soft dry wood at normal temperature.

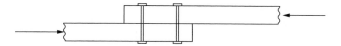

FIGURE P8.8   A single shear connection.

8.9 For the connection shown in Figure P8.9, adjust the reference design value which is 1000 lb. Use dry wood under normal temperature conditions.

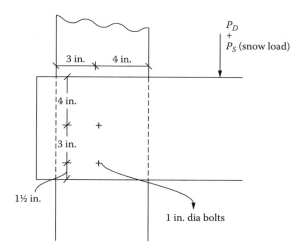

FIGURE P8.9   A perpendicular-to-grain bolted connection.

8.10 Toenails of $50d$ pennyweight (0.244 in. diameter, 51/2 in. length) are used to connect a beam to the top plate of a stud wall, as shown in Figure P8.10. It is subjected to the dead and live loads. The reference design value is 250 lb. Determine the adjusted reference design value for soft wood under normal temperature and dry conditions. Show the connection.

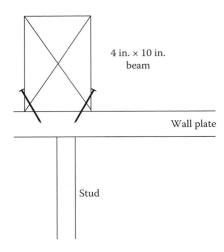

4 in. × 10 in.
beam

Wall plate

Stud

FIGURE P8.10   Toenails connection to a top plate.

8.11 Design a nail connection to transfer tensile service dead and live loads of 400 and 600 lb, respectively, acting along the axis of a 2 in. × 6 in. diagonal member connected to a 4 in. × 4 in. vertical member. Use No. 1 Southern Pine soft dry wood. Assume two rows of $30d$ nails.

8.12 Determine the number and show the placement of $20d$ nails to transfer the factored dead and snow load of 1.5 k from a 2 in. × 8 in. diagonal member attached at an angle of 30° to a 4 in. × 6 in. vertical member. The loads act vertically downward. Use No. 1 Douglas Fir-Larch dry wood. [*Hint*: The reference design value requires the adjustment for a load that acts at an angle to wood fibers. However, ignore this effect.]

8.13 Determine the tensile capacity of a spliced connection acted upon by the dead and snow loads. The joint connects two 2 in. × 6 in. No. 1 Southern Pine members together by $10d$ nails through two side plates of 1 in. thickness, as shown in Figure P8.13.

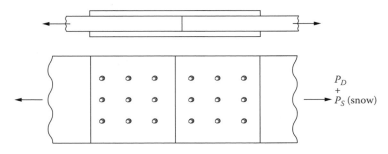

P_D
+
P_S (snow)

FIGURE P8.13   A spliced nailed connection.

8.14 Two 2 in. × 8 in. members of No. 1 Douglas Fir-Larch are to be spliced connected with a single 1½ in. thick plate on the top with two rows of nine size screws. The service loads comprise of 200 lb of dead load and 500 lb of live load that act normal to the fibers. Design the connection.

8.15 Southern Pine #1, 10 ft long 2 in. × 4 in. wall studs, spaced at 16 in. on center are nailed on to Southern Pine #1 top and bottom plates with two 10*d* nails at each end. The horizontal service wind load of 30 psf acts on the toenails. Is the connection adequate?

8.16 The service dead load and live load in Problem 8.11 are increased five folds. Design a lag screw connection using three rows of ¾ in. lag screws. Assume the edge distance, end distance and bolt spacing of 4 in. each.

8.17 Design a ½ in. lag screw connection to transfer 1.5 k of unfactored snow load. A 2 in. × 6 in. is connected to a 4 in. × 6 in. member, as shown in Figure P8.17. The wood is soft Hem Fir-Larch No. 1 in dry condition at normal temperature.

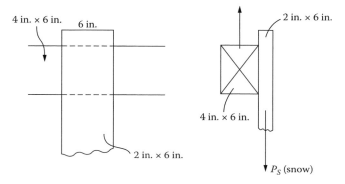

FIGURE P8.17 A beam–column single shear connection.

8.18 Determine the number and placement of ⅝ in. bolts to transfer a service snow load of 3 k through a joint, as shown in Figure P8.18. Use a reference design value of 830 lbs.

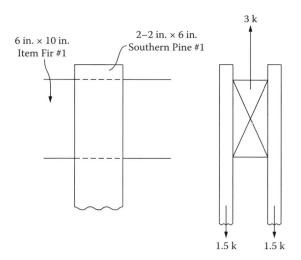

FIGURE P8.18 A beam–column double shear connection.

8.19 The controlling load on the structural member of Problem 8.17 is an unfactored wind load of 2 k that acts horizontally. Design the ½ in. bolted connection.

8.20 A bolted spliced joint consists of 3 in. × 10 in. main member and one 2 in. × 10 in. side member of Southern Pine #1 soft dry wood, connected by 6 1 in. bolts in two rows in each splice. Determine the joint capacity for dead and live loads. The edge distance, end distance and bolt spacing are 3 in. each.

Part III

Steel Structures

9 Tension Steel Members

PROPERTIES OF STEEL

Steel structures commonly consist of frames made of column and beam elements. The bracing in the form of diagonal members are provided to resist the lateral loads. For steel elements, generally, the standard shapes that are specified according to the American Society of Testing Materials (ASTM) standards are used and the properties of these elements are listed in the beginning of the manual of the American Institute of Steel Construction (AISC) under the "Dimensions and Properties" section. A common element is an I-shaped section having horizontal flanges that are connected at the top and bottom of a vertical web. This type of section is classified into W, M, S, and HP shapes, the difference in these shapes essentially being in the width and thickness of flanges. A typical designation "W14×68" means a wide flange section having a nominal depth of 14 in. and a weight of 68 lb/ft of length. The other standard shapes are channels (C and MC), angles (L), and tees (WT, MT, and ST).

Tubular shapes are common for compression members. The rectangular and square sections are designated by letters HSS along with the outer dimensions and the wall thickness. The round tubing is designated as HSS round (for Grade 42) and pipes (for Grade 35) along with the outer diameter and the wall thickness. The geometric properties of the frequently used wide flange sections are given in Appendix C.1, channel sections in Appendix C.2, angle sections in Appendix C.3, rectangular tubing in Appendix C.4, square tubing in Appendix C.5, round tubing in Appendix C.6, and pipes in Appendix C.7.

The structural shapes are available in many grades of steel classified according to the ASTM specifications. The commonly used grades of steel for various structural shapes are listed in Table 9.1.

The yield strength is a very important property of steel because so many design procedures are based on this value. For all grades of steel, the modulus of elasticity is practically the same at a level of 29×10^3 ksi, which means the stress–strain relation of all grades of steel is similar.

A distinguished property that makes steel a very desirable structural material is its ductility—a property that indicates that a structure will withstand extensive amount of deformation under very high level of stresses without failure.

THE 2005 UNIFIED DESIGN SPECIFICATIONS

A major unification of the codes and specifications has been done by the American Institute of Steel Construction (AISC). Formerly, the AISC provided four design publications, one separately for the allowable stress design (ASD) method, the load resistance factor design (LRFD) method, the single-angle members and the hollow tubular structural sections. The 13th edition of the *Steel Construction Manual* of AISC 2005 combined and updated all these provisions in a single volume. In addition, the 2005 AISC specifications have established the common requirements of these methods for the analyses and designs of the structural elements and to apply the same sets of specifications for the ASD and LRFD.

TABLE 9.1
The Common Steel Grades

ASTM Classification	Yield Strength, F_y (ksi)	Ultimate Strength, F_u (ksi)	Applicable Shapes
A36	36	58	W, M, S, HP, L, C, MC, WT
A572 Grade 50	50	65	Same
A992 Grade 50	50	65	Same
A500 Grade B	46	58	HSS—rectangular and square
A500 Grade B	42	58	HSS—round
A53 Grade B	35	60	Pipe—round

The factors unifying the two approaches are as follows:

1. The nominal strength is the limiting state for failing of a steel member under different modes like compression, tension, or bending. It is the capacity of the member. The same limits must be considered under both philosophies of design.
2. For the ASD, the available strength is the allowable strength, which is the nominal strength divided by a factor of safety. The available strength for the LRFD is the design strength, which is the nominal strength multiplied by a resistance (uncertainty) factor.
3. The required strength for a member is given by the total of the service loads that act on the structure for the ASD method. The required strength for the LRFD method is given by the total of the factored (magnified) loads.
4. The required strength for loads should not exceed the available strength of the material.

Since the allowable strength of the ASD and the design strength of the LRFD are both connected with the nominal strength as indicated in item 2 above, there can be a direct relationship between the factor of safety of the ASD and the resistance factor of the LRFD. This has been discussed in the "Working Stress Design, Strength Design, and Unified Design of Structures" section in Chapter 1.

LIMIT STATES OF DESIGN

All designs are based on checking of the limit states. For each member type (tensile, column, beam), the AISC specifications identify the limit states that should be checked. The limit states consider all possible modes of failures like yielding, rupture, and buckling, and also consider the serviceability limit states like deflection and slenderness.

The limit states design process consists of the following:

1. Determine all applicable limit states (modes of failures) for the type of member to be designed.
2. Determine the expression for the nominal strength (and the available strength) with respect to each limit state.
3. Determine the required strength for the loads and their combinations.
4. Configure the member size equating items 2 and 3 above of this section.

In the ASD, the safety is established through a safety factor, which is independent of the types of loading. In the LRFD, the safety is established through a resistance factor and a load factor that varies with load types and load combinations.

DESIGN OF TENSION MEMBERS

In the *AISC Manual* (2005), Chapter D of Part 16 applies to the members that are subject to axial tension and Section J4 of Chapter J applies to the connections and the connecting elements like gusset plates that are in tension.

The limiting states for the tensile members and the connecting elements are controlled by the following modes:

1. Tensile strength
2. Shear strength of connection
3. Block shear strength of connection along the shear/tension failure path

The shear strength of connection (item 2) will be discussed in Chapter 13 on steel connection.

TENSILE STRENGTH OF ELEMENTS

The design tensile strength of a member shall be lower of the values obtained for the limit states of (1) the *tensile yielding* at the gross area and (2) the *tensile rupture* at the net area.

Thus, the strength is lower of the following two values.

- Based on the limit state of yielding in the gross section

$$P_u = 0.9 F_y A_g \tag{9.1}$$

- Based on the limit state of rupture in the net section

$$P_u = 0.75 F_u A_e \tag{9.2}$$

where
  P_u is the factored design tensile load
  F_y is the yield strength of steel
  F_u is the ultimate strength of steel
  A_g is the gross area of member
  A_e is the effective net area

In connecting members, if a portion of a member is not fully connected like a leg of an angle section, the unconnected part is not subjected to the full stress. This is referred to as a *shear leg*. A factor is used to account for the shear lag. Thus,

$$A_e = A_n U \tag{9.3}$$

where
  A_n is the net area
  U is the shear lag factor

The serviceability limit state of the slenderness ratio L/r* to be less than 300 for members in tension has been relaxed in the new specifications although it is still followed in the practice.

---

* L is the length of the member and r is the radius of gyration given by $\sqrt{I/A}$.

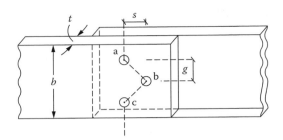

FIGURE 9.1 Zigzag pattern of holes.

NET AREA, A_N

The net area is the product of the thickness and the net width of a member. To compute net width, the sum of widths of the holes for bolts is subtracted from the gross width. The hole width is taken as 1/8 in. greater than the bolt diameter.

For a chain of holes in a zigzag line shown as a-b-c in Figure 9.1, a quantity $s^2/4g$ is added to the net width for each zigzag of the gage space, g in the chain, e.g., for a-b and b-c in Figure 9.1. Thus,

$$A_n = bt - \sum ht + \sum \left(\frac{s^2}{4g}\right) t \qquad (9.4)$$

where
  s is the longitudinal (in the direction of loading) spacing between two consecutive holes (pitch)
  g is the transverse (perpendicular to force) spacing between the same two holes (gage)

For angles, the gage for holes in the opposite legs, as shown in Figure 9.2, is $g = g_1 + g_2 - t$.

Example 9.1

An angle ∟ 5 × 5 × 1/2* has a staggered bolt pattern, as shown in Figure 9.3. The holes are for bolts of 7/8 in. diameter. Determine the net area.

Solution

1. $A_g = 4.75$ in.², $t = 0.5$ in.
2. $h = d + (1/8) = (7/8) + (1/8) = 1$ in.
3. $g = g_1 + g_2 - t = 3 + 2 - 0.5 = 4.5$ in.
4. Section through line a-b-d-e: deducting for two holes

$$A_n = A_g - \sum ht$$

$$= 4.75 - 2(1)(0.5) = 3.75 \text{ in.}^2$$

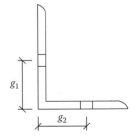

FIGURE 9.2 Gage for holes in angle section.

5. Section through line a-b-c-d-e: deducting for three holes and adding $s^2/4g$ for b-c and c-d

---

* Properties of this section not included in the appendix.

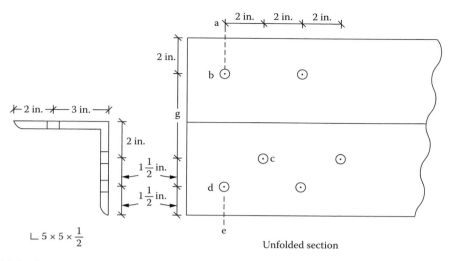

FIGURE 9.3 Bolts pattern for Example 9.1.

$$A_n = A_g - 3ht + \left(\frac{s^2}{4g}\right)_{bc} t + \left(\frac{s^2}{4g}\right)_{cd}$$

$$= 4.75 - 3(1)(0.5) + \left[\frac{2^2}{4(4.5)}\right]0.5 + \left[\frac{2^2}{4(1.5)}\right]0.5$$

$$= 3.69 \,\text{in.}^2 \leftarrow \text{Controls}$$

SHEAR LAG FACTOR FOR UNATTACHED ELEMENTS

1. For plates, $U = 1$ except for item 2.
2. For plates with longitudinal welds only (in the direction of loading) and no transferred weld (Figure 9.4),
   $U = 1.0$ for $L \geq 2w$
   $U = 0.87$ for $L < 2w$ and $\geq 1.5w$
   $U = 0.75$ for $L < 1.5w$
3. For single angles,
   $U = 0.8$ for four or more bolts in the direction loading
   $U = 0.6$ for two to three bolts in the direction of loading
4. For W, M, S, HP, WT,
   $U = 0.9$ for flange connection with three or more bolts in the direction of loading where width of section $\geq 2/3$ depth

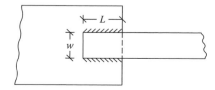

FIGURE 9.4 Welded members—definition sketch.

$U=0.85$ for flange connection with three or more bolts in the direction of loading where width of section $<2/3$ depth

$U=0.7$ for web connection through four or more bolts in the direction of loading.

In the older AISC manuals, a factor of 0.75 was recommended for all shapes and categories not covered above.

Example 9.2

Determine the effective net area for the single-angle member in Example 9.1.

Solution

1. Since the number of bolts in the direction of loading are 2, $U=0.6$.
2. From Example 9.1, $A_n=3.69$ in.^{2}
3. $A_e=A_nU=(3.69)(0.6)=2.21$ in.^{2}

Example 9.3

What is the design strength of element of Example 9.1 for A36 steel?

Solution

1. $A_g=4.75$ in.^{2}
2. $A_e=2.21$ in.^{2} (from Example 9.2)
3. From Equation 9.1,

$$P_u = 0.9F_yA_g = 0.9(36)(4.75) = 153.9 \text{ k.}$$

4. From Equation 9.2,

$$P_u = 0.75F_uA_e = 0.75(58)(2.21) = 96.14 \text{ k} \leftarrow \text{Controls.}$$

BLOCK SHEAR STRENGTH

In certain connections, a *block* of material at the end of the member may tear out. In a single-angle member shown in Figure 9.5, the block shear failure may occur along the plane abc. The shaded block will fail by shear along the plane ab and tension in the section bc.

Figure 9.6 shows a tensile plate connected to a gusset plate. In this case, the block shear failure could occur in both the gusset plate and the main tensile member. The tensile failure occurs along the section ad and the shear failure along the planes ab and cd.

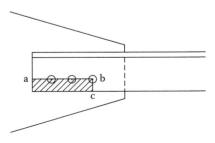

FIGURE 9.5 Block shear in a single angle member.

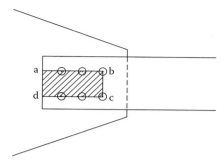

FIGURE 9.6   Block shear in a plate member.

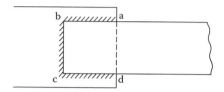

FIGURE 9.7   Block shear in a welded member.

A welded member shown in Figure 9.7 experiences the block shear failure along the welded planes abcd. It has the tensile area along bc and the shear area along ab and cd.

Both the tensile area and shear area contribute to the strength. The resistance to shear block will be the sum of the strengths of the two surfaces.

The resistance (strength) to shear block is given by a single two parts equation

$$R_u = \phi R_n = \phi\left(0.6 F_u A_{nv} + U_{bs} F_u A_{nt}\right) \le \phi\left(0.6 F_y A_{gv} + U_{bs} F_u A_{nt}\right) \qquad (9.5)$$

where

ϕ is the resistance factor $= 0.75$
A_{nv} is the net area subjected to shear
A_{nt} is the net area subjected to tension
A_{gv} is the gross area along the shear surface
$U_{bs} = 1.0$ when the tensile stress is uniform (most cases)
$\quad = 0.5$ when the tensile is nonuniform

Example 9.4

An L6 × 4 × 1/2* tensile member of A36 steel is connected by three 7/8 in. bolts, as shown in Figure 9.8. Determine the strength of the member.

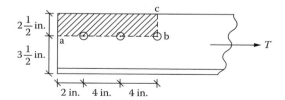

FIGURE 9.8   A three-bolt connection—Example 9.4.

---

* Section properties not included in the appendix.

Solution

I. Tensile strength of member
   A. Yielding in gross area
      1. $A_g = 4.75$ in.^{2}
      2. $h = (7/8) + (1/8) = 1$ in.
      3. From Equation 9.1,

      $$P_u = 0.9(36)(4.75) = 153.9 \text{ k}$$

   B. Rupture in net area
      1. $A_n = A_g -$ one hole area
         $= 4.75 - (1)(1)(1/2) = 4.25$ in.^{2}
      2. $U = 0.6$ for three bolts in a line
      3. $A_e = UA_n = 0.6\,(4.25) = 2.55$ in.^{2}
      4. From Equation 9.2

      $$P_u = 0.75(58)(2.55) = 110.9 \text{ k} \leftarrow \text{Controls}$$

II. Block shear strength
   A. Gross shear area along ab

   $$A_{gv} = 10\left(\frac{1}{2}\right) = 5 \text{ in.}^2$$

   B. Net shear area along ab

   $$A_{nv} = A_{gv} - 2\,\tfrac{1}{2}\text{ holes area}$$

   $$= 5 - 2.5(1)\left(\frac{1}{2}\right) = 3.75 \text{ in.}^2$$

   C. Net tensile area along bc

   $$A_{nt} = 2.5t - 1/2 \text{ hole}$$

   $$= 2.5\left(\frac{1}{2}\right) - \frac{1}{2}(1)\left(\frac{1}{2}\right) = 1.0 \text{ in.}^2$$

   D. $U_{bs} = 1.0$
   E. From Equation 9.4

   $$\phi(0.6F_u A_{nv} + U_{bs}F_u A_{nt}) = 0.75[0.6(58)(3.75) + (1)(58)(1.0)] = 141.4 \text{ k}$$
   $$\phi(0.6F_y A_{gv} + U_{bs}F_u A_{nt}) = 0.75[0.6(36)(5) + (1)(58)(1.0)] = 124.5 \text{ k}$$

The strength is 110.9 k controlled by rupture of the net section.

DESIGN PROCEDURE FOR TENSION MEMBERS

The type of connection used for a structure affects the choice of the tensile member. The bolt-type connections are convenient for members consisting of the angles, channels W and S shapes. The welded connection suits the plates, channels, and structural tees.

The procedure to design a tensile member consists of the following:

1. Determine the critical combination(s) of factored loads.
2. For each critical load combination, determine the gross area required by Equation 9.1 and select a section.
3. Make provision for the holes or for weld based on the connection requirements, and determine the effective net area.
4. Compute the loading capacity of the effective net area of the selected section by Equation 9.2. This capacity should be more than the design load(s) of step 1. If it is not, revise the selection.
5. Check for the block shear strength by Equation 9.5. If it is not adequate, either revise the connection or revise the member size.
6. The limitation of the maximum slenderness ratio of 300 has been removed in the AISC 2005. However, it is still a preferred practice except for rods and hangers.

Although rigid frames are common in steel structures, roof trusses having nonrigid connections are used for industrial or mill buildings. The members in the bottom chord of a truss are commonly in tension. Some of the web members are in tension and the others are in compression. With changing of the wind direction, the forces in the web members alternate between tension and compression. Accordingly, the web members have to be designed to function both as tensile as well as compression elements.

Example 9.5

A roof system consists of a warren type roof truss, as shown in Figure 9.9. The trusses are spaced 25 ft apart. The following loads are passed on to the truss through the purlins. Design the bottom chord members consisting of the two angles section separated by a 3/8 in. gusset plate. Assume one line of two 3/4 in. diameter bolts at each joint. Use A572 steel.

Dead load (deck, roofing, insulation) = 10 psf
Snow = 25 psf
Roof LL = 20 psf
Wind (vertical) = 18 psf

Solution
A. Computation of loads
    1. Adding 20% to dead load for the truss weight, $D = 12$ psf.
    2. Consider the following load combinations:
       (a) $1.2D + 1.6(L_r$ or $S) + 0.8W = 1.2(12) + 1.6(25) + 0.8(18)$
                                       $= 68.8$ psf ← Controls
       (b) $1.2D + 1.6W + 0.5(L_r$ or $S) = 1.2(12) + 1.6(18) + 0.5(25) = 55.7$ psf

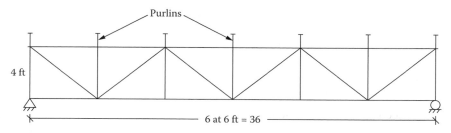

FIGURE 9.9 A warren roof truss.

3. Tributary area of an entire truss $= 36 \times 25 = 900\,\text{ft}^2$.
4. Total factored load on the truss $= 68.8 \times 900 = 61,920\,\text{lb}$ or $61.92\,\text{k}$.
5. This load is distributed through 6 purlins comprising of 5 interior purlins and one-half on each ends since the exterior joint tributary is one-half that of the interior joints. Thus, the joint loads are

$$\text{Interior joints} = \frac{61.92}{6} = 10.32\,\text{k}$$

$$\text{Exterior joints} = \frac{10.32}{2} = 5.16\,\text{k}.$$

B. Analysis of truss
   1. The loaded truss is shown in Figure 9.10.
   2. Reaction @ L_0 and $L_6 = 61.62/2 = 30.96\,\text{k}$
   3. The bottom chord members L_2L_3 and L_3L_4 are subjected to the maximum force. A free-body diagram of the left of section a-a is shown in Figure 9.11.
   4. M @ $U_2 = 0$
      $-30.96\,(12) + 5.16\,(12) + 10.32\,(6) + F_{L2L3}\,(4) = 0$
      $F_{L2L3} = 61.92\,\text{k} \leftarrow P_u$
C. Design of member
   1. From Equation 9.1

$$A_g = \frac{P_u}{0.9F_y} = \frac{61.92}{0.9(50)} = 1.38\,\text{in.}^2$$

Try $2\,\llcorner 3 \times 2 \times 1/4\; A_g = 2.4\,\text{in.}^2$

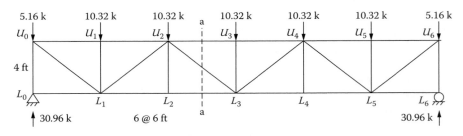

FIGURE 9.10   Analysis of truss.

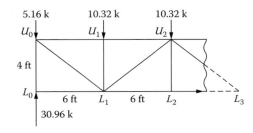

FIGURE 9.11   Free body diagram of truss.

2. $h = (3/4) + (1/8) = (7/8)$ in.

$A_n = A_g$ − one hole area

$$= 2.40 - (1)\left(\frac{7}{8}\right)\left(\frac{1}{4}\right) = 2.18\,in.^2$$

$U = 0.6$ for 2 bolts per line
$A_e = 0.6\,(2.18) = 1.31$ in.^{2}

3. From Equation 9.2

$P_u = 0.75 F_u A_e$

$$= 0.75(65)(1.31) = 63.86\,k > 61.92\,k\ \mathbf{OK}$$

D. Check for block shear strength
    Similar to Example 9.4

PROBLEMS

9.1 A 1/2 in. × 10 in. plate is attached to another plate by means of six 3/4 in. diameter bolts, as shown in Figure P9.1. Determine the net area of the plate.

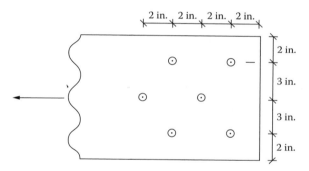

FIGURE P9.1   Plate to plate connection.

9.2 A 3/4 in. × 10 in. plate is connected to a gusset plate by 7/8 in. diameter bolts, as shown in Figure P9.2. Determine the net area of the plate.

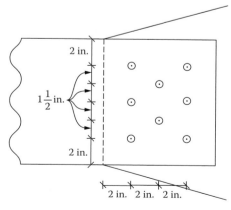

FIGURE P9.2   Plate to gusset plate connection.

9.3 An ⌐ 5×5×1/2 has staggered holes for 3/4 in. diameter bolts, as shown in Figure P9.3. Determine the net area for the angle. (A_g = 4.75 in.²)

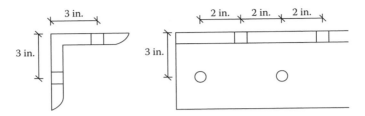

FIGURE P9.3 Staggered angle connection.

9.4 An ⌐ 8×4×1/2 has staggered holes for 7/8 in. diameter bolts, as shown in Figure P9.4. Determine the net area. (A_g = 5.75 in.²)

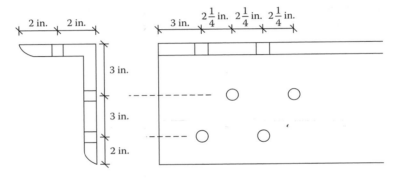

FIGURE P9.4 Staggered long leg angle connection.

9.5 A channel section C9×20 has the bolt pattern shown in Figure P9.5. Determine the net area for 3/4 in. bolts.

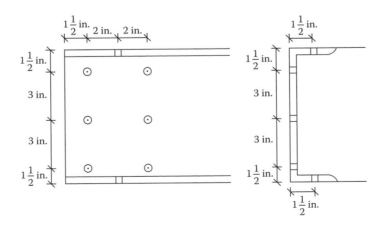

FIGURE P9.5 Staggered channel connection.

9.6 Determine the effective net area for Problem 9.2.

9.7 Determine the effective net area for Problem 9.3.

9.8 Determine the effective net area for Problem 9.4.

9.9 Determine the effective net area for the connection shown in Figure P9.9 for an ∟ 5×5×½.

9.10 For Problem 9.9 with welding both in the longitudinal and transverse directions, determine the effective net area.

9.11 Determine the tensile strength of the plate in Problem 9.1 for A36 steel.

9.12 A tensile member of Problem 9.4 is subjected to a dead load of 30 k and a live load of 60 k. Is the member adequate? Use A572 steel.

9.13 Is the member of Problem 9.9 adequate to support the following loads all acting in tension? Use A992 steel.

Dead load = 25 k
Live load = 50 k
Snow load = 40 k
Wind load = 35 k

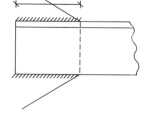

FIGURE P9.9 Welded connection.

9.14 An angle of A36 steel is connected to a gusset plate with six 3/4 in. bolts, as shown in Figure P9.14. The member is subjected to a dead load of 50 k and a live load of 110 k. Design a 6 in. size (6×?) member. (6×6×1, A_g=11 in.²; 6×4×⅞, A_g=7.98; 6×3½×½, A_g=4.5)

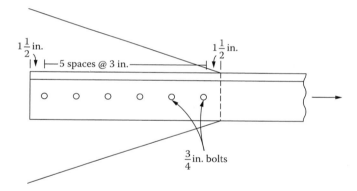

FIGURE P9.14 A connection for Problem 9.14.

9.15 An angle of A36 steel is connected by 7/8 in. bolts, as shown in Figure P9.15. It is exposed to a dead load of 60 k, a live load of 100 k, and a wind load of 40 k. Design a 8 in. size (8×?) member. (8×6×⅞, A_g=11.5; 8×4×¾, A_g=8.44; 8×8×½, A_g=7.75)

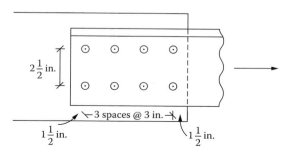

FIGURE P9.15 Two rows connection for Problem 9.15.

9.16 Compute the strength including the block shear capacity of a member comprising ∟ 3½×3½×1/2 as shown in Figure P9.16. The bolts are 3/4 in. Steel is A36.

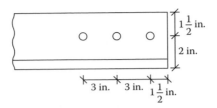

FIGURE P9.16   A tensile member for Problem 9.16.

9.17 A tensile member comprises of W 12×30 section of A36 steel, as shown in Figure P9.17 with each side of flanges having three holes for 7/8 in. bolts. Determine the strength of the member including the block shear strength.

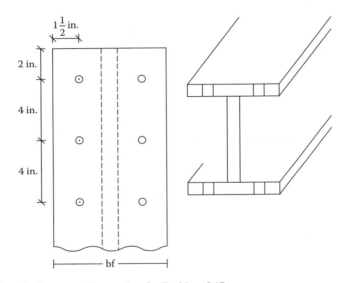

FIGURE P9.17   A wide flange tensile member for Problem 9.17.

9.18 Determine the strength of the welded member shown in Figure P9.18 including the block shear capacity. Steel is A572.

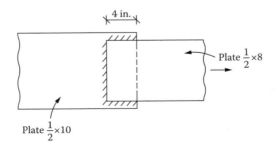

FIGURE P9.18   A welded member for Problem 9.18.

10 Compression Steel Members

STRENGTH OF COMPRESSION MEMBERS OR COLUMNS

The basic strength requirement or compression in the LRFD format is

$$P_u \leq \phi P_n \tag{10.1}$$

where

P_u is the factored axial load

$\phi = 0.9$, resistance factor for compression

P_n is the nominal compressive strength of column

For a compression member that fails by yielding, $P_n = F_y A_g$, similar to a tensile member. However, the steel columns are leaner, i.e., the length dimension is much larger than the cross-sectional dimension. Accordingly, the compression capacity is more often controlled by the rigidity of the column against buckling instead of the yielding. There are two common modes of failure in this respect.

1. **Local instability**: If the parts (elements) comprising a column are relatively very thin, a localized buckling or winkling of one or more of these elements may occur prior to the instability of the entire column. Based on the ratio of width to thickness of the element, a section is classified as a *slender* or a *non-slender* section for the purpose of local instability.
2. **Overall instability**: Instead of an individual element getting winkled, the entire column may bend or buckle lengthwise under the action of the axial compression force. This can occur in three different ways.
2a. **Flexural buckling**: A deflection occurs by bending about the weak axis, as shown in Figure 10.1. The slenderness ratio is a measure of the flexure buckling of a member. When the buckling occurs at a stress level within the proportionality limit of steel, it is called the *elastic buckling*. When the stress at buckling is beyond the proportionality limit, it is an *inelastic buckling*. The columns of any shape can fail in this mode by either the elastic or inelastic buckling.
2b. **Torsional buckling**: This type of failure is caused by the twisting of the member longitudinally, as shown in Figure 10.2. The standard hot-rolled shapes are not susceptible to this mode of buckling. A thinly built-up section may be exposed to torsional buckling.
2c. **Flexural–torsional buckling**: This failure occurs by the combination of flexure and torsional buckling when a member twists while bending, as shown in Figure 10.3. Only the section with a single axis of symmetry or an nonsymmetric section such as a channel, tee, and angle is subjected to this mode of buckling.

The nominal compressive strength, P_n in Equation 10.1 is the lowest value obtained according to the limit states of flexural buckling, torsional buckling, and flexural–torsional buckling.

The flexural buckling limit state is applicable to all sections. For singly symmetric, nonsymmetric, and certain double symmetric cruciform and built-up sections, torsional buckling or flexural–torsional buckling are also applicable.

The limit states are considered separately for the categories of the non-slender or the slender sections according to the local instability criteria.

The torsional buckling alone or in combination with the flexural buckling can be too complex to evaluate. It is desirable to prevent it when feasible. This could be done by bracing the member to prevent twisting.

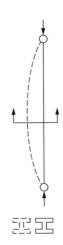

FIGURE 10.1 Flexural buckling.

LOCAL BUCKLING CRITERIA

In the context of local buckling, the elements of a structural section are classified in two categories:

1. *Unstiffened element*: It has an unsupported edge (end) parallel to (along) the direction of the load, like an angle section.
2. *Stiffened element*: It is supported along both of its edges, like the web of a wide flange section.

The two types of elements are illustrated in Figure 10.4.

When the ratio of width to thickness of an element of a section is greater than the specified limit λ_r, as shown in Table 10.1, it is classified as a slender shape. The cross section of a slender element is not fully effective in resisting a compressive force. Such elements should be avoided or else their strength should be reduced, as discussed in the "Slender Compression Members" section. The separate provisions for strength reduction are made in the AISC manual for stiffened and unstiffened sections. The terms are explained in Figure 10.4.

FLEXURAL BUCKLING CRITERIA

The term (KL/r), known as the *slenderness ratio* is important in column design. Not only the compression capacity of a column depends upon the slenderness ratio, the ratio sets a limit between the elastic and nonelastic buckling of the column. When the slenderness ratio exceeds a value of $4.71\sqrt{E/F_y}$, the column acts as an elastic column and the limiting (failure) stress level is within the elastic range.

According to the classic Euler formula, the critical load is inversely proportional to $(KL/r)^2$, where K is the effective length factor (coefficient) discussed in the "Effective Length Factor for Slenderness Ratio" section, L is the length of the column, and r is the radius of gyration given by $\sqrt{I/A}$.

Although it is not a mandatory requirement in the *AISC Manual 2005*, the AISC recommends that the slenderness ratio for a column should not exceed a value of 200.

EFFECTIVE LENGTH FACTOR FOR SLENDERNESS RATIO

The original flexural buckling or Euler formulation considered that the column is pinned at both ends. The term K was introduced to account for the other end conditions because the end condition will make a column to buckle differently. For example, if a column is fixed at both

FIGURE 10.2 Torsional buckling.

ends, it will buckle at the points of inflection about $L/4$ distance away from the ends, with an effective length of one-half of the column length. Thus, the effective length of a column is the distance at which the column is assumed to buckle in the shape of an elastic curve. The length between the supports, L is multiplied by a factor to calculate the effective length.

When columns are part of a frame, they are constrained at the ends by their connection to beams and to other columns. The effective length factor for such columns is evaluated by the use of the alignment charts or nomographs given in Figures 10.5 and 10.6; the former is for the braced frames where the sidesway is prevented and the latter is for the moment frames where the sidesway is permitted.

In the nomographs, the subscripts A and B refer to two ends of a column for which K is desired. The term G is the ratio of the column stiffness to the girder stiffness expressed as

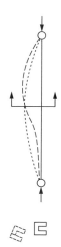

FIGURE 10.3 Flexural–torsional buckling.

$$G = \frac{\sum I_c/L_c}{\sum I_g/L_g}$$

(10.2)

where
   I_c is the moment of inertia of column section
   L_c is the length of column
   I_g is the moment of inertia of girder beam meeting the column
   L_g is the length of girder
   Σ is the summation of all members meeting at joint A for G_A and at joint B for G_B

The values of I_c and I_g are taken about the axis of bending of the frame. For a column base connected to the footing by a hinge, G is taken as 10 and when the column is connected rigidly (fixed) to the base, G is taken as 1.

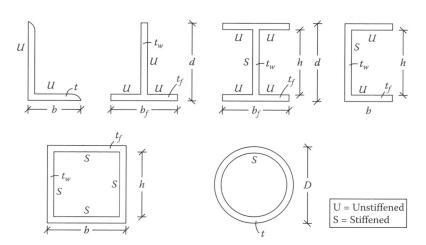

FIGURE 10.4   Stiffened and unstiffened elements.

TABLE 10.1
Slenderness Limit for Compression Member

Element	Width:Thickness Ratio	λ_r
W, S, M, H	$b_f/2t_f$	$0.56\sqrt{E/F_y}$
	h/t_w	$1.49\sqrt{E/F_y}$
C	b_f/t_f	$0.56\sqrt{E/F_y}$
	h/t_w	$1.49\sqrt{E/F_y}$
T	$b_f/2t_f$	$0.56\sqrt{E/F_y}$
	d/t_w	$0.75\sqrt{E/F_y}$
Single L or double L with separation	b/t	$0.45\sqrt{E/F_y}$
Box, tubing	b/t	$1.4\sqrt{E/F_y}$
Circular	D/t	$0.11E/F_y$

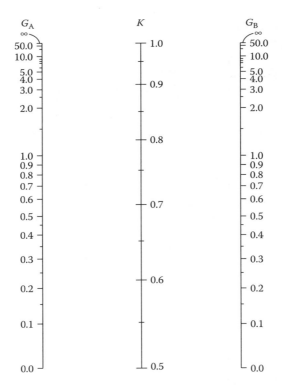

FIGURE 10.5 Alignment chart, sidesway prevented. (Courtesy of American Institute of Steel Construction, Chicago, IL.)

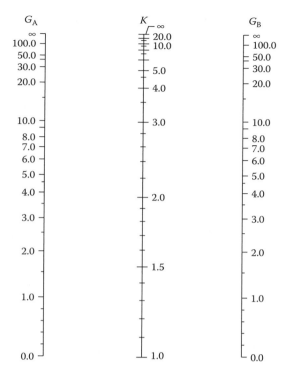

FIGURE 10.6 Alignment chart, sideway not prevented. (Courtesy of American Institute of Steel Construction, Chicago, IL.)

After determining G_A and G_B for a column, K is obtained by connecting straight line between points G_A and G_B on the nomograph. Since the values of I (moment of inertia) of the columns and beams at the joint are required to determine G, the factor K can not be determined unless the size of the columns and the beams are known. On the other hand, the factor K is required to determine the column size. Thus, these nomographs need some preliminary assessments of the value of K and the dimensions of the columns and girders.

One of the conditions for the use of the nomographs or the alignment charts is that all columns should buckle elastically, i.e., $KL > 4.71\sqrt{E/F_y}$. If a column buckles inelastically, a stiffness reduction factor, τ_a has to be applied. The factor τ_a is the ratio of the tangent modulus of elasticity to the modulus of elasticity of steel. The value has been tabulated in the AISC manual as a function of P_u/A_g. Without τ_a, the value of K is on a conservative side.

However, in lieu of applying the monographs, in a simplified method the factors (coefficients) listed in Figure 7.6 are used to ascertain the effective length. Figure 7.6 is used for isolated columns also. When Figure 7.6 is used for the unbraced frame columns, the lowest story (base) columns are approximated by the condition (f) with $K=2$, and the upper story columns are approximated by the condition (c) with $K=1.2$. For braced frames, the condition (a) with $K=0.65$ is a good approximation.

Example 10.1

A rigid unbraced moment frame is shown in Figure 10.7. Determine the effective length factors with respect to weak axis for members AB and BC.

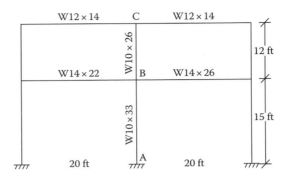

FIGURE 10.7   An unbraced frame.

Solution

1. The section properties and G ratios are arranged in the table below:

Joint	Column				Girder				G
	Section	I, in.⁴	L, ft	I/L	Section	I, in.⁴	L, ft	I/L	
A	Fixed								1
B	W10×33	171	15	11.40[a]	W14×22	199	20	9.95	
	W10×26	144	12	12.00	W14×26	245	20	12.25	
	Σ			23.40				22.20	23.4/22.20 = 1.05
C	W10×26	144	12	12.00	W12×14	88.6	20	4.43	
					W12×14	88.6	20	4.43	
	Σ			12.00				8.86	12.00/8.86 = 1.35

[a] Mixed units (I in in.⁴ and L in ft) can be used since the ratio is being used.

2. Column AB
   From Figure 10.6, the alignment chart for unbraced frame (side sway permitted) connecting a line from $G_A = 1$ to $G_B = 1.05$, $K = 1.3$.
3. Column BC
   From the alignment chart with $G_A = 1.05$ (point B) and $G_B = 1.35$ (point C), $K = 1.38$.

LIMIT STATES FOR COMPRESSIVE STRENGTH

The limit states of design of a compression member depends upon the category to which the compression member belongs, as described in the "Strength of Compression Members or Columns" section. The limit states applicable to all categories of columns are summarized in Table 10.2. The strength requirements for each of the limiting states of the table are discussed below.

NON-SLENDER MEMBERS

FLEXURAL BUCKLING OF NON-SLENDER MEMBERS IN ELASTIC AND INELASTIC REGIONS

Based on the limit state for flexural buckling, the nominal compressive strength P_n is given by

$$P_n = F_{cr}A_g \tag{10.3}$$

TABLE 10.2
Applicable Limit States for Compressive Strength

Type of Column	Local Buckling (Local Instability)	
	Non-Slender Column, $\lambda \leq \lambda_r$	Slender Column, $\lambda > \lambda_r$
Overall Instability		
1. Hot rolled members	Flexure buckling in elastic or inelastic region	Flexure buckling in elastic or inelastic region incorporating the reduction factors for slender element
2. Built-up or cruciform doubly symmetric members	Lowest of the following two limits:	Lowest of the following two limits, incorporating the reduction factors for slender element:
	1. Flexure buckling in elastic or inelastic region	1. Flexure buckling in elastic or inelastic region
	2. Torsional buckling	2. Torsional buckling
3. Singly symmetric and unsymmetric members	Lowest of the following three limits:	Lowest of the following three limits, incorporating the reduction factors for slender element:
	1. Flexure buckling in elastic or inelastic region	1. Flexure buckling in elastic or inelastic region
	2. Torsional buckling	2. Torsional buckling
	3. Flexure–torsional buckling	3. Flexure–torsional buckling

where

F_{cr} is the flexure buckling state (stress)

A_g is the gross cross-sectional area

Including the nominal strength in Equation 10.1, the strength requirement of a column can be expressed as

$$P_u = \phi F_{cr} A_g \tag{10.4}$$

The flexural buckling stress, F_{cr} is determined as follows:

INELASTIC BUCKLING

When $KL/r \leq 4.71\sqrt{E/F_y}$, it is an inelastic buckling, for which

$$F_{cr} = (0.658^{F_y/F_e}) F_y \tag{10.5}$$

where F_e is the elastic critical buckling or Euler stress calculated according to Equation 10.6.

$$F_e = \frac{\pi^2 E}{(KL/r)^2} \tag{10.6}$$

ELASTIC BUCKLING

When $KL/r > 4.71\sqrt{E/F_y}$, it is an elastic buckling, for which

$$F_{cr} = 0.877 F_e \tag{10.7}$$

The value of $4.71\sqrt{E/F_y}$, at the threshold of inelastic and elastic buckling, is given in Table 10.3 for various types of steel.

The available critical stress ϕF_{cr} in Equation 10.4 for both the inelastic and elastic regions are given in Table 10.4 in terms of KL/r, adapted from the *AISC Manual 2005*.

TORSIONAL AND FLEXURAL–TORSIONAL BUCKLING OF NON-SLENDER MEMBERS

TABLE 10.3
Numerical Limits of Inelastic–Elastic Buckling

Type of Steel	$4.71\sqrt{E/F_y}$
A36	133.7
A992	113.43
A572	113.43

This section applies to the singly symmetric, nonsymmetric, and certain doubly symmetric members, such as cruciform and built-up sections. The nominal strength is governed by Equation 10.3. Also the F_{cr} value is determined according to Equations 10.5 and 10.7 above, except for two-angle members and tee-shaped members. However, for determining the Euler stress F_e, instead of Equation 10.6, a different set of formulas are used that involve the warping and the torsional constants for the section. For two-angle and tee-shaped members, the F_{cr} is determined directly by a different type of equation. For the set of equations for torsional-related buckling, a reference is made to Section E4 of Chapter 16 of the AISC manual.

SLENDER COMPRESSION MEMBERS

The approach to design slender members having $\lambda > \lambda_r$ is similar to the non-slender members in all categories except that a slenderness reduction factor, Q is included in the expression $4.7\sqrt{E/F_y}$ to classify the inelastic and elastic regions and Q is also included in the equations for F_{cr}. The slenderness reduction factor Q has two components: Q_s for the slender unstiffened elements and Q_a for the slender stiffened elements. These are given by a set of formulas for different shapes of columns. A reference is made to the Section E7 of Chapter 16 of the AISC manual.

All W shapes have non-slender flanges for A992 steel. All W shapes listed for the columns in the AISC manual have non-slender web (except for W14×43). However, many W shapes meant to be used as beams have slender webs in the compression.

This chapter considers only the doubly symmetric non-slender members covered in the "Non-Slender Members" section. By proper selection of a section, this condition, i.e., $\lambda \le \lambda_r$ could be satisfied.

USE OF THE COMPRESSION TABLES

Section 4 of the *AISC Manual 2005* contains the tables "available strength in axial compression, in kips" for various shapes and sizes. These tables directly give the capacity as a function of effective length (KL) with respect to least radius of gyration for various sections. Design of columns is a direct procedure from these tables. An abridged table for $F_y = 50\,ksi$ is given in Appendix C.8.

When the values of K and/or L are different in the two directions, both $K_x L_x$ and $K_y L_y$ are computed. If $K_x L_x$ is bigger, it is adjusted as $K_x L_x/(r_x/r_y)$. The higher of the adjusted $K_x L_x/(r_x/r_y)$ and $K_y L_y$ value is entered in the table to pick up a section that matches to the factored design load P_u.

Designing for a case when $K_x L_x$ is bigger, the adjustment of $K_x L_x/(r_x/r_y)$ is not straight forward because the values of r_x and r_y are not known. The initial selection could be made based upon the $K_y L_y$ value and then the adjusted value of $K_x L_x/(r_x/r_y)$ is determined based on the initially selected section, to make a revised selection.

Example 10.2

A 25 ft long column is rigidly fixed to the foundation. The other end is braced in the weak axis and free to translate in the strong axis. It is subjected to a dead load to 120 k and a live load of 220 k. Design the column of A992 steel.

TABLE 10.4
Available Critical Stress ϕF_{cr} for Compression Members ($F_y = 50$ ksi and $\phi = 0.90$)

KL/r	ϕF_{cr}, ksi	KL/r	ϕF_{cr}, ksi	KL/r	ϕF_{cr}, ksi	KL/r	ϕF_{cr}, ksi	KL/r	ϕF_{cr}, ksi
1	45.0	41	39.8	81	27.9	121	15.4	161	8.72
2	45.0	42	39.5	82	27.5	122	15.2	162	8.61
3	45.0	43	39.3	83	27.2	123	14.9	163	8.50
4	44.9	44	39.1	84	26.9	124	14.7	164	8.40
5	44.9	45	38.8	85	26.5	125	14.5	165	8.30
6	44.9	46	38.5	86	26.2	126	14.2	166	8.20
7	44.8	47	38.3	87	25.9	127	14.0	167	8.10
8	44.8	48	38.0	88	25.5	128	13.8	168	8.00
9	44.7	49	37.7	89	25.2	129	13.6	169	7.89
10	44.7	50	37.5	90	24.9	130	13.4	170	7.82
11	44.6	51	37.2	91	24.6	131	13.2	171	7.73
12	44.5	52	36.9	92	24.2	132	13.0	172	7.64
13	44.4	53	36.7	93	23.9	133	12.8	173	7.55
14	44.4	54	36.4	94	23.6	134	12.6	174	7.46
15	44.3	55	36.1	95	23.3	135	12.4	175	7.38
16	44.2	56	35.8	96	22.9	136	12.2	176	7.29
17	44.1	57	35.5	97	22.6	137	12.0	177	7.21
18	43.9	58	35.2	98	22.3	138	11.9	178	7.13
19	43.8	59	34.9	99	22.0	139	11.7	179	7.05
20	43.7	60	34.6	100	21.7	140	11.5	180	6.97
21	43.6	61	34.3	101	21.3	141	11.4	181	6.90
22	43.4	62	34.0	102	21.0	142	11.2	182	6.82
23	43.3	63	33.7	103	20.7	143	11.0	183	6.75
24	43.1	64	33.4	104	20.4	144	10.9	184	6.67
25	43.0	65	33.0	105	20.1	145	10.7	185	6.60
26	42.8	66	32.7	106	19.8	146	10.6	186	6.53
27	42.7	67	32.4	107	19.5	147	10.5	187	6.46
28	42.5	68	32.1	108	19.2	148	10.3	188	6.39
29	42.3	69	31.8	109	18.9	149	10.2	189	6.32
30	42.1	70	31.4	110	18.6	150	10.0	190	6.26
31	41.9	71	31.1	111	18.3	151	9.91	191	6.19
32	41.8	72	30.8	112	18.0	152	9.78	192	6.13
33	41.6	73	30.5	113	17.7	153	9.65	193	6.06
34	41.4	74	30.2	114	17.4	154	9.53	194	6.00
35	41.2	75	29.8	115	17.1	155	9.40	195	5.94
36	40.9	76	29.5	116	16.8	156	9.28	196	5.88
37	40.7	77	29.2	117	16.5	157	9.17	197	5.82
38	40.5	78	28.8	118	16.2	158	9.05	198	5.76
39	40.3	79	28.5	119	16.0	159	8.94	199	5.70
40	40.0	80	28.2	120	15.7	160	8.82	200	5.65

Source: Courtesy of American Institute of Steel Construction, Chicago, IL.

Solution

A. Analytical solution
1. Assume a dead load of 100 lb/ft

    Weight of column $= 25(0.1) = 2.5$ k

2. Factored design load

    $P_u = 1.2(120 + 2.5) + 1.6(220) = 499$ k

3. For yield limit state

    $$A_g = \frac{P_u}{\phi F_y} = \frac{499}{0.9(50)} = 11.1 \text{ in.}^2$$

4. The size will be much larger than the step 3 for the buckling mode of failure

    Select a section W14×61 $A = 17.9$ in.^{2}

    $r_x = 5.98$ in.

    $r_y = 2.45$ in.

    $\dfrac{b_f}{2t_f} = 7.75$

    $\dfrac{h}{t_w} = 30.4$

    $0.56\sqrt{\dfrac{E}{F_y}} = 0.56\sqrt{\dfrac{29{,}000}{50}} = 13.49$

    $1.49\sqrt{\dfrac{E}{F_y}} = 1.49\sqrt{\dfrac{29{,}000}{50}} = 35.88$

5. Since $b_f/2t_f < 0.56\sqrt{E/F_y}$ and $h/t_w < 1.49\sqrt{E/F_y}$, it is a non-slender section
6. $K_x = 1.2$ from Figure 7.6

    $K_y = 0.65$

    $\dfrac{K_x L_x}{r_x} = \dfrac{1.2(25 \times 12)}{5.98} = 60.2$

    $\dfrac{K_y L_x}{r_y} = \dfrac{0.65(25 \times 12)}{2.45} = 79.59 \leftarrow$ Controls

    Since $79.59 < 200$ **OK**

7. From Table 10.3, $4.71\sqrt{E/F_y} = 113.43$

    Since $79.59 < 113.43$, inelastic buckling

8. $F_e = \dfrac{\pi^2 E}{\left(KL/r\right)^2}$

$= \dfrac{\pi^2 (29,000)}{(79.59)^2} = 45.14\,\text{ksi}$

9. $F_{cr} = (0.658^{50/45.14})\,50 = 31.45\,\text{ksi}$
10. $\phi P_n = (0.9)(31.45)(17.9) = 507\,\text{k}$ **OK**

B. Use of Appendix C.8
1. $K_x L_x = 1.2(25) = 30\,\text{ft}$
    $K_y L_y = 0.65(25) = 16.25\,\text{ft}$
2. Select preliminary section
    Based on $K_y L_y = 16.25\,\text{ft}$, section W14 × 61, capacity = 507 k, from Appendix C.8.
3. For section W14 × 61, $r_x = 5.98\,\text{in.}$, $r_y = 2.45\,\text{in.}$

$$\text{Adjusted}\,\frac{K_x L_x}{r_x/r_y} = \frac{1.2(25)}{5.98/2.45} = 12.29$$

Use the larger value of $K_y L_y$ of 16.25 ft
4. Section from Appendix C.8 W14 × 61 with capacity = 507 k

Example 10.3

A column as shown in Figure 10.8 is fabricated from Grade 50 steel. Determine the limit state that will control the design of the column.

Solution

1. The doubly symmetric built-up section will be subjected to the flexure and the torsional buckling.
2. $\dfrac{b}{t} = \dfrac{10}{0.25} = 40$

$1.4\sqrt{\dfrac{E}{F_y}} = 1.4\sqrt{\dfrac{29,000}{50}} = 33.72$

Since 40 > 33.72, it is a slender column; the reduction factors have to be applied

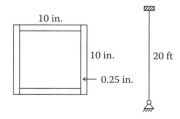

FIGURE 10.8 Built-up column.

3. $I = I_{out} - I_{inside}$

$$= \frac{1}{12}(10)(10)^3 - \frac{1}{12}(9.5)(9.5)^3$$

$$= 154.58 \text{ in.}^4$$

$A = (10)(10) - (9.5)(9.5)$

$$= 9.75 \text{ in.}^2$$

$$r = \sqrt{\frac{I}{A}} = \sqrt{\frac{154.58}{9.75}} = 3.98 \text{ in.}$$

4. $K = 2.0$

$$\frac{KL}{r} = \frac{2.0(20 \times 12)}{3.98} = 120.6$$

From Table 10.2, $4.71\sqrt{\frac{E}{F_y}} = 113.43$

$$\frac{KL}{r} > 4.71\sqrt{\frac{E}{F_y}}, \text{ elastic flexural buckling}$$

5. The lowest of the following two limit states will control
   a. Elastic flexure buckling with reduction factors
   b. Torsional buckling with reduction factors

PROBLEMS

10.1 A W8×31 column of A36 steel is 20 ft long. Along Y-axis it is hinged at both ends. Along X-axis, it is hinged at one end and free to translate at the other end. In which direction it is likely to buckle? ($r_x = 3.47$ in., $r_y = 2.02$ in.)

10.2 An HSS 5×2½×¼ column is supported, as shown in Figure P10.2. Determine the controlling (higher) slenderness ratio.

10.3 A single-story single-bay frame has the relative I values, as shown in Figure P10.3. Determine the effective length of the columns along X-axis. The sway is permitted in X-direction.

10.4 The frame of Figure P10.4 is braced and bends about X-axis. All beams are W18×35 and all columns are W10×54. Determine the effective length factors for AB and BC.

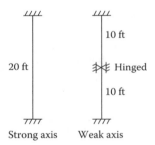

FIGURE P10.2 An HSS column.

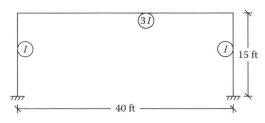

FIGURE P10.3 Frame for Problem 10.3.

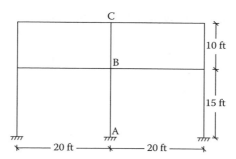

FIGURE P10.4   Frame of Problem 10.4.

10.5   An unbraced frame of Figure P10.5 bends along *X*-axis. Determine the effective length factors for AB and BC.

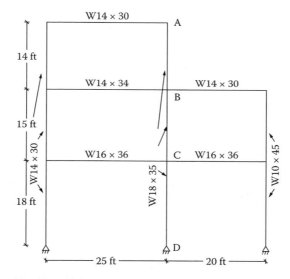

FIGURE P10.5   Frame of Problem 10.5.

10.6   Determine the effective length factors for AB and BC of the frame of Figure P10.5 for bending along *Y*-axis. Whether the factors of Problems 10.5 or 10.6 control?

10.7   Determine the strength of column of Figure P10.7 of A992 steel, when (a) the length is 15 ft and (b) the length is 30 ft.

FIGURE P10.7   Column of Problem 10.7.

10.8 Compute the strength of the member shown in Figure P10.8 of A36 steel.

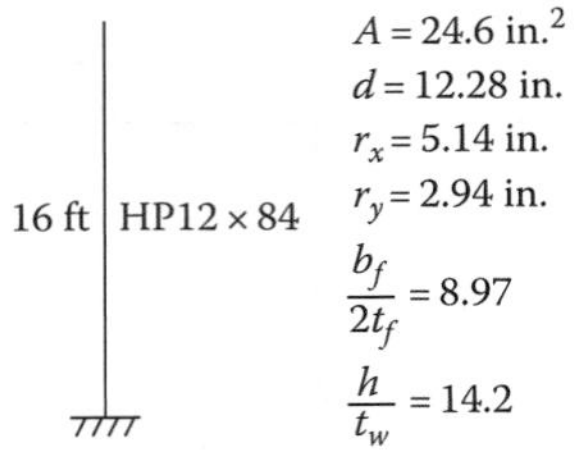

$A = 24.6$ in.^{2}
$d = 12.28$ in.
$r_x = 5.14$ in.
$r_y = 2.94$ in.
$\dfrac{b_f}{2t_f} = 8.97$
$\dfrac{h}{t_w} = 14.2$

16 ft | HP12 × 84

FIGURE P10.8 Column of Problem 10.8.

10.9 Compute the strength of the member shown in Figure P10.9 of A500 Grade B steel.

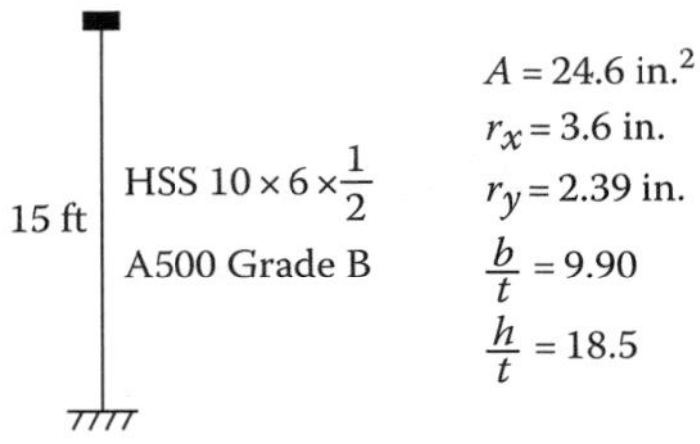

HSS $10 \times 6 \times \dfrac{1}{2}$
A500 Grade B
15 ft

$A = 24.6$ in.^{2}
$r_x = 3.6$ in.
$r_y = 2.39$ in.
$\dfrac{b}{t} = 9.90$
$\dfrac{h}{t} = 18.5$

FIGURE P10.9 Column of Problem 10.9.

10.10 A W18 × 130 section is used as a column with one end pinned and the other end is fixed against rotation but is free to translate. The length is 12 ft. Determine the strength of the A992 steel column.

10.11 Determine the maximum dead and live loads that can be supported by the compression member shown in Figure P10.11. The live load is twice of the dead load.

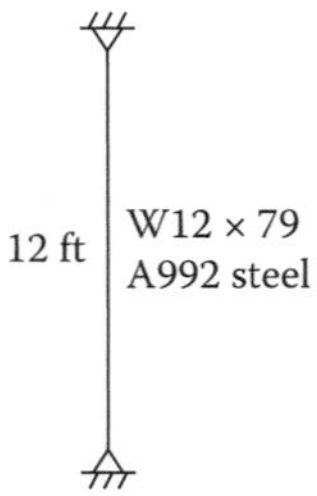

12 ft | W12 × 79
A992 steel

FIGURE P10.11 Column of Problem 10.11.

10.12 Determine the maximum dead and live loads to be supported by the column of Figure P10.12. The live load is one-and-a-half times of the dead load.

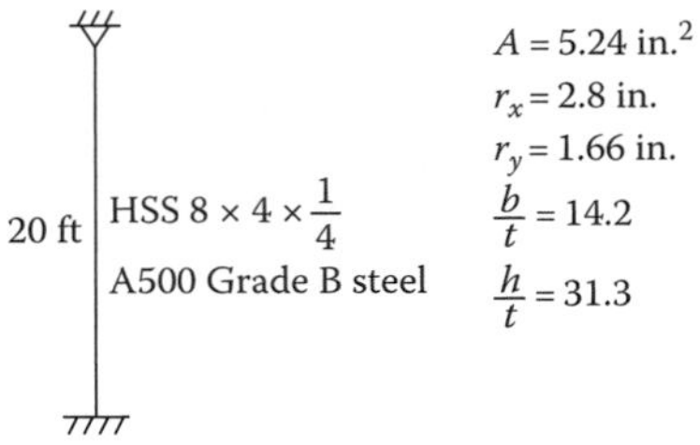

20 ft | HSS $8 \times 4 \times \dfrac{1}{4}$
A500 Grade B steel

$A = 5.24$ in.^{2}
$r_x = 2.8$ in.
$r_y = 1.66$ in.
$\dfrac{b}{t} = 14.2$
$\dfrac{h}{t} = 31.3$

FIGURE P10.12 Column of Problem 10.12.

10.13 Determine whether the member in Figure P10.13 of A992 steel is adequate to support the loads as indicated.

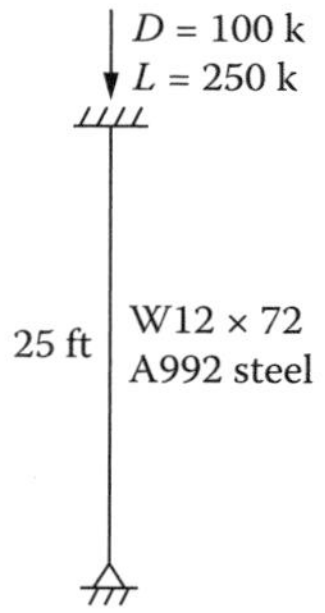

$D = 100\ k$
$L = 250\ k$

25 ft | W12 × 72
A992 steel

FIGURE P10.13   Column of Problem 10.13.

10.14 Check whether the member of Figure P10.14 of A36 steel is adequate for the indicated loads.

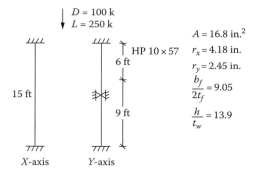

$D = 100\ k$
$L = 250\ k$

$A = 16.8\ \text{in.}^2$
HP 10 × 57 $r_x = 4.18\ \text{in.}$
6 ft
$r_y = 2.45\ \text{in.}$
15 ft
$\dfrac{b_f}{2t_f} = 9.05$
9 ft
$\dfrac{h}{t_w} = 13.9$

X-axis   Y-axis

FIGURE P10.14   Column of Problem 10.14.

10.15 An HSS6×4×5/16 section (46 ksi steel) shown in Figure P10.15 is applied by a dead load of 40 k and a live load of 50 k. Check the column adequacy.

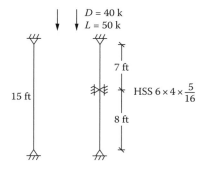

$D = 40\ k$
$L = 50\ k$

7 ft
15 ft
$\text{HSS } 6 \times 4 \times \dfrac{5}{16}$
8 ft

FIGURE P10.15   Column of Problem 10.15.

10.16 Select an HSS section for the column shown in Figure P10.16.

D = 50 k
L = 150 k
12 ft

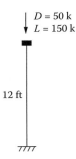

FIGURE P10.16 Column of Problem 10.16.

10.17 Design a standard pipe section of A53 Grade B steel for the column shown in Figure P10.17.

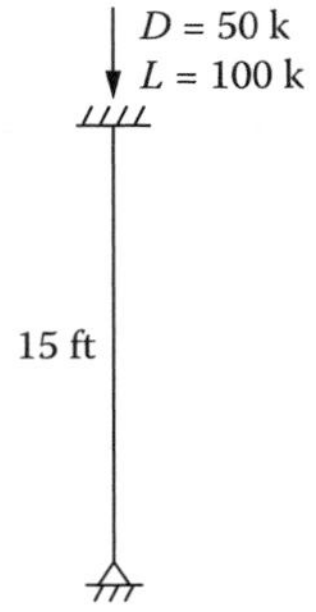

D = 50 k
L = 100 k
15 ft

FIGURE P10.17 Column of Problem 10.17.

10.18 Select a W14 shape of A992 steel for a column of 25 ft length shown in Figure P10.18. Both ends are fixed. There are bracings at 10 ft from top and bottom in the weaker direction.

D = 350 k
L = 1000 k
25 ft
10 ft
5 ft
10 ft
X-axis
Y-axis

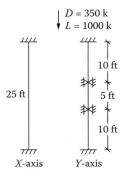

FIGURE P10.18 Column of Problem 10.18.

10.19 Design a W14 section column AB of the frame shown in Figure P10.19. It is unbraced along *X*-axis and braced in the weak direction. The loads on the column are dead load = 200 k and live load = 600 k. First determine the effective length factor using Figure 7.6. After selecting the preliminary section, use the alignment chart with the same size for section BC to revise the selection. Use W16 × 100 for the beam sections meeting at B.

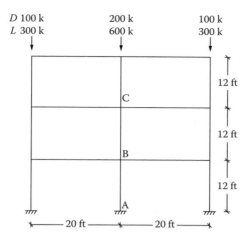

FIGURE P10.19   Braced frame of Problem 10.19.

10.20 Design the column AB of Problem 10.19 for the frame braced in both directions.

10.21 A WT12×34 column of 18 ft length is pinned at both ends. Show what limiting states will determine the strength of the column. Use A992 steel.

10.22 An A572 steel column in Figure P10.22 is fixed at one end and hinged at the other end. Indicate the limit states that will control the strength of the column.

FIGURE P10.22   Column of Problem 10.22.

10.23 A double-angle section with a separation 3/8 in. is subjected to the loads shown in Figure P10.23. Determine the limit states that will govern the design of the column. Use Grade 50 steel.

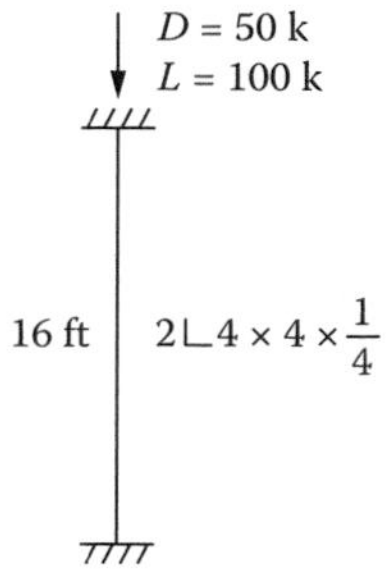

$$D = 50 \text{ k}$$
$$L = 100 \text{ k}$$

16 ft $2\llcorner 4 \times 4 \times \dfrac{1}{4}$

FIGURE P10.23   Column of Problem 10.23.

10.24 A cruciform column is fabricated from the Grade 50 steel, as shown in Figure P10.24. Determine the limit states that will control the design.

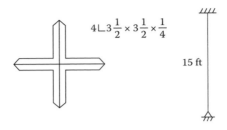

FIGURE P10.24   A cruciform column of Problem 10.24.

10.25 For the column section and the loading shown in Figure P10.25, determine the limit states for which the column should be designed. Use A992 steel.

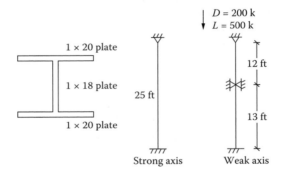

FIGURE P10.25   A built-in column of Problem 10.25.

11 Flexural Steel Members

THE BASIS OF DESIGN

Beams are the structural members that support transverse loads on them and are subjected to flexure and shear. An *I* shape is a common cross section for a steel beam where the material in the flanges at the top and bottom is most effective in resisting the bending moment and the web provides for most of the shear resistance. As discussed in the context of wood beams, the design process involves selection of a beam section on the basis of the maximum bending moment to be resisted. The selection is, then, checked for the shear capacity. In addition, the serviceability requirement imposes the deflection criteria for which the selected section should be checked.

The basis of design for bending or flexure is as follows:

$$M_u \leq \phi M_n \tag{11.1}$$

where
M_u is the factored design (imposed) moment
ϕ is the resistance factor for bending $= 0.9$
M_n is the nominal moment strength of steel

NOMINAL STRENGTH OF STEEL IN FLEXURE

Steel is a ductile material. As discussed in the "Elastic and Plastic Designs" section in Chapter 1, steel can acquire the plastic moment capacity M_p wherein the stress distribution above and below the neutral axis will be represented by the rectangular blocks corresponding to the yield strength of steel, i.e., $M_p = F_y Z$; Z being the plastic moment of inertia of the section.

However, there are certain other factors that undermine the plastic moment capacity. One such factor relates to the unsupported (unbraced) length of the beam and another factor relates to the slender dimensions of the beam section. The design capacity is determined considering both of these. The effect of the unsupported length on the strength has been discussed first in the "Lateral Unsupported Length" section. The beam's slender dimensions effect the strength similar to the local instability of the compression members. This is described in the "Slender Beam" section.

LATERAL UNSUPPORTED LENGTH

As a beam bends, it develops the compression stress in one part and the tensile stress in the other part of the cross section of the beam. The compression region acts analogous to a column. If the entire member is slender, it will buckle outward similar to a column. However, in this case, the compression portion is restrained by the tensile portion. As a result, a twist will occur in the section. This form of instability, as shown the Figure 11.1, is called the *lateral torsional buckling* (LTB).

The lateral torsional buckling could be prevented in two ways:

FIGURE 11.1 Buckling and twisting effect in a beam.

1. Lateral bracings could be applied to the compression flange at close intervals that would prevent the lateral translation (buckling) of the beam, as shown in Figure 11.2. This support could be provided by a floor member securely attached to the beam.

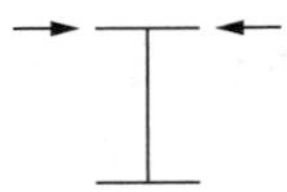

FIGURE 11.2 Lateral bracing of compression flange.

2. The cross bracings or diaphragm could be provided between adjacent beams, as shown in Figure 11.3, which will directly prevent the twisting of the sections.

Depending upon the lateral support condition on the compression side, the strength of the limit state of a beam is either due to the plastic yielding of the section or the lateral torsional buckling (LTB) of the section. The latter condition has two further divisions: the inelastic LTB or the elastic LTB. These three zones of the limit states are shown in Figure 11.4 and described in the following section.

In Figure 11.4, the first threshold value for the unsupported or the unbraced length is L_p given by the following relation:

$$L_p = 1.76 r_y \sqrt{\frac{E}{F_y}} \tag{11.2}$$

where
L_p is the first threshold limit unsupported length (in.)
r_y is the radius of gyration about Y-axis, listed in the Appendices C.1 through C.7

The second threshold value is L_r which is conservatively given by the following relation:

$$L_r = \pi r_{ts} \sqrt{\frac{E}{0.7F_y}} \tag{11.3}$$

FIGURE 11.3 Cross bracing or diaphragm.

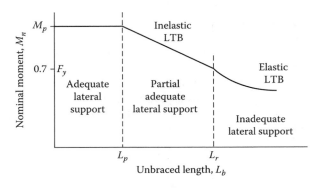

FIGURE 11.4 Nominal moment strength as a function of unbraced length.

where

L_r is the second threshold unsupported length (in.)

r_{ts} is the special radius of gyration for L_r, listed in the Appendices C.1 through C.7

FULLY PLASTIC ZONE WITH ADEQUATE LATERAL SUPPORT

When the lateral support is continuous or closely spaced so that the unbraced (unsupported) length of a beam, L_b is less than or equal to L_p from Equation 11.2, the beam could be loaded to reach the plastic moment capacity throughout the section.

The limit state in this case is the yield strength given as follows:

$$M_u = \phi F_y Z_x \text{ with } \phi = 0.9 \tag{11.4}$$

The LTB does not apply in this zone.

INELASTIC LATERAL TORSIONAL BUCKLING (I-LTB) ZONE

When the lateral unsupported (unbraced) length, L_b is more than L_p but less than or equal to L_r, the section will not have a sufficient capacity to develop the for full yield stress, F_y in the entire section. Some fibers will be stressed to F_y before the buckling occurs. This will lead to the inelastic lateral torsional buckling.

At $L_b = L_p$, the moment capacity is the plastic capacity M_p. As the length L_b increases beyond the L_p value, the moment capacity becomes less. At the L_r value of the unbraced length, the section buckles elastically attaining the yield stress only at the top or the bottom fiber. Accounting for the residual stress in the section during the manufacturing, the effective yield stress is $(F_y - F_r)$, where F_r is the residual stress. The residual stress is taken as 30% of the yield stress. Thus, at $L_b = L_r$, the moment capacity is $(F_y - F_r)S_x$ or $0.7F_yS$.

When the unbraced length, L_b is between the L_p and L_r values, the moment capacity is interpolated linearly between the magnitudes of M_p and $0.7F_yS$ as follows:

$$M_u = \phi \left[M_p - \left(M_p - 0.7F_yS \right) \left(\frac{L_b - L_p}{L_r - L_p} \right) \right] C_b \tag{11.5}$$

A factor C_b has been introduced in Equation 11.5 to account for a situation when the moment within the unbraced length is not uniform (constant). On a conservative side, $C_b = 1$.

ELASTIC LATERAL TORSIONAL BUCKLING (E-LTB) ZONE

When the unbraced length, L_b exceeds the threshold value of L_r, the beam buckles before the effective yield stress $(0.7F_y)$ is reached anywhere in the cross section. This is the elastic buckling. The moment capacity is made up of the torsional resistance and the warping resistance of the section.

$$M_u < 0.7\phi F_yS \tag{11.6}$$

SLENDER BEAM SECTIONS

The above discussion of the beam strength did not account for the shape of a beam, i.e., it assumed that the beam section is robust enough so as not to create any "localized" problem. However, if the

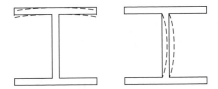

FIGURE 11.5 Local buckling of section.

TABLE 11.1
Shape Classification Limits

Element	λ	λ_p	λ_r
Flange	$b_f/2t_f{}^a$	$0.38\sqrt{\dfrac{E}{F_y}}$	$1.0\sqrt{\dfrac{E}{F_y}}$
Web	h/t_w	$3.76\sqrt{\dfrac{E}{F_y}}$	$5.70\sqrt{\dfrac{E}{F_y}}$

^{a} For channel shape, this is b_f/t_f.

flange and the web of a section are relatively thin, they might get buckled, as shown in Figure 11.5, even before the lateral torsional buckling happens. This mode of failure is called the *flange local buckling* (FLB) or the *web local buckling* (WLB).

The sections are divided in three classes based on the width-to-thickness ratios of the flange and the web. The threshold values of classification are given in Table 11.1.

When $\lambda \le \lambda_p$, the shape is compact
When $\lambda > \lambda_p$, but $\lambda \le \lambda_r$, the shape is noncompact
When $\lambda > \lambda_r$, the shape is slender

Both the flanges and web are evaluated by the above criteria. Based on the above limits, the flange of a section might fall into one category whereas the web of the same section might fall into the other category.

The values of λ_p and λ_r for various types of steel are listed in Table 11.2.

In addition to the unsupported length, the bending moment capacity of a beam also depends on the compactness or width:thickness ratio, as shown in Figure 11.6.

This localized buckling effect could be the FLB or the WLB depending upon which one falls into the noncompact or slender category. All W, S, M, HP, C, and MC shapes listed in the *AISC Manual 2005* have the compact webs at $F_y \le 65$ ksi. Thus, only the flange criteria need to be applied. Fortunately, most of the shapes satisfy the flange compactness requirements also.

Without accounting for the lateral unsupported length effect, i.e., assuming a fully laterally supported beam, the following strength limits are applicable based on the compactness criteria.

COMPACT FULL PLASTIC LIMIT

As long as $\lambda \le \lambda_p$, the beam moment capacity is equal to M_p and the limit state of the moment is given by the yield strength expressed by Equation 11.4.

TABLE 11.2
Magnitude of the Classification Limits

Element	Limits	A36 $F_y = 36\,\text{ksi}$	A572 A992 $F_y = 50\,\text{ksi}$
Flange	$0.38\sqrt{\dfrac{E}{F_y}}$	10.79	9.15
	$1.0\sqrt{\dfrac{E}{F_y}}$	28.38	24.08
Web	$3.76\sqrt{\dfrac{E}{F_y}}$	106.72	90.55
	$5.70\sqrt{\dfrac{E}{F_y}}$	161.78	137.27

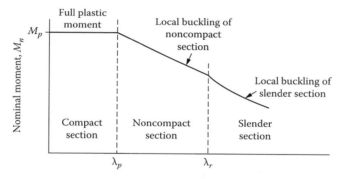

FIGURE 11.6 Nominal moment strength as a function of compactness.

NONCOMPACT FLANGE LOCAL BUCKLING (N-FLB)*

For sections having the value of λ between λ_p and λ_r limits shown in Table 11.1, the moment capacity is interpolated between M_p and $0.7F_yS$ as a gradient of the λ values on the same line like Equation 11.5, expressed as follows:

$$M_u = \phi\left[M_p - (M_p - 0.7F_yS)\left(\frac{\lambda - \lambda_p}{\lambda_r - \lambda_p}\right)\right]. \tag{11.7}$$

SLENDER FLANGE LOCAL BUCKLING (S-FLB)

For sections with $\lambda > \lambda_r$, the moment-resisting capacity is inversely proportional to the square of slenderness ratio as follows:

$$M_u = \frac{0.9\phi Ek_c S}{\lambda^2} \tag{11.8}$$

* All webs are compact for $F_y = 36$ ksi and 50 ksi.

where, $k_c = \dfrac{4}{\sqrt{h/t_w}}$; k_c being ≥ 0.35 and ≤ 0.76.

SUMMARY OF BEAM RELATIONS

Considering both the lateral support and the compactness criteria, the flexural strength (the moment capacity) is taken to be the lowest value obtained according to the limit states of the LTB and the compression FLB. The applicable limits and the corresponding equations are shown in Table 11.3. Most of the beam sections fall in the full plastic zone where Equation 11.4 could be applied. In this chapter, it has been assumed that the condition of the adequate lateral support will be satisfied, if necessary, by providing the bracings at intervals less than the distance L_p and also the condition of the flange and the web compactness is fulfilled.

As stated earlier, all W, S, M, HP, C, and MC shapes have the compact webs for F_y of 36, 50, and 65 ksi. All W, S, M, C, and MC shapes have the compact flanges for F_y of 36 and 50 ksi, except for the sections listed in Table 11.4. Thus, a beam will be compact if the sections listed in Table 11.4 are avoided.

DESIGN AIDS

The *AISC Manual 2005* provides the design tables. A beam can be selected by entering the table either with the required section modulus or with the design bending moment.

These tables are applicable to adequately support compact beams for which yield limit state is applicable. For simply supported beams with uniform load over the entire span, tables are provided, which show the allowable uniform loads corresponding to various spans. These tables are also for the adequately supported beams but extend to the noncompact members.

Also included in the manual are more comprehensive charts that plot the total moment capacity against the unbraced length starting at spans less than L_p and continuing to spans greater than

---

TABLE 11.3
Applicable Limiting States of Beam Design

		Flange Local Buckling (FLB)[a]		
Unbraced Length, L_b	**Compact $\lambda < \lambda_p$**	**Noncompact (Inelastic) $\lambda > \lambda_p$ and $\leq \lambda_r$**	**Slender (Elastic) $\lambda > \lambda_r$**
Adequate lateral support $L_b \leq L_p$	Limit state: Yield strength: Equation 11.4[b] LTB does not apply	Limit state: Inelastic FLB: Equation 11.7 LTB does not apply	Limit state: Elastic FLB: Equation 11.8 LTB does not apply
Lateral torsional buckling (LTB) — Partial inadequate support $L_b > L_p$ and $L_b \leq L_r$	Limit state: Inelastic LTB: Equation 11.5	Limit states: Lower of the following two: 1. Inelastic LTB: Equation 11.5 2. Noncompact FLB: Equation 11.7	Limit states: Lower of the following two: 1. Inelastic LTB: Equation 11.5 2. Slender FLB: Equation 11.8
Inadequate support, $L_b > L_r$	Limit state: Elastic LTB: Equation 11.6	Limit states: Lower of the following two: 1. Elastic LTB: Equation 11.6 2. Noncompact FLB: Equation 11.7	Limit states: Lower of the following two: 1. Elastic LTB: Equation 11.6 2. Slender FLB: Equation 11.8

[a] The WLB are not included since all I-shaped and C-shaped sections have compact web. In the case of a WLB member, the LTB and FLB categories of formulas are similar to the above equations but are modified for (1) the web plastification factor (R_{pc}) and (2) the bending strength reduction factor (R_{pg}).

[b] Most beams fall into the adequate laterally supported compact category. This chapter considers only this state of design.

L_r covering compact as well as noncompacting members. These charts are applicable to the condition when the bending moments within the unbraced length is greater than that at both ends, i.e., $C_b = 1$. The charts can be directly used to select a beam section.

A typical chart is given at Appendix C.9. Enter the chart with given unbraced length on the bottom scale, proceed upward to meet the horizontal line corresponding to the design moment on the left-hand scale. Any beam listed above and to the right of the intersection point will meet the design requirement. The section listed at the first solid line after the intersection represents the most economical section.

TABLE 11.4
List of Noncompact Flange Sections
W21×48
W14×99
W14×90
W12×65
W10×12
W8×31
W8×10
W6×15
W6×9
W6×8.5
M4×6

Example 11.1

A floor system is supported by steel beams, as shown in Figure 11.7. The live load is 100 psf. Design the beam. Determine the maximum unbraced length of beam to satisfy the requirement of the adequate lateral support. $F_y = 50$ ksi

Solution

A. Analytical
   1. Tributary area of beam/ft = $10 \times 1 = 10$ ft²/ft
   2. Weight of slab/ft = $1 \times 10 \times \dfrac{6}{12} \times 150 = 750$ lb/ft
   3. Estimated weight of beam/ft = 30 lb/ft
   4. Dead load/ft = 780 lb/ft
   5. Live load/ft = $100 \times 10 = 1000$ lb/ft
   6. Design load/ft

$$w_u = 1.2(780) + 1.6(1000) = 2536 \text{ lb/ft or } 2.54 \text{ k/ft}$$

   7. Design moment

$$M_u = \frac{w_u L^2}{8} = \frac{2.54(25)^2}{8} = 198.44 \text{ ft-k}$$

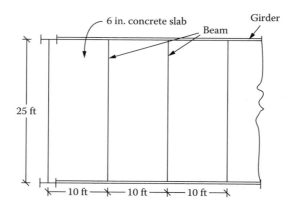

FIGURE 11.7   A floor system supported by beams.

8. From Equation 10.4:

$$Z = \frac{198.44(12)}{(0.9)(50)} = 52.91 \, \text{in.}^3$$

9. Select W16 × 31
   $Z = 54 \, \text{in.}^3$
   $r_x = 6.41 \, \text{in.}$
   $r_y = 1.17 \, \text{in.}$
   $\dfrac{b_f}{2t_f} = 6.28$
   $\dfrac{h}{t_w} = 51.6$

10. $\dfrac{b_f}{2t_f} = 6.28 < 9.15$ from Table 11.2, it is a compact flange

   $\dfrac{h}{t_w} = 51.6 < 90.55$ from Table 11.2, it is a compact web

   Equation 11.4 applies, selection **OK**
11. Unbraced length from Equation 11.2

$$L_p = 1.76 r_y \sqrt{\frac{E}{F_y}}$$

$$= 1.76(1.17)\sqrt{\frac{29000}{50}}$$

$$= 49.59 \, \text{in. or } 4.13 \, \text{ft}$$

B. Use of chart
   1. From Appendix C.9
      For an unbraced length of 4 ft and a design moment of 198 ft-k, the most suitable section is W16 × 31 (same as above).
   2. If the beam is supported at mid-span, i.e., the unbraced length is 12.5 ft, from Appendix C.9, most suitable section is W16 × 40.
   3. If the beam is supported only at end, i.e., the unbraced length is 25 ft, from Table 3-10 in the *AISC Manual 2005* (not included in the text), the most suitable section is W12 × 53, which accounts for all necessary adjustments for partial or inadequate support and compactness and noncompactness of the selected section.

SHEAR STRENGTH OF STEEL

The section of beam selected for the moment capacity is checked for its shear strength capacity. The design relationship for shear strength is

$$V_u = \phi_v V_n \tag{11.9}$$

where
   V_u is the factored shear force applied
   ϕ_v is the resistance factor for shear
   V_n is the nominal shear strength

The shear in an I-shaped section is primarily resisted by the web. As such, the web area, A_w is taken to be equal to the depth of beam (d) times the web thickness (t_w). The *AISC Manual 2005* assigns the shear yield strength F_v to be 60% of the tensile yield strength of steel, F_y.

Thus, the limit state for shear yielding is expressed as

$$V_u = \phi_v 0.6 F_y A_w \qquad (11.10)$$

The above relation is for a compact web section when $\phi_v = 1$. However, as the ratio of depth to thickness of web, h/t_w exceeds $2.24\sqrt{E/F_y}$ for W, S, HP sections or $2.46\sqrt{E/F_y}$ for other type of shapes (except the round shapes), the inelastic web buckling occurs whereby ϕ_v is reduced to 0.9 and Equation 11.10 is further multiplied by a reduction factor C_v.

At h/t_w exceeding $3.06\sqrt{E/F_y}$, the elastic web buckling condition sets in and the factor C_v is further reduced. However, most of the sections of $F_y < 50$ ksi steel have the compact shapes that satisfy Equation 11.10.

Example 11.2

Check the beam of Example 11.1 for shear strength.

Solution

1. $V_u = \dfrac{w_u L}{2}$

   $= \dfrac{2.54(25)}{2} = 31.75 \, \text{k}$

2. For W16 × 31

   $A_w = d t_w = 15.9(0.275) = 4.37 \, \text{in.}^2$

3. From Equaiton 11.10

   $\phi_v 0.6 F_y A_w = (1)(0.6)(50)(4.37) = 131.1 \, k > 31.75$ **OK**

BEAM DEFLECTION LIMITATIONS

Deflection is a service requirement. A limit on deflection is imposed so that the serviceability of a floor or a roof is not impaired due to the cracking of plastic, or concrete slab, the distortion of the partitions or any other kind of undesirable occurrence. There are no standard limits because such values depend upon the function of a structure. For cracking of plaster, usually a live load deflection limit of span/360 is observed. It is imperative to note that being a serviceability consideration, the deflections are always computed with the service (unfactored) loads and moments.

For a common case of the uniformly distributed load on a simple beam, the deflection is given by the following formula:

$$\delta = \frac{5}{384} \frac{wL^4 *}{EI} \qquad (11.11)$$

---

* In FPS units, the numerator is multiplied by $(12)^3$ to convert δ in inch unit when w is is k/ft, L is in ft, E is in ksi, and I is in in.^{4} Similarly, Equation 11.12 is also multiplied by $(12)^3$ when M is in ft-k.

However, depending upon the loading condition, the theoretical derivation of the expression for deflection might be quite involved. For some commonly encountered load conditions, when the expression of the bending moment is substituted in the deflection expression, a generalized from of deflection can be expressed as follows:

$$\delta = \frac{ML^2}{CEI} \tag{11.12}$$

where
   w is the service loads combination
   M is the moment due to the service loads

The values of constant C are indicated in Table 11.5 for different load cases.

In a simplified form, the designed factored moment, M_u can be converted to the service moment dividing by a factor of 1.5 (i.e., $M = M_u/1.5$). The service live load moment, M_L is approximately 2/3 of the total moment M (i.e., $M_L = 2M_u/4.5$). The factor C from Table 11.5 can be used in Equation 11.12 to compute the expected deflection, which should be checked against the permissible deflection, Δ to satisfy the deflection limitation.

Example 11.3

Check the beam of Example 10.1 for deflection limitation. The maximum permissible live load deflection is $L/360$. Use (1) the conventional method, (2) the simplified procedure.

Solution

   (a) Convention method
      1.   Service live load = 1000 lb/ft or 1 k/ft
      2.   For W16 × 31, $I = 375$ in.[4]

TABLE 11.5
Deflection Loading Constants

Diagram of Load Condition	Constant C for Equation 11.2
	9.6
	12
	9.39
	10.13
	4
	3

---

3. $\delta = \dfrac{5}{384} \dfrac{wL^4}{EI} \times (12)^3$

$\qquad = \dfrac{5}{384} \dfrac{(1.0)(25)^4(12)^3}{(29,000)(375)}$

$\qquad = 0.81\,\text{in.}$

4. $\Delta = \dfrac{L \times 12}{360}$

$\qquad = \dfrac{25 \times 12}{360}$

$\qquad = 0.83\,\text{in.}$

(b) Simplified procedure

1. $M_L = \dfrac{2M_u}{4.5} = \dfrac{2(198.44)}{4.5} = 88.20 \text{ ft-k}$

2. From Equation 11.12

$\delta = \dfrac{ML^2 \times (12)^3}{CEI}$

$\qquad = \dfrac{(88.20)(25)^2(12)^3}{(10)(290,000)(375)}$

$\qquad = 0.88\,\text{in.}$

3. $\Delta < \delta$ **NG** (border case)

PROBLEMS

11.1 Design a beam of A992 steel for the loads in Figure P11.1. Determine the maximum unbraced length of beam to satisfy the requirement of the adequate lateral support.

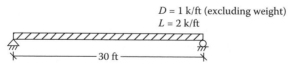

D = 1 k/ft (excluding weight)
L = 2 k/ft

30 ft

FIGURE P11.1   Beam of Problem 11.1.

11.2 Design a simply supported 20 ft span beam having the following concentrated loads at the mid-span. Determine the maximum unbraced length of beam to satisfy the requirement of the adequate lateral support.

Service dead load = 10 k
Service live load = 25 k

11.3 Design a beam of A992 steel for the loading shown in Figure P11.3. The compression flange bracing is provided at each concentrated load. The selected section should be such that the full lateral support condition is satisfied. Determine the maximum unbraced length of beam to satisfy the requirement of the adequate lateral support.

11.4 Design a cantilever beam of A992 steel for the loading shown in Figure P11.4. The compression flange bracing is provided at each concentrated load. The selected section should be such that the full lateral support condition is satisfied. Determine the maximum unbraced length of beam to satisfy the requirement of the adequate lateral support.

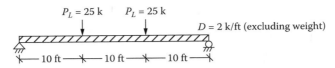

FIGURE P11.3   Beam of Problem 11.3.

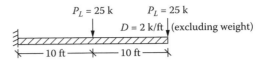

FIGURE P11.4   Beam of Problem 11.4.

11.5   A floor system supporting a 6 in. concrete slab is shown in Figure P11.5. The live load is 100 psf. Design the beam of A36 steel. Recommend the compression flange bracing so that the beam has a full lateral support.

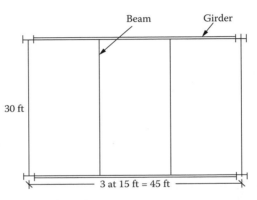

FIGURE P11.5   Floor system of Problem 11.5.

11.6   Design an A992 steel girder of Problem 11.5. Recommend the compression flange bracing so that the beam has a full lateral support.

11.7   From the sections listed, sort out which of the sections of A992 steel are compact, noncompact, and slender?

   (1) W21×93, (2) W18×97, (3) W14×99, (4) W12×65, (5) W10×68, (6) W8×31, (7) W6×15.

11.8   A grade 50 W21×62 section is used for a simple span of 20 ft. The only dead load is the weight of the beam. The beam is fully laterally braced. What is the largest service concentrated load that can be placed at the center of the beam? What is the maximum unbraced length?

11.9   A W18×97 beam of A992 steel is selected to span 20 ft. If the compression flange is supported at the end and at midpoint. Which formula do you recommend to solve for the moment capacity? Determine the maximum unbraced length of beam to satisfy the requirement of the adequate lateral support.

11.10   A W18×97 beam of A992 steel is selected to span 20 ft. It is supported at the ends only. Which formula do you recommend to solve for the moment capacity?

11.11   A W21×48 section is used to span 20 ft and is supported at the ends only. Which formula do you recommend to solve for the moment capacity?

11.12   A W21×48 section is used to span 20 ft and is supported at the ends and at the center. Which formula do you recommend to solve for the moment capacity?

11.13 Check the selected beam section of Problem 11.1 for the shear strength capacity.

11.14 Check the selected beam section of Problem 11.2 for the shear strength capacity.

11.15 Check the selected beam section of Problem 11.3 for the shear strength capacity.

11.16 What is the shear strength of the beam of Problem 11.9?

11.17 What is the shear strength of the beam of Problem 11.11?

11.18 Compute the total load and the live load deflections for the beam of Problem 11.1 by (1) the conventional method and (2) the simplified procedure. The permissible deflection for total load is $L/240$ and for live load $L/360$.

11.19 Compute the total load and the live load deflections for the beam of Problem 11.2 by (1) the conventional method and (2) the simplified procedure. The permissible deflection for total load is $L/240$ and for live load $L/360$.

11.20 Compute the total load and the live load deflections for the beam of Problem 11.3 by (1) the conventional method, and (2) the simplified procedure. The permissible deflection for total load is $L/240$ and for live load $L/360$.

11.21 Check the total load and the live load deflections for the beam of Problem 11.5 by (1) the conventional method and (2) the simplified procedure. The permissible deflection for total load is $L/240$ and for live load $L/360$.

11.22 Check the total load and the live load deflections for the beam of Problem 11.6 by (1) the conventional method and (2) the simplified procedure. The permissible deflection for total load is $L/240$ and for live load $L/360$.

12 Combined Forces on Steel Members

DESIGN APPROACH TO THE COMBINED FORCES

Design of tensile, compression, and bending members was separately treated in Chapters 9, 10, and 11, respectively. In actual structures, the axial and the bending forces generally act together, specifically the compression due to gravity loads and the bending due to lateral loads. An interaction formula is a simplest way for such cases wherein the sum of the ratios of the factored design load to the limiting axial strength and the factored design moment to the limiting moment strength should not exceed 1.

The test results show that assigning of an equal weight to the axial force ratio and the moment ratio in the interaction equation provides sections that are too large. Accordingly, the AISC suggested the following modifications to the interaction equations in which the moment ratio is reduced when the axial force is high and the axial force ratio is reduced when the bending moment is high.

1. For $\dfrac{P_u}{\phi P_n} \geq 0.2$

$$\frac{P_u}{\phi P_n} + \frac{8}{9}\left(\frac{M_{ux}}{\phi_b M_{nx}} + \frac{M_{uy}}{\phi_b M_{ny}}\right) \leq 1 \tag{12.1}$$

2. For $\dfrac{P_u}{\phi P_n} < 0.2$

$$\frac{1}{2}\frac{P_u}{\phi P_n} + \left(\frac{M_{ux}}{\phi_b M_{nx}} + \frac{M_{uy}}{\phi_b M_{ny}}\right) \leq 1 \tag{12.2}$$

where
  ϕ is the resistance factor for axial force (0.9 or 0.75 for tensile member, 0.9 for compression member)
  ϕ_b is the resistance factor for bending (0.9)
  P_u is the factored design load determined by structural analysis (required force)
  P_n is the nominal axial capacity, determined according to Chapter 9 or 10.
  M_{ux}, M_{uy} are factored design moments about X- and Y-axes as determined by structural analysis including the second-order effects (required moments)
  M_{nx}, M_{ny} are nominal bending capacities along X- and Y-axes, if only bending moments were present, which are determined by different methods mentioned in Chapter 11.

COMBINATION OF TENSILE AND FLEXURE FORCES

Some members of a structural system are subjected to axial tension as well as bending. An example is the bottom chord of a trussed bridge. The hanger type of structures acted upon by transverse loads is another example.

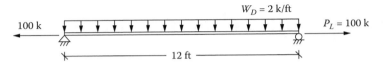

FIGURE 12.1   A tensile and flexure forces member.

The analysis, where a member size is known and the adequacy of the member to handle a certain magnitude of force is to be checked, is a direct procedure by Equations 12.1 and 12.2. However, the design of a member that involves selecting of a suitable size for a known magnitude of load is a trial-and-error procedure by the interaction equations, Equations 12.1 and 12.2. The *AISC Manual 2005* presents a simplified procedure to make an initial selection of a member size. This procedure, however, necessitates the application of the factors that are available from the special tables in the manual. Since the manual is not a precondition for this text, that procedure is not used here.

Example 12.1

Design a member to support the load shown in Figure 12.1. It has one line of four holes for 7/8 in. bolt in the web for the connection. The beam has adequate lateral support. Use Grade 50 steel.

Solution

A. Analysis of structure
   1. Assume a beam weight of 50 lb/ft
   2. $W_u = 1.2(2.05) = 2.46$ k/ft
   3. $M_u = \dfrac{W_u L^2}{8} = \dfrac{(2.46)(12)^2}{8} = 44.28$ ft-k or 531.4 in.-k
   4. $P_u = 1.6(100) = 160$ k
B. Design
   1. Try W 10×26 section*
   2. $A_g = 7.61$ in.^{2}
   3. $I_x = 144$ in.^{4}
   4. $Z_x = 31.3$ in.^{3}
   5. $t_w = 0.26$ in.
   6. $b_f/2t_f = 5.70$
   7. $h/t_w = 29.5$
C. Axial (tensile) strength
   1. $U = 0.7$ from "Shear Lag" section of Chapter 9, $h = 7/8 + 1/8 = 1$, $A_h = 1(0.26) = 0.26$ in.^{2}
   2. $A_n = A_g - A_h = 7.61 - 0.26 = 7.35$ in.^{2}
   3. $A_e = 0.7(7.35) = 5.15$ in.^{2}
   4. Tensile strength
      $\phi F_y A_g = 0.9(50)(7.62) = 342.9$ k
      $\phi F_u A_e = 0.75(65)(5.15) = 251.06$ k $\leftarrow$ Controls
D. Moment strength
   1. $0.38\sqrt{\dfrac{E}{F_y}} = 9.15 > 5.7$, it is a compact flange

   $3.76\sqrt{\dfrac{E}{F_y}} = 90.55 > 29.5$, it is a compact web

---

* As a guess, the minimum area for axial load alone should be $A_u = \dfrac{P_u}{\phi F_y} = \dfrac{160}{0.9(50)} = 3.55$ in.^{2} The selected section is twice this size because a moment M_u is also acting.

2. Adequate lateral support (given)
3. Moment strength
$$\phi_b F_y Z = 0.9(50)(313) = 1400.5 \text{ in.-k}$$
E. Interaction equation

1. $\dfrac{P_u}{\phi P_n} = \dfrac{160}{251.06} = 0.64 > 0.2$, Use Equation 12.1

2. $\dfrac{P_u}{\phi P_n} + \dfrac{8}{9}\left(\dfrac{M_{ux}}{\phi_b M_{nx}}\right) = (0.64) + \dfrac{8}{9}\left(\dfrac{531.4}{1400.5}\right) = 0.97 < 1$ **OK**

COMBINATION OF COMPRESSION AND FLEXURE FORCES: THE BEAM-COLUMN MEMBERS

Instead of axial tension, when an axial compression acts together with a bending moment, which is a more frequent case, a secondary effect sets in. The member bends due to the moment. This causes the axial compression force to act off center resulting in an additional moment equal to the axial force times the lateral displacement. This additional moment causes further deflection, which in turn produces more moment, and so on until an equilibrium is reached. This additional moment known as the *second-order moment* is not as much of a problem with the axial tension, which tends to reduce the deflection.

There are two kinds of the second-order moments, as discussed below.

MEMBERS WITHOUT SIDESWAY

Consider an isolated beam-column member AB of a frame in Figure 12.2. Due to load w_u on the member itself a moment M_{nt} results assuming that the top joint B does not deflect with respect to the bottom joint A (i.e., there is no sway). This causes the member to bend, as shown in Figure 12.3. The total moment consists of the primary (the first order) moment, M_{nt} and the second-order moment, $P_u \delta$. Thus,

$$M_{u1} = M_{nt} + P_u \delta \tag{12.3}$$

where M_{nt} is the first-order moment in a member assuming no lateral movement (no translation).

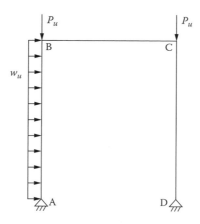

FIGURE 12.2 Second-order effects on a frame.

MEMBERS WITH SIDESWAY

Now consider that the frame is subject to a sidesway where the ends of the column can move with respect to each other, as shown in Figure 12.4. M_{lt} is the primary (first order) moment caused by the lateral translation of the frame only. Since the end B is moved by Δ with respect to A, the second-order moment is $P_u\Delta$.

Therefore, the total moment is

$$M_{u2} = M_{lt} + P_u\Delta \qquad (12.4)$$

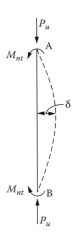

where M_{lt} is the first-order moment caused by the lateral translation.

It should be understood that the moment M_{u1} (Equation 12.3) is the property of the member and M_{u2} (Equation 12.4) is a characteristic of a frame. When a frame is braced against sidesway, M_{u2} does not exist. For an unbraced frame, the total moment is the sum of M_{u1} and M_{u2}. Thus,

$$M_u = \left(M_{nt} + P_u\delta\right) + \left(M_{lt} + P\Delta\right) \qquad (12.5)$$

FIGURE 12.3   Second-order moment within a member.

The second-order moments are evaluated directly or through the factors that magnify the primary moments. In the second case,

$$M_u = B_1 M_{nt} + B_2 M_{lt} \qquad (12.6)$$

where B_1 and B_2 are the magnification factors when the first-order moment analysis is used.

For braced frames, only the factor B_1 is applied. For unbraced frames, the factors B_1 and B_2 both are applied.

FIGURE 12.4   Second-order moment with sideway.

MAGNIFICATION FACTOR, B_1

This factor is determined assuming the braced (no sway) condition. It can be demonstrated that for a sine curve, the magnified moment directly depends on the ratio of the applied axial load to the elastic (Euler) load of the column. The factor is expressed as follows:

$$B_1 = \frac{C_m}{1 - \left(P_u/P_{e1}\right)} \geq 1 \qquad (12.7)$$

where C_m is an expression that accounts for the nonuniform distribution of the bending moment within a member as discussed in the "Moment Modification Factor, C_m," section below.

$$P_{e1} = \text{Euler buckling strength}$$

$$= \frac{\pi^2 EA}{\left(KL/r\right)^2}$$

The slenderness ratio $((KL)/r)$ is along the axis in which the bending occurs and P_u is the applied factored axial compression load.

Equation 12.7 suggests that B_1 should be greater or equal to 1; it is a magnification factor.

Moment Modification Factor, C_m

In Equation 12.7 for B_1, a factor C_m appears. Without this factor, B_1 may be over-magnified. When a column is bent in a single curvature with equal end moments, the deflection occurs, as shown in Figure 12.5a. In this case, $C_m=1$. When the end moments bend a member in a reverse curvature, as shown in Figure 12.5b, the max deflection that occurs at some distance out from the center is smaller than the first case; using $C_m=1$ will overdo the magnification. The purpose of the modifier C_m is to reduce the magnified moment when the variation of the moment within a member requires that B_1 should be reduced. The modification factor depends on the rotational restraint placed at the member ends. There are two types of loading for C_m.

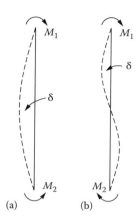

FIGURE 12.5 Deflection of a column under different end moment conditions.

1. When there is no transverse loading between the two ends of a member:
The modification factor is given by

$$C_m = 0.6 - 0.4\left(\frac{M_1}{M_2}\right) \le 1 \tag{12.8}$$

where
M_1 is the smaller end moment
M_2 is the larger of the end moments

The ratio (M_1/M_2) is negative when the end moments have the opposite directions causing the member to bend in a single curvature. (This is opposite of the sign convention for concrete column in the "Short Columns with Combined Loads" section in Chapter 16.) The ratio is taken to be positive when the end moments have the same directions causing the member to bend in a reverse curvature.
2. When there is a transverse loading between the two ends of a member:
The C_m value is as follows:
   a. $C_m=0.85$ for a member with the restrained (fixed) ends
   b. $C_m=1.0$ for a member with the unrestrained ends

Example 12.2

The service loads on a W12×72 braced frame member of A572 steel are shown in Figure 12.6. The bending is about the strong axis. Determine the magnification factor B_1. Assume the pinned end condition.

Solution

A. Design loads
   1. $P_u = 1.2(100) + 1.6(200) = 440\,\text{k}$
   2. $(M_u)_B = 1.2(15) + 1.6(40) = 82\,\text{ft-k}$
   3. $(M_u)_A = 1.2(20) + 1.6(50) = 104\,\text{ft-k}$

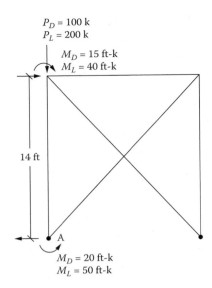

$P_D = 100$ k
$P_L = 200$ k

$M_D = 15$ ft-k
$M_L = 40$ ft-k

14 ft

A

$M_D = 20$ ft-k
$M_L = 50$ ft-k

FIGURE 12.6   A braced frame for Example 12.2.

B. Modification factor
  1. $\dfrac{M_1}{M_2} = \dfrac{-82}{104} = -0.788$
  2. $C_m = 0.6 - 0.4(-0.788) = 0.915$
C. Euler buckling strength
  1. $K = 1$
  2. For W12 × 72, $A = 21.1$ in.^{2}
     $r_x = 5.31$ in., bending in X-direction
  3. $\dfrac{KL}{r_x} = \dfrac{(1)(14 \times 12)}{5.31} = 31.64$
  4. $P_{e1} = \dfrac{\pi^2 EA}{(KL/r)^2}$

     $= \dfrac{\pi^2 (29,000)(21.2)}{(31.64)^2} = 6,055 \text{k}$

  5. $B_1 = \dfrac{C_m}{1 - (P_u/P_{e1})}$

     $= \dfrac{0.915}{1 - \left(\dfrac{440}{6,055}\right)} = 0.99 < 1$

     Use $B_1 = 1$

BRACED FRAME DESIGN

To braced frames, only the magnification factor B_1 is applied. As stated earlier, the use of the interaction equation, Equations 12.1 or 12.2, is direct in analysis when the member size is known. However, it is a trial-and-error procedure for designing of a member.

Instead of making a blind guess, design aids are available to make a feasible selection prior to application of the interaction equation. The procedure presented in the *AISC Manual 2005* for initial selection needs an intensive input of the data from the special tables included in the manual. In a previous version of the AISC manual, a different approach was suggested, which was less data intensive. This approach is described below.

The interaction equations can be expressed in terms of an equivalent axial load. With respect to Equation 12.1, this modification has been demonstrated below:

$$\frac{P_u}{\phi P_n} + \frac{8}{9}\left(\frac{M_{ux}}{\phi_b M_{nx}} + \frac{M_{uy}}{\phi_b M_{ny}}\right) = 1$$

Multiply the both sides by ϕP_n,

$$P_u + \frac{8}{9}\frac{\phi P_n}{\phi_b}\left(\frac{M_{ux}}{M_{nx}} + \frac{M_{uy}}{M_{ny}}\right) = \phi P_n \tag{12.9}$$

Treating ϕP_n as P_{eff}, this can be expressed as

$$P_{eff} = P_u + m\,M_{ux} + m\,U\,M_{uy} \tag{12.10}$$

where
P_u is the factored axial load
M_{ux} is the magnified factored moment about X-axis
M_{uy} is the magnified factored moment about Y-axis

The values of the coefficient m, reproduced from the AISC manual, are given in Table 12.1.

The manual makes an iterative application of Equation 12.10 to determine the equivalent axial compressive load, P_{eff} for which a member could be picked up as an axially loaded column only. However, this also requires the use of an additional table to select the value of U.

This chapter suggests an application of Equation 12.10 just only to make an educated guess for a preliminary section. The initially selected section will then be checked by the interaction equations. The procedure is as follows:

1. With the known value of the effective length, KL select a value of m from Table 12.1 for the shape category proposed to be used, i.e., W12. Assume $U=3$.
2. From Equation 12.10, solve for P_{eff}
3. Pick up a section having the cross-sectional area larger than the following:

$$A_g = \frac{P_{eff}}{\phi F_y}$$

4. Confirm the selection by the appropriate interaction equation, Equation 12.1 or Equation 12.2

Example 12.3

For a braced frame, the axial load and the end moments obtained from structural analysis are shown in Figure 12.7. Design a W14 member of A992 steel. Use $K=1$ for the braced frame.

TABLE 12.1
Values of Factor m

F_y	36 ksi							50 ksi						
KL (ft)	10	12	14	16	18	20	22 and over	10	12	14	16	18	20	22 and over
					First approximation									
All shapes	2.4	2.3	2.2	2.2	2.1	2.0	1.9	2.4	2.3	2.2	2.0	1.9	1.8	1.7
					Subsequent approximations									
W, S 4	3.6	2.6	1.9	1.6	—	—	—	2.7	1.9	1.6	1.6	—	—	—
W, S 5	3.9	3.2	2.4	1.9	1.5	1.4	—	3.3	2.4	1.8	1.6	1.4	1.4	—
W, S 6	3.2	2.7	2.3	2.0	1.9	1.6	1.5	3.0	2.5	2.2	1.9	1.8	1.5	1.5
W 8	3.0	2.9	2.8	2.6	2.3	2.0	2.0	3.0	2.8	2.5	2.2	1.9	1.6	1.6
W 10	2.6	2.5	2.5	2.4	2.3	2.1	2.0	2.5	2.5	2.4	2.3	2.1	1.9	1.7
W 12	2.1	2.1	2.0	2.0	2.0	2.0	2.0	2.0	2.0	2.0	1.9	1.9	1.8	1.7
W 14	1.8	1.7	1.7	1.7	1.7	1.7	1.7	1.8	1.7	1.7	1.7	1.7	1.7	1.7

Note: Values of m are for $C_m = 0.85$. When C_m is other than 0.85, multiply the tabular value of m by $C_m/0.85$.

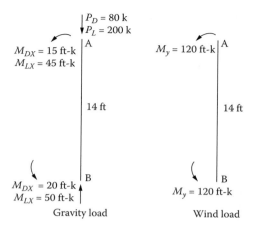

$P_D = 80$ k
$P_L = 200$ k

$M_{DX} = 15$ ft-k
$M_{LX} = 45$ ft-k

A

$M_y = 120$ ft-k

A

14 ft

14 ft

$M_{DX} = 20$ ft-k
$M_{LX} = 50$ ft-k

B

$M_y = 120$ ft-k

B

Gravity load

Wind load

FIGURE 12.7   A column member of a braced frame.

Solution

A. Critical load combinations
   (a) $1.2D + 1.6L$
   1. Assume a member weight of 100 lb/ft, total weight $= 100(14) = 1400$ lb or 1.4 k
   2. $P_u = 1.2(81.4) + 1.6(200) = 417.7$ k
   3. $(M_{nt})_x$ at A $= 1.2(15) + 1.6(45) = 90$ ft-k
   4. $(M_{nt})_x$ at B $= 1.2(20) + 1.6(50) = 104$ ft-k
   (b) $1.2D + L + 1.6W$
   1. $P_u = 1.2(84) + 200 = 297.7$ k
   2. $(M_{nt})_x$ at A $= 1.2(15) + 45 = 63$ ft-k
   3. $(M_{nt})_x$ at B $= 1.2(20) + 50 = 74$ ft-k
   4. $(M_{nt})_y = 1.6(120) = 192$ ft-k
B. Trial selection
   1. For the load combination (a)
       From Table 12.1 for $KL = 14$ ft, $m = 1.7$
       $P_{eff} = 417.7 + 1.7(104) = 594.5$ k
   2. For load combination (b), Let $U = 3$
       $P_{eff} = 297.7 + 1.7(74) + 1.7(3)(192) = 1402.7$ k ← Controls
   3. $A_g = \dfrac{P_{eff}}{\phi F_y} = \dfrac{1402.7}{(0.9)(50)} = 31.17$ in.^{2}
   4. Select W14 × 109 $A = 32.0$ in.^{2}
       $Z_x = 192$ in.^{3}
       $Z_y = 92.7$ in.^{3}
       $r_x = 6.22$ in.
       $r_y = 3.73$ in.
       $b_f/2t_f = 8.49$
       $h/t_w = 21.7$

   Checking of the trial section for the load combination (b)
C. Along the strong axis
   1. Moment strength
       $\phi M_{nx} = \phi F_y Z_x = 0.9(50)(192) = 8640$ in.-k or 720 ft-k
   2. Modification factor: reverse curvature

   $$\frac{(M_{nt})_x \, @A}{(M_{nt})_x \, @B} = \frac{63}{74} = 0.85$$

   $$C_{mx} = 0.6 - 0.4(0.85) = 0.26$$

3. Magnification factor

$$K = 1$$

$$\frac{KL}{r_x} = \frac{(1)(14 \times 12)}{6.22} = 27.0$$

$$(P_{e1})_x = \frac{\pi^2 EA}{(KL/r_x)^2} = \frac{\pi^2 (29,000)(32)}{(27.0)^2} = 12,551$$

4. $(B_1)_x = \dfrac{C_m}{1 - (P_u/P_{e1})}$

$$= \frac{0.26}{1 - (297.7/12,551)} = 0.256 < 1; \text{ use } 1$$

5. $(M_u)_x = B_1(M_{nt})_x$
$= 1(74) = 74 \text{ ft-k}$

D. Along the minor axis
   1. Moment strength
      $\phi M_{ny} = \phi F_y Z_y = 0.9(50)(92.7) = 4171.5 \text{ in.-k or } 347.63 \text{ ft-k}$
   2. Modification factor: reverse curvature

$$\frac{(M_{nt})_y \, @A}{(M_{nt})_y \, @B} = \frac{208}{208} = 1$$

$$C_{mx} = 0.6 - 0.4(1) = 0.2$$

   3. Magnification factor

$$K = 1$$

$$\frac{KL}{r_y} = \frac{(1)(14 \times 12)}{3.73} = 45.0$$

$$(P_{e1})_y = \frac{\pi^2 EA}{(KL/r_y)^2} = \frac{\pi^2 (29,000)(32)}{(45.0)^2} = 4518.4$$

4. $(B_1)_y = \dfrac{C_m}{1 - (P_u/P_{e1})}$

$$= \frac{0.2}{1 - (297.7/4518.4)} = 0.21 < 1; \text{ use } 1$$

5. $(M_u)_y = (B_1)_y(M_{nt})_y$
$= 1(192) = 192 \text{ ft-k}$

E. Compression strength
   1. $\dfrac{KL}{r_x} = \dfrac{(1)(14 \times 12)}{6.22} = 27.0$

   2. $\dfrac{KL}{r_y} = \dfrac{(1)(14 \times 12)}{3.73} = 45.0 \leftarrow \text{Controls}$

3. $4.71\sqrt{\dfrac{E}{F_y}} = 4.71\sqrt{\dfrac{29{,}000}{50}} = 113.43 > 45$ Inelastic buckling

4. $F_e = \dfrac{\pi^2 E}{\left(KL/r_y\right)^2} = \dfrac{\pi^2 (29{,}000)}{(45.0)^2} = 141.2$

5. $F_{cr} = (0.658^{50/141.2})50 = 43.11$

6. $\phi P_n = 0.9 F_{cr} A_g$
   $= 0.9(43.11)(32) = 1241.6\,\text{k}$

F. Interaction equation

$$\dfrac{P_u}{\phi P_n} = \dfrac{297.7}{1241.6} = 0.24 > 0.2$$

$$\dfrac{P_u}{\phi P_n} + \dfrac{8}{9}\left(\dfrac{M_{ux}}{\phi_b M_{nx}} + \dfrac{M_{uy}}{\phi_b M_{ny}}\right)$$

$$0.24 + \dfrac{8}{9}\left(\dfrac{74}{720} + \dfrac{192}{347.63}\right)$$

$$= 0.82 < 1 \quad \textbf{OK}$$

MAGNIFICATION FACTOR FOR SWAY, B_2

The term B_2 is used to magnify the column moments under sidesway condition. For sidesway to occur in a column on a floor, it is necessary that all of the columns on that floor should sway simultaneously. Hence the total load acting on all columns on a floor appears in the expression for B_2. The *AISC Manual 2005* presents the following two relations for B_2:

$$B_2 = \dfrac{1}{1 - \dfrac{\Sigma P_u}{\Sigma H}\left(\dfrac{\Delta H}{L}\right)} \tag{12.11}$$

or

$$B_2 = \dfrac{1}{1 - \dfrac{\Sigma P_u}{\Sigma P_{e2}}} \tag{12.12}$$

where
  ΔH is the lateral deflection of the floor (story) in question
  L is the story height
  ΣH is the sum of horizontal forces on the floor in question
  ΣP_u is the total design axial force on all the columns on the floor in question
  ΣP_{e2} is the summation of the elastic (Euler) capacity of all columns on the floor in question, given by

$$\Sigma P_{e2} = \Sigma \dfrac{\pi^2 EA}{\left(KL/r\right)^2} \tag{12.13}$$

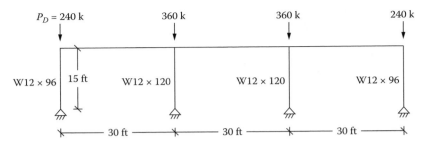

FIGURE 12.8 An unbraced frame.

The term P_{e2} is similar to the term P_{e1}, except that the factor K is determined in the plane of bending for an unbraced condition (K in P_{e1}, was for the braced condition).

The designer can use either of the Equation 12.11 or Equation 12.12; the choice is a matter of convenience. In Equation 12.11, initial size of the members is not necessary since A and r as a part of P_{e2} are not required unlike in Equation 12.12. Further, a limit on $\Delta H/L$, known as the *drift index*, can be set by the designer to control the sway. This is limited to 0.004 with factored loads.

Example 12.4

An unbraced frame of A992 steel at the base floor level is shown in Figure 12.8. The loads are factored dead loads. Determine the magnification factor for sway for the column member (1) bending in Y-axis.

Solution

A. Exterior columns
   1. Factored weight of column $= 1.2(0.096 \times 15) = 1.7\,k$
   2. $P_u = 240 + 1.7 = 241.7\,k$
   3. $K = 2$
   4. For W12 × 96, $A = 28.2\,in.^2$
      $r_y = 3.09\,in.$
   5. $\dfrac{KL}{r_y} = \dfrac{2(15 \times 12)}{3.09} = 116.50$
   6. $P_{e2} = \dfrac{\pi^2 EA}{\left(KL/r_y\right)^2} = \dfrac{\pi^2 (29{,}000)(28.2)}{(116.5)^2} = 594.1k$

B. Interior columns
   1. Factored weight of column $= 1.2(0.12 \times 15) = 2.2\,k$
   2. $P_u = 360 + 2.2 = 362.2\,k$
   3. $K = 2$
   4. For W12 × 120, $A = 35.3\,in.^2$
      $r_y = 3.13\,in.$
   5. $\dfrac{KL}{r_y} = \dfrac{2(15 \times 12)}{3.13} = 115.0$
   6. $P_{e2} = \dfrac{\pi^2 EA}{\left(KL/r_y\right)^2} = \dfrac{\pi^2 (29{,}000)(35.3)}{(115.0)} = 763.2k$

C. For the entire story,
   1. $\Sigma P_u = 2(241.7) + 2(362.2) = 1208\,k$
   2. $\Sigma P_{e2} = 2(594.1) + 2(763.2) = 2714.6$

3. From Equation 12.12,

$$B_2 = \cfrac{1}{1-\left(\cfrac{\Sigma P_u}{\Sigma P_{e2}}\right)}$$

$$= \cfrac{1}{1-\left(\cfrac{1208}{2714.6}\right)} = 1.80$$

Example 12.5

In Example 12.4, the total factored horizontal force on the floor is 200 k and the allowable drift index is 0.002. Determine the magnification factor for sway.

Solution
From Equation 12.11,

$$B_2 = \cfrac{1}{1-\cfrac{\Sigma P_u}{\Sigma H}\left(\cfrac{\Delta H}{L}\right)}$$

$$= \cfrac{1}{1-\left(\cfrac{1208}{200}\right)(0.002)} = 1.01$$

UNBRACED FRAME DESIGN

The interaction equations, Equations 12.1 and 12.2 are used for the unbraced frame design as well. M_{ux} and M_{uy} in the equations are computed by Equation 12.6 magnified for both B_1 and B_2.

The trial size can be determined from Equation 12.10 following the procedure stated in the "Braced Frame Design" section. When an unbraced frame is subjected to the symmetrical vertical (gravity) loads along with a lateral load, as shown in Figure 12.9, the moment M_{nt} in member AB is computed for the gravity loads. This moment is amplified by the factor B_1 to account for the P–δ effect. The moment M_{lt} is computed due to the horizontal load H. It is then magnified by the factor B_2 for P–Δ effect.

When an unbraced frame supports an asymmetric loading, as shown in Figure 12.10, the eccentric loading causes it to deflect sideways. First, the frame is considered braced by a fictitious support called an *artificial joint restraint* (AJR). The moment M_{nt} and the deflection δ are computed, which is amplified by the factor B_1.

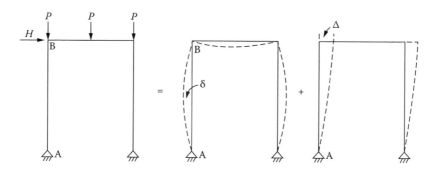

FIGURE 12.9 Symmetrical vertical loads and a lateral load on an unbraced frame.

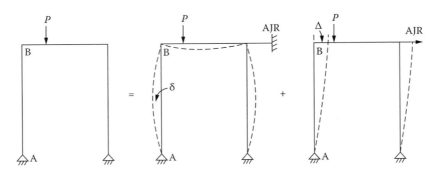

FIGURE 12.10 Assymetric load on an unbraced frame.

To compute M_{lt}, a force equal to AJR but opposite in direction is then applied. This moment is magnified by the factor B_2 for P–Δ effect.

When both the asymmetric gravity loads and the lateral loads are present, the above two cases are combined, i.e., AJR force is added to the lateral loads to compute M_{lt} for the P–Δ effect.

Alternatively, two structural analyses are performed. The first analysis is performed as a braced frame; the resulting moment is M_{nt}. The second analysis is done as an unbraced frame. The results of the first analysis are subtracted from the second analysis to obtain the M_{lt} moment.

Example 12.6

An unbraced frame of A992 steel is subjected to the dead load, the live load, and the wind load. The structural analysis provides the axial forces and the moments on the column along X-axis, as shown in Figure 12.11. Design for a maximum drift of 0.5 in.

Solution

A. Critical load combinations
  (a) $1.2D + 1.6L$
      1. Assume a member weight of 100 lb/ft, total weight = 100(15) = 1500 lb or 1.5 k
      2. $P_u = 1.2(81.5) + 1.6(210) = 433.8$ k
      3. $(M_{nt})_x$ at A = 1.2(15) + 1.6(45) = 90 ft-k
      4. $(M_{nt})_x$ at B = 1.2(20) + 1.6(50) = 104 ft-k
      5. $(M_{lt}) = 0$ since the wind load is not in this combination
  (b) $1.2D + L + 1.6W$
      1. $P_u = 1.2(81.5) + 210 = 307.8$ k
      2. $(M_{nt})_x$ at A = 1.2(15) + 45 = 63 ft-k
      3. $(M_{nt})_x$ at B = 1.2(20) + 50 = 74 ft-k

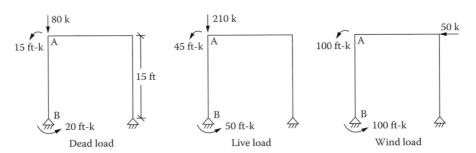

FIGURE 12.11 Loads on an unbraced frame.

    4. $(M_{lt})_x$ at A= 1.6(100) = 160 ft-k
    5. $(M_{lt})_x$ at B = 1.6(100) = 160 ft-k
B. Trial selection
  1. For the load combination (a)

$$K = 1.2, \ KL = 1.2(15) = 18 \text{ ft}$$

From Table 12.1 for W12 section, $m = 1.9$

$$P_{eff} = 433.8 + 1.9(104) = 631.4 \text{ k}$$

  2. For load combination (b)

$$P_{eff} = 307.8 + 1.9(74) + 1.9(160) = 752.4 \text{ k} \leftarrow \text{Controls}$$

  3. $A_g = \dfrac{P_{eff}}{\phi F_y} = \dfrac{751.4}{(0.9)(50)} = 16.72 \text{ in.}^2$

  4. Select W12 × 74 (W12 × 69 has the non-compact flange)
     $A = 21.1$ in.^{2}
     $Z_x = 108$ in.^{3}
     $r_x = 5.31$ in.
     $r_y = 3.04$ in.
     $b_f/2t_f = 8.99$
     $h/t_w = 22.6$

Checking of the trial section for the critical load combination (b)

C. Moment strength
  1. $0.38\sqrt{\dfrac{E}{F_y}} = 9.15 > \dfrac{b_f}{2t_f}$, compact

  2. $3.76\sqrt{\dfrac{E}{F_y}} = 90.55 > \dfrac{h}{t_w}$, compact

  3. $\phi M_{nx} = \phi F_y Z_x = 0.9(50)(108) = 4860$ in.-k or 405 ft-k
D. Modification factor: reverse curvature
  1. $\dfrac{(M_{nt})_x @ A}{(M_{nt})_x @ B} = \dfrac{63}{74} = 0.85$
  2. $C_{mx} = 0.6 - 0.4(0.85) = 0.26$
E. Magnification factor, B_1
  1. $K = 1$ for braced condition
  2. $\dfrac{KL}{r_x} = \dfrac{(1)(15 \times 12)}{5.31} = 33.9$

  3. $(P_{e1})_x = \dfrac{\pi^2 EA}{(KL/r_x)^2} = \dfrac{\pi^2 (29{,}000)(21.1)}{(33.9)^2} = 5250.3$

  4. $(B_1)_x = \dfrac{C_m}{1 - \dfrac{P_u}{(P_{e1})_x}}$

$$= \dfrac{0.26}{1 - \left(\dfrac{307.8}{5250.3}\right)} = 0.28 < 1; \text{ use } 1$$

F. Magnification factor for sway, B_2
  1. $K = 1.2$ for unbraced condition
  2. $\dfrac{KL}{r_x} = \dfrac{(12)(15 \times 12)}{5.31} = 40.68$

3. $(P_{e2})_x = \dfrac{\pi^2 EA}{(KL/r_x)^2} = \dfrac{\pi^2 (29{,}000)(21.1)}{(40.68)^2} = 3645.7$

4. $\Sigma P_u = 2(307.8) = 615.6\,\text{k}$, since there are two columns in the frame

5. $\Sigma(P_{e2})_x = 2(3645.7) = 7291.4\,\text{k}$

6. $\dfrac{\Delta H}{L} = \dfrac{0.5}{15 \times 12} = 0.00278$

7. From Equation 12.11,

$$B_2 = \dfrac{1}{1 - \dfrac{\Sigma P_u}{\Sigma H}\left(\dfrac{\Delta H}{L}\right)}$$

$$= \dfrac{1}{1 - \left(\dfrac{615.6}{50}\right)(0.00278)} = 1.035$$

8. From Equation 12.12

$$B_2 = \dfrac{1}{1 - \dfrac{\Sigma P_u}{\Sigma P_{e2}}}$$

$$= \dfrac{1}{1 - \left(\dfrac{615.6}{7291.4}\right)} = 1.09 \leftarrow \text{Controls}$$

G. Design moment

$(M_u)_x = B_1(M_{nt})_x + B_2(M_{lt})_x$

$= 1(74) + 1.09(160) = 248.4\ \text{ft-k}$

H. Compression strength

1. $\dfrac{KL}{r_x} = \dfrac{(1.2)(15 \times 12)}{5.31} = 40.7$

2. $\dfrac{KL}{r_y} = \dfrac{(1.2)(15 \times 12)}{3.04} = 71.05 \leftarrow \text{Controls}$

3. $4.71\sqrt{\dfrac{E}{F_y}} = 4.71\sqrt{\dfrac{29{,}000}{50}} = 113.43 > 71.05$ Inelastic buckling

4. $F_e = \dfrac{\pi^2 E}{(KL/r_y)^2} = \dfrac{\pi^2 (29{,}000)}{(71.05)^2} = 56.64$

5. $F_{cr} = (0.658^{50/56.64})50 = 34.55\,\text{ksi}$

6. $\phi P_n = 0.9 F_{cr} A_g$
   $= 0.9(34.55)(21.1) = 656.2\,\text{k}$

I. Interaction equation

1. $\dfrac{P_u}{\phi P_n} = \dfrac{307.8}{656.2} = 0.47 > 0.2$,   apply Equation 12.1

2. $\dfrac{P_u}{\phi P_n} + \dfrac{8}{9}\left(\dfrac{M_{ux}}{\phi_b M_{nx}}\right) = 0.47 + \dfrac{8}{9}\left(\dfrac{248.4}{405}\right)$

$$= 1.0 \quad \textbf{OK} \text{ (border case)}$$

Select a W12 × 72 section

OPEN-WEB STEEL JOISTS

A common type of floor system for small to medium size steel frame building consists of the open-web steel joists with or without joist girders. The joist girders, when used, are designed to support the open-web steel joists. The floor and roof slabs are supported by the open-web joists. A typical plan is shown in Figure 12.12.

The open-web joists are parallel chord truss where web members are made from steel bars or small angles. A section is shown in Figure 12.13. The open-web joists are pre-engineered systems that can be quickly erected. The open spaces in the web can accommodate ducts and piping.

The AISC specifications do not the cover open-web joists. A separate organization, the Steel Joist Institute (SJI) is responsible for the specifications relating to the open-web steel joists and joist girders. The SJI's publication titled "*Standard Specifications*" deals with all aspects of the open-web joists including their design, manufacturing, application, erection, stability, and the handling.

Three categories of joists are presented in the standard specifications:

1. Open-web joists, K-series
   For span range 8–60 ft, depth 8–30 in., chords $F_y = 50$ ksi and web $F_y = 36$ or 50 ksi
2. Long span steel joists, LH-series
   For span range 21–96 ft, depth 18–48 in., chords $F_y = 36$ or 50 ksi and web $F_y = 36$ or 50 ksi
3. Deep long span joists, DLH-series
   For span range 61–144 ft, depth 52–72 in., chords $F_y = 36$ or 50 ksi, and web $F_y = 36$ or 50 ksi

Open-web joist uses a standardized designation; for example, "18 K 6" means that the depth of the joist is 18 in., it is a K-series joist that has a relative strength of 6. The higher the strength number, the stronger is the joist. Different manufactures of 18 K 6 joists will have different member cross sections but they all must have a depth of 18 in. and a load capacity as tabulated by the SJI.

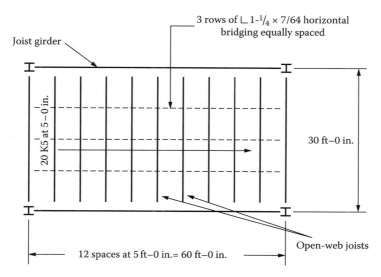

FIGURE 12.12 An open-web joist floor system.

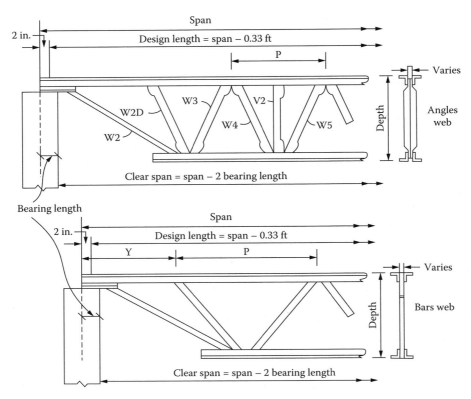

FIGURE 12.13   Open-web steel joist.

The joists are designed as a simply supported uniformly loaded trusses supporting a floor or a roof deck. They are constructed as so that the top chord of the joists are braced against the lateral buckling.

The SJI specifications stipulate the following basis of design:

1. The bottom chord is designed as an axially loaded tensile member. The design standards and the limiting states of Chapter 9 for tensile member are applied.
2. The top chord is designed for the axial compression force only when the panel length, l does not exceed 24 in., which is taken as the spacing between lines of bridging. The design is done according to the standards of Chapter 10 on columns. When the panel length exceeds 24 in., the top chord is designed as a continuous member subject to the combined axial compression and bending, as discussed in this chapter.
3. The web is designed for the vertical shear force determined from a full uniform loading but it should not be less than ¼ of the end reaction. The combined axial compression and bending are investigated for the compression web members.
4. The bridging comprising of a cross connection between the adjoining joists is required for the top and the bottom chords. This consists of one or both of the following types:
   a. The horizontal bridging by a continuous horizontal steel member. The ratio of the length between the attachment to the least radius of gyration l/r should not exceed 300.
   b. The diagonal bridging by cross bracing between the joists with the l/r ratio not exceeding 200.

The number of rows of the top chord and the bottom chord bridging should not be less than those prescribed in the bridging tables of the SJI standards. The spacing should be such that the radius of

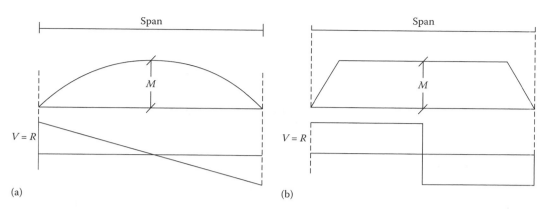

FIGURE 12.14 Shear and moment envelopes. (a) Standard joists shear and bending moment diagrams. (b) KCS joists shear and bending moment diagrams.

gyration of the top chord about its vertical axis should not be less than $l/145$, where l is the spacing in inches between the lines of bridging.

For design convenience, the SJI in their standard specification, have included, the standard load tables that can be directly used to determine the joist size. Tables for K-series joists are included in Appendix C.10. The loads in the tables represent the uniformly distributed loads. The joist are designed for a simple span uniform loading, which produces a parabolic moment diagram for the chord members and a linearly sloped (triangular shaped) shear diagram for the web members, as shown in Figure 12.14a.

To address the problem of supporting of the uniform loads together with the concentrated loads, the special K-series joists, known as KCS joists, are designed. The KCS joists are designed for a flat moment and rectangular shear envelopes, as shown in Figure 12.14b.

When we enter the table at Appendix C.10, under the column "18 K 6," across a row corresponding to the joist span, the first figure is the total lb/ft load that a 18 K 6 joist can support and the second light-faced figure is the unfactored live load from the consideration of $L/360$ deflection. For a live load deflection of $L/240$, multiply the load figure by the ratio 360/240, i.e., 1.5.

The following example demonstrates the use of the joist table.

Example 12.7

Select an open-web steel joist for a span of 30 ft to support a dead load of 35 psf and a live load of 40 psf. The joist spacing is 4 ft. The maximum live load deflection is $L/240$.

Solution

    A. Design loads
1. Tributary area/ft = 4 ft²/ft
2. Dead load/ft = $35 \times 4 = 140$ lb/ft
3. Weight of joist/ft = 10 lb/ft
4. Total dead load = 150 lb/ft
5. Factored dead load = 1.2(150) = 180 lb/ft
6. Live load/ft = $40 \times 4 = 160$ lb/ft
7. Factored live load = 1.6(160) = 256 lb/ft
8. Total factored load = 436 lb/ft

    B. Standard load table at Appendix C.10
1. Enter the row corresponding to span 30. The section suitable for total factored load of 436 lb/ft is $18 K \times 6$, which has a capacity of 451 lb/ft.

2. Live load capacity for $L/240$ deflection

$$= \frac{360}{240}(175) = 262.5\,\text{lb/ft} > 256\,\text{lb/ft}\quad \textbf{OK}$$

3. The joists of different depths might be selected using the other standard load tables of the SJI. In fact, the SJI include an economy table for the lightest joist selection.

JOIST GIRDERS

The loads on a joist girder are applied through open-web joists that the girder supports. This load is equal in magnitude and evenly spaced along the top chord of the girder applied through the panel points.

The bottom chord is designed as an axially loaded tension member. The radius of gyration of the bottom chord about its vertical axis should not be less than $l/240$, where l is the distance between the lines of bracing.

The top chord is designed as an axially loaded compression member. The radius of gyration of the top chord about the vertical axis should not be less than the span/575.

The web is designed for the vertical shear for full loading but should not be less than ¼ of the end reaction. The tensile web members are designed to resist at least 25% of the axial force in compression.

The SJI, in the standard specifications, have included the girder tables that are used to design girders. Selected tables have been included at Appendix C.11. The following are the design parameters of a joist girder:

1. Span of the girder
2. Number and spacing of the open-web joists on the girder: when the spacing is known, number equals the span/spacing; for the known number of joists, spacing equals the span/number
3. The point load on the panel points in kips: Total factored unit load in psf is multiplied by the joist spacing and the length of joist (bay length) converted to kips
4. Depth of girder

For any of the above three known parameters, the fourth one can be determined from the girder tables. In addition, the table gives the weight of the girder in lb/ft to confirm that it had been adequately included in the design loads.

Usually, the first three parameters are known and the depth of the girder is determined. A rule of thumb is about an inch of depth for each feet of span for an economic section. Each joist girder uses a standardized designation; for example, "36G 8N 15F" means that the depth of the girder 36 in., it provides for 8 equal joist spaces and it supports a factored load of 15 k at each panel location (a symbol K at the end, in place of F, is used for the service load capacity at each location).

Example 12.8

Specify the size of the joist girder for the floor system shown in Figure 12.15.

Solution

A. Design loads
1. Including 1 psf for the weight of girder, total factored load

$$= 1.2(15+1)+1.6(30) = 67.2\,\text{psf}$$

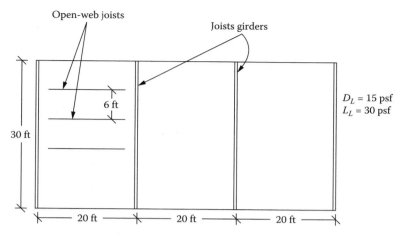

FIGURE 12.15   Floor system for Example 12.8.

2. Panel area $= 6 \times 20 = 120\,\text{ft}^2$
3. Factored concentrated load/panel point

$$= 67.2 \times 120 = 8064\,\text{lb} \text{ or } 8.1\,\text{k, use } 9\,\text{k}$$

B. Joist details
   1. Space $= 6\,\text{ft}$
   2. Numbering joists $= \dfrac{30}{6} = 5$

C. Girder depth selection
   1. Refer to Appendix C.11. For 30 ft span, 5 N, and 9 k load, the range of depth is 24 in. to 36 in.
      Select 28G 5N 9F
   2. From Appendix C.11, weight per ft girder $= 17\,\text{lb/ft}$
      Unit weight $= \dfrac{17}{20} = 0.85\,\text{psf} <$ assumed 1 psf   **OK**
   3. The information shown in Figure 12.16 will be specified to the manufacturer

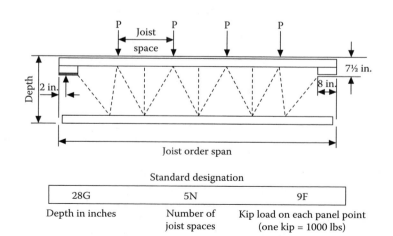

28G	5N	9F
Depth in inches	Number of joist spaces	Kip load on each panel point (one kip = 1000 lbs)

Standard designation

FIGURE 12.16   Selection of joist girder.

PROBLEMS

In all problems assume the full lateral support conditions.

12.1 A W12×35 section of A992 steel with a single line (along the tensile force) of four 3/4 in. bolts in the web is subjected to a tensile live load of 65 k and a bending moment due to the dead load along the weak axis of 20 ft-k. Is this member satisfactory?

12.2 A W10×33 member is to support a factored tensile force of 100 k and a factored moment along X-axis of 100 ft-k. It is a fully welded member of Grade 50 steel. Is the member adequate for the loads?

12.3 A 12 ft long hanger supports a tensile dead load of 50 k and a live load of 100 k at an eccentricity of 4 in. with respect to the X-axis. Design a W10 section of A992 steel. There is one line of three bolts of 3/4 in. diameter on one side of the top flange and one line of three bolts of the same size on the other side of the top flange. The bottom flange has a bolt pattern similar to the top flange.

12.4 Design a W8 or W10 member to support the loads shown in Figure P12.4. It has a single line of three holes for 7/8 in. bolts in the web. The member consists of A992 steel.

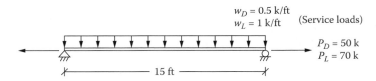

$$w_D = 0.5 \text{ k/ft}$$
$$w_L = 1 \text{ k/ft} \quad \text{(Service loads)}$$

$$P_D = 50 \text{ k}$$
$$P_L = 70 \text{ k}$$

15 ft

FIGURE P12.4   A tensile and flexure member.

12.5 The member of Problem #12.4 is a factored bending moment of along Y-axis of 40 ft-k in addition to the above loading along X-axis. Design the member. [*Hint*: Since a sizeable bending along Y-axis is involved, initially select a section at least four times of that required for axial load alone.]

12.6 A horizontal beam section W10×26 of A992 steel is subjected to the service live loads shown in Figure P12.6. The member is bent about X-axis. Determine the magnitude of the magnification factor, B_1.

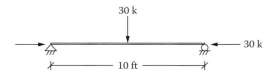

30 k

30 k

10 ft

FIGURE P12.6   A compression flexure member.

12.7 A braced frame member W12×58 of A992 steel is subjected to the loads shown in Figure P12.7. The member is bent about X-axis. Determine the magnitude of the magnification factor, B_1. Assume the pin-end conditions.

12.8 In Problem #12.7, the moments at the ends A and B are both clockwise. The ends are restrained (fixed). Determine the magnification factor, B_1.

12.9 In Problem #12.7, in addition to the loads shown, a uniformly distributed wind load of 1 k/ft acts laterally between A and B. Determine the magnification factor, B_1.

12.10 In Problem #12.7, in addition to the shown X-axis moments, the moments in Y-axis at A and B are as follows. Determine the magnification factor, B_1.

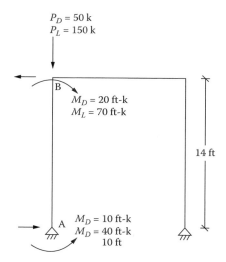

$P_D = 50$ k
$P_L = 150$ k

B

$M_D = 20$ ft-k
$M_L = 70$ ft-k

14 ft

A $M_D = 10$ ft-k
     $M_D = 40$ ft-k
     10 ft

FIGURE P12.7   An unbraced frame member.

At B $(M_D)_y = 10$ ft-k $(M_L)_y = 20$ ft-k, clockwise

At A $(M_D)_y = 8$ ft-k $(M_L)_y = 15$ ft-k, counterclockwise

12.11 The member of a A572 steel section, as shown in Figure P12.11, is used as a beam-column in a braced frame. It is bent about the strong axis. Is the member adequate?

12.12 A horizontal component of a braced frame is shown in Figure P12.12. It is bent about the strong axis. Is the member adequate? Use A992 steel.

12.13 The member of a A572 steel section, as shown in Figure P12.13, is used as a beam-column in a braced frame. It has restrained ends. Is the member adequate?

12.14 A W12×74 section of A572 steel is part of a braced frame. It is subjected to the service, dead, live, and seismic loads, as shown in Figure P12.14. The bending is along the strong axis. It has pinned ends. Is the section satisfactory?

12.15 For a braced frame, the service axial load and the moments obtained by structural analysis are shown in Figure P12.15. Design a W14 section of A992 steel. The ends are restrained.

12.16 In Problem #12.15, the wind load moments act along Y-axis (instead of the X-axis). Design the member.

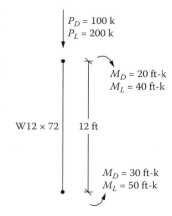

$P_D = 100$ k
$P_L = 200$ k

$M_D = 20$ ft-k
$M_L = 40$ ft-k

W12 × 72   12 ft

$M_D = 30$ ft-k
$M_L = 50$ ft-k

FIGURE P12.11 A beam–column member.

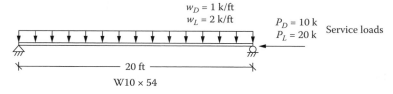

$w_D = 1$ k/ft
$w_L = 2$ k/ft

$P_D = 10$ k
$P_L = 20$ k   Service loads

20 ft

W10 × 54

FIGURE P12.12   A horizontal component of a braced frame.

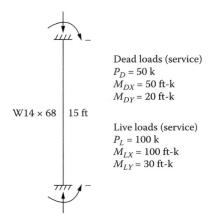

FIGURE P12.13 A restrained braced frame member.

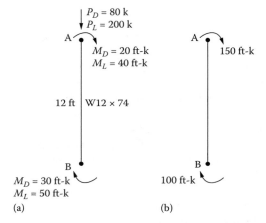

FIGURE P12.14 (a) Gravity and (b) seismic loads on a braced frame member.

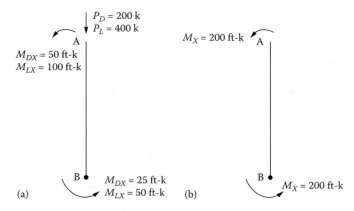

FIGURE P12.15 (a) Gravity and (b) wind loads on a braced frame member.

12.17 For a 12-ft high beam column in an unbraced A36 steel frame, a section W10×88 has been selected for $P_u=500$ k. There are five columns of the same size bearing the same load and have the same buckling strength. Assume that the members are fixed at the support and are free to sway (rotation is fixed) at the other end in both directions. Determine the magnifications factors in both directions.

12.18 In Problem #12.17, the drift along X-axis is 0.3 in as a result of a factored lateral load of 300 k. Determine the magnification factor B_2.

12.19 An unbraced frame of A992 steel is shown in Figure P12.19. Determine the magnification factors along both axes.

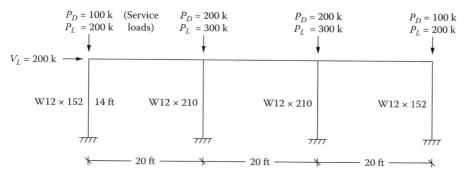

FIGURE P12.19 An unbraced frame magnification factors.

12.20 The allowable story drift in Problem 2.19 is 0.5 in. in X-direction. Determine the magnification factor B_2.

12.21 A 10-ft long W12×96 column of A992 steel in an unbraced frame is subjected to the following loading conditions. Is the section satisfactory?

1. $P_u=240$ k $M_{ntx}=50$ ft-k $M_{nty}=30$ ft-k $M_{ltx}=100$ ft-k $M_{lty}=70$ ft-k (all loads factored)
2. It is bent in reverse curvature with equal and opposite end moments
3. There are five similar columns in a story
4. The column is fixed at the base and is free to translate without rotation at the other end

12.22 Select a W12 column member of A992 steel of an unbraced frame for the following conditions:

1. $K=1.2$
2. $L=12$ ft
3. $P_u=350$ k
4. $M_{ntx}=75$ ft-k
5. $M_{nty}=40$ ft-k
6. $M_{ltx}=150$ ft-k
7. $M_{lty}=80$ ft-k (all loads factored)
8. Allowable drift=0.3 in.
9. It has intermediate transverse loading between the ends
10. The ends are restrained against rotation
11. There are four similar in a story

12.23 An unbraced frame of A992 steel is subjected to the dead load, live load, and wind load in X-axis. The structural analysis provided the results, as shown in Figure P12.23. Design a W14 section for a maximum drift of 0.5 in. Each column is subjected to the same axial force and moment.

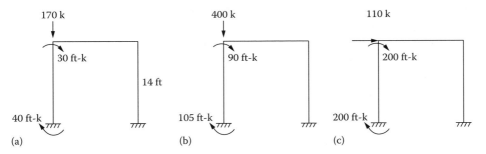

170 k   400 k   110 k

30 ft-k   90 ft-k   200 ft-k

14 ft

40 ft-k   105 ft-k   200 ft-k

(a)   (b)   (c)

FIGURE P12.23   (a) Dead, (b) live and (c) wind loads on an unbraced frame section.

12.24 One story unbraced frame of A992 steel is subjected to the dead load, roof live load, and wind load. The bending is in X-axis. The structural analysis provided the results, as shown in Figure P12.24. The moments at the base are 0. Design a W12 section for a maximum drift of 0.5 in.

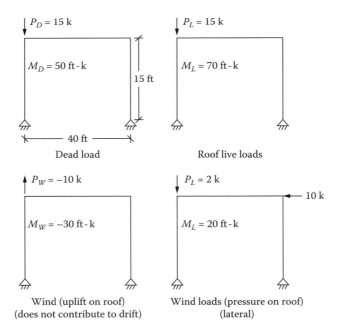

$P_D = 15$ k $P_L = 15$ k

$M_D = 50$ ft-k $M_L = 70$ ft-k   15 ft

40 ft

Dead load   Roof live loads

$P_W = -10$ k $P_L = 2$ k   10 k

$M_W = -30$ ft-k $M_L = 20$ ft-k

Wind (uplift on roof)   Wind loads (pressure on roof)
(does not contribute to drift)   (lateral)

FIGURE P12.24   Dead, roof live loads and wind loads on an unbraced frame section.

12.25 Select a K-series open-web steel joist spanning 25 ft to support a dead load of 30 psf and a live load of 50 psf. The joist spacing is 3.5 ft. The maximum live load deflection is $L/360$.

12.26 Select an open-web steel joist for the following flooring system:

1. Joist spacing 3 ft
2. Span length 20 ft
3. Floor slab 3 in. concrete
4. Other dead load 30 psf
5. Live load 60 psf
6. Maximum live load deflection $L/240$

12.27 On a 18 K 10 joist spanning 30 ft, how much total unit load and unfactored live load in psf can be imposed? The joists spacing is 4 ft. The maximum live load deflection is $L/200$.

12.28 The service dead load in psf on a 18 K 6 joist is one-half of the live load. What are the magnitudes of these loads on the joist loaded to the capacity at a span of 20 ft, spaced 4 ft on center?

12.29 Indicate the joist girder designation for the flooring system shown in Figure P12.29.

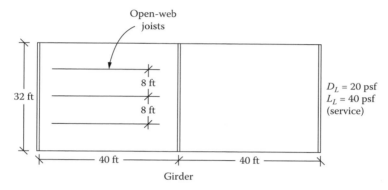

FIGURE P12.29   An open-web joists and joist girder floor system.

12.30 For a 30 ft × 50 ft bay, joists spaced 3.75 ft on center, indicate the designation of the joist girders to be used for a dead load of 20 psf and a live load of 30 psf.

13 Steel Connections

TYPES OF CONNECTIONS AND JOINTS

Most structures' failure occur at a connection. Accordingly, the AISC has placed lots of emphasis on connections and has brought out separate detailed design specifications related to connections in the *2005 Steel Design Manual*. Steel connections are made by bolting and welding; riveting is obsolete now. Bolting of steel structures is rapid and requires less skilled labor. On the other hand, welding is simple and many complex connections with bolts become very simple when welds are used. But the requirements of skilled workers and inspections make welding difficult and costly, which can be partially overcome by shop welding instead of field welding. When a combination is used, welding can be done in shop and bolting in field.

Based on the mode of load transfer, the connections are categorized as

1. Simple or axially loaded connection when the resultant of the applied forces passes through the center of gravity of the connection
2. Eccentrically loaded connection when the line of action of the resultant of the forces does not pass through the center of gravity of the connection

The following types of joints are formed by the two connecting members.

1. *Lap joint*: As shown in Figure 13.1, the line of action of the force in one member and the line of action of the force in the other connecting member have a gap between them. This causes a bending within the connection, as shown by the dashed lines. For this reason, the lap joint is used for minor connections only.
2. *Butt joint*: It provides a more symmetrical loading, as shown in Figure 13.2, that eliminates the bending condition.

The connectors (bolts or welds) are subjected to the following types of forces (and stresses).

1. *Shear*: The forces acting on the splices shown in Figure 13.3 can shear the shank of the bolt. Similarly, the weld in Figure 13.4 resists the shear.
2. *Tension*: The hanger-type connection shown in Figures 13.5 and 13.6 imposes tension in bolts and welds.
3. *Shear and tension combination*: The column-to-beam connections shown in Figures 13.7 and 13.8 cause both shear and tension in bolts and welds. The welds are weak in shear and are usually assumed to fail in shear regardless of the direction of the loading.

BOLTED CONNECTIONS

The ordinary or common bolts, also known as *unfinished bolts*, are classified as A307 bolts. The characteristics of A307 steel is very similar to A36 steel. Their strength is considerably less than those of high-strength bolts. Their use is recommended for structures subjected to static loads and for the secondary members like purlins, girts, and bracings. With the advent of high-strength bolts, the use of the ordinary bolts has been neglected though for ordinary construction, the common bolts are quite satisfactory.

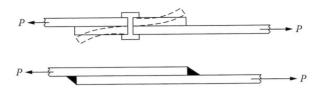

FIGURE 13.1 Lap joint.

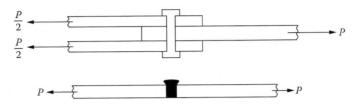

FIGURE 13.2 Butt joint.

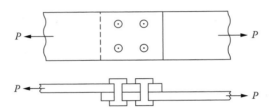

FIGURE 13.3 Bolts in shear.

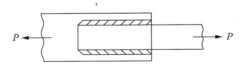

FIGURE 13.4 Welds in shear.

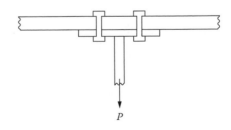

FIGURE 13.5 Bolts in tension.

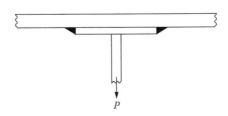

FIGURE 13.6 Welds in tension.

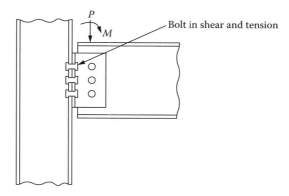

FIGURE 13.7 Bolts in shear and tension.

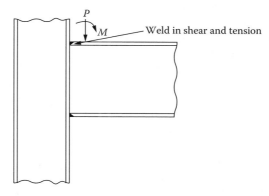

FIGURE 13.8 Welds in shear and tension.

High-strength bolts have strength that is twice or more of the ordinary bolts. There are two types of high-strength bolts: A325 type and the higher strength A490 type. High-strength bolts are used in two types of connections: the bearing-type connections and the slip-critical or friction-type connections.

In the bearing-type connection, in which the common bolts can also be used, no fictional resistance in the faying (contact) surfaces is assumed and a slip between the connecting members occurs as the load is applied. This brings the bolt in contact with the connecting member and the bolt bears the load. Thus, the load transfer takes place through the bolt.

In a slip-critical connection, the bolts are torqued to a high tensile stress in the shank. This develops a clamping force on the connected parts. The shear resistance to the applied load is provided by the clamping force, as shown in Figure 13.9.

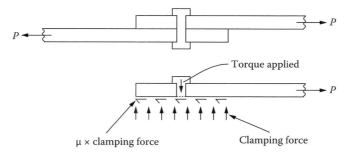

FIGURE 13.9 Frictional resistance in a slip-critical connection.

TABLE 13.1
Minimum Pretension on Bolts, k

Bolt Diameter, in.	Area, in.^{2}	A325 Bolts	A490 Bolts
1/2	0.0196	12	15
5/8	0.307	19	24
3/4	0.442	28	35
7/8	0.601	39	49
1	0.785	51	64

Thus, in a slip-critical connection, the bolts themselves are not stressed since the entire force is resisted by the friction developed on the contact surfaces. For this purpose, the high-strength bolts are tightened to a very high degree. The minimum pretension applied to bolts is 0.7 times the tensile strength of steel. These are given in Table 13.1.

The methods available to tighten the bolts comprise (1) the turn of the nut method, (2) the calibrated wrench method, and (3) the direct tension indicator method or the alternate designed twist-off type bolts are used whose tips are sheared off at a predetermined tension level.

The slip critical is a costly process subject to inspections. It is used for structures subjected to dynamic loading such as bridges and where the stress reversals and the fatigued loading take place.

For most situations, the bearing-type connection should be used where the bolts can be tighten to the snug-tight condition which means the tightness that could be obtained by the full effort of a person using a spud wrench or the pneumatic wrench.

SPECIFICATIONS FOR SPACING OF BOLTS AND EDGE DISTANCE

1. *Definitions*: The following definitions are given with respect to Figure 13.10.
   Gage, *g*: It is the center-to-center distance between two successive lines of bolts, perpendicular to the axis of a member (perpendicular to the load).
   Pitch, *p*: It is the center-to-center distance between two successive bolts along the axis of a member (in line with the force).
   Edge distance, L_e: It is the distance from the center of the outer most bolt to the edge of a member.
2. *Minimum spacing*: The minimum center-to-center distance for standard, oversized, and slotted holes should not be less than 3*d*, *d* being the bolt diameter.
3. *Maximum spacing*: The maximum spacing of bolts of the painted members or the unpainted members not subject to corrosion should not exceed 24 times the thickness of thinner member or 12 in.
4. *Minimum edge distance*: The minimum edge distance in any direction are tabulated by the AISC. It is generally 1¾ times the bolt diameter for the sheared edges and 1¼ times the bolt diameter for the rolled or gas cut edges.
5. *Maximum edge distance*: The maximum edge distance should not exceed 12 times the thickness if thinner member or 6 in.

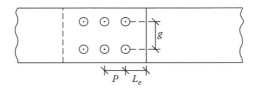

FIGURE 13.10 Definition sketch.

BEARING-TYPE CONNECTIONS

The design basis of connection is as follows:

$$P_u \leq \phi R_n \tag{13.1}$$

where
P_u is the applied factored load on connection
ϕ is the resistance factor $= 0.75$ for connection
R_n is the nominal strength of connection

In terms of the nominal unit strength (stress), Equation 13.1 can be expressed as

$$P_u \leq \phi F_n A \tag{13.2}$$

For bearing-type connections, F_n refers to the nominal unit strength (stress) for different limit states or modes of failure and A to the relevant area of failure.

The failure of a bolted joint in bearing-type connection can occur by the following modes:

1. Shearing of the bolt across the plane between the members: In single shear in the lap joint and in double shear in the butt joint, as shown in Figure 13.11.
For a single shear

$$A = \frac{\pi}{4} d^2$$

and for a double shear

$$A = \frac{\pi}{2} d^2$$

2. Bearing failure on the contact area between the bolt and the plate, as shown in Figure 13.12.

$$A = d \cdot t$$

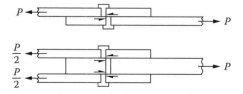

FIGURE 13.11   Shear failure.

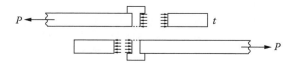

FIGURE 13.12   Bearing failure.

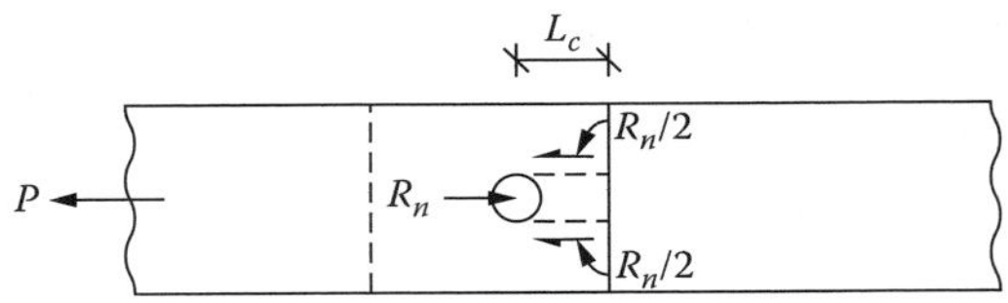

FIGURE 13.13   Tearing out of plate.

3. Tearing out of the plate from the bolt, as shown in
Figure 13.13.

$$\text{The tearing area} = 2L_c t$$

FIGURE 13.14   Tensile failure
of plate.

4. Tensile failure of plate as shown in Figure 13.14. This condition has been discussed in Chapter 9 on tension members. It
is not a part of the connection.

For the shearing type of the limiting state, F_n in Equation 13.2 is the nominal unit shear strength
of bolts, F_{nv} which is taken as 50% of the ultimate strength of bolts. The cross-sectional area, A_b is
taken as the area of the unthreaded part or the body area of bolt. If the threads are in the plane of
shear or are not excluded from shear plane, a factor 0.8 is applied to reduce the area. This factor is
incorporated in the strength, F_{nv}.

Thus, for the shear limit state, the design strength is given by

$$P_u \le 0.75 F_{nv} A_b n_b \qquad (13.3)$$

where

F_{nv} is as given in Table 13.2
$A_b = (\pi/4)d^2$ for single shear and $A = (\pi/2)d^2$ double shear
n_b is the number of bolts in the connection

In Table 13.2, threads not excluded from shear plane is referred to as the N-type connection, like
$32 - N$ and threads excluded from shear plane as the X-type connection, like $325 - X$.

The other two modes of failure, i.e., the bearing and the tearing out of a member are based not
on the strength of bolts but upon the parts being connected. The areas for bearing and tearing are
described above. The nominal unit strengths in the bearing and the (shear) tear out depend upon the
deformation around the holes that can be tolerated and on the types of holes. The bearing strength
is very high because the tests have shown that the bolts and the connected member actually do not
fail in bearing but the strength of the connected parts is impaired. The AISC expressions combine
the bearing and (shear) tear state limits together as follows:

TABLE 13.2
Nominal Unit Shear Strength, F_{nv}

Bolt Type	F_{nv}, ksi
A307	24
A325 threads not excluded from shear plane	48
A325 threads excluded from shear plane	60
A490 threads not excluded from shear plane	60
A490 threads excluded from shear plane	75

1. For standard, oversized, short-slotted holes and long-slotted holes with slot parallel to the force where deformation can be ≤0.25 in.

$$P_u = 1.2\phi L_c t F_u n_b \le 2.4\phi dt F_u n_b \qquad (13.4)$$

where F_u is the ultimate strength of the connected member.

2. For standard, oversized, short-slotted holes and long-slotted holes parallel to the force where deformations can be >0.25 in.

$$P_u = 1.5\phi L_c t F_u n_b \le 3\phi dt F_u n_b \qquad (13.5)$$

3. For long-slotted holes, slots being perpendicular to the force

$$P_u = 1.0\phi L_c t F_u n_b \le 2.0\phi dt F_u n_b \qquad (13.6)$$

The distance L_c in the above expressions has been illustrated in Figure 13.15.
   For the edge bolt #1

$$L_c = L_e - \frac{h}{2} \qquad (13.7)$$

For the interior bolt #2

$$L_c = s - h \qquad (13.8)$$

where h = hole diameter = $(d + 1/8)$ in.*

Example 13.1

A A36 channel section C9 × 15 is connected to a 3/8 in. steel gusset plate, with 7/8 in. diameter A325 bolts. A service dead load of 20 k and live load of 50 k is applied to the connection. Design the connection. The slip of the connection is permissible. The threads are excluded from the shear plane.

Solution

   A. The factored load

$P_u = 1.2(20) + 1.6(50) = 104$ k

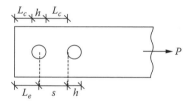

FIGURE 13.15   Definition of L_c.

---

* The AISC stipulates $d + 1/16$ but 1/8 has been used conservatively.

B. Shear limit state
   1. $A_b = (\pi/4)(7/8)^2 = 0.601$ in.^{2}
   2. For A325-X, $F_{nv} = 60$ ksi
   3. From Equation 13.3

$$\text{No. of bolts} = \frac{P_u}{0.75 F_{nv} A_b}$$

$$= \frac{104}{0.75(60)(0.601)} = 3.85 \text{ or } 4 \text{ bolts}$$

C. Bearing limit state
   1. Minimum edge distance

$$L_e = 1\tfrac{3}{4}\left(\frac{7}{8}\right) = 1.53 \text{ in., use 2 in.}$$

   2. Minimum spacing

$$s = 3\left(\frac{7}{8}\right) = 2.63 \text{ in., use 3 in.}$$

   3. $h = d + 1/8 = 1$ in.
   4. For holes near edge

$$L_e = L_e - \frac{h}{2}$$

$$= 2 - \frac{1}{2} = 1.5 \text{in.}$$

$t = 5/16$ in. for the web of the channel section
For a standard size hole of deformation <0.25 in.

Strength/bolt $= 1.2\phi L_c t F_u$

$$= 1.2(0.75)(1.5)\left(\frac{5}{16}\right)(58) = 24.5\text{k} \leftarrow \text{Controls}$$

Upper limit $= 2.4\phi d t F_u$

$$= 2.4(0.75)\left(\frac{7}{8}\right)\left(\frac{5}{16}\right)(58) = 28.55\text{k}$$

   5. For interior holes

$$L_c = s - h = 3 - 1 = 2 \text{ in.}$$

   6. Strength/bolt $= 1.2(0.75)(2)\left(\frac{5}{16}\right)(58) = 32.63\text{k}$

Upper limit $= 2.4(0.75)\left(\frac{7}{8}\right)\left(\frac{5}{16}\right)(58) = 28.55\text{k} \leftarrow \text{Controls}$

7. Suppose there are n lines of holes with two bolts in each, then

$$P_u = 104 = n(24.5) + n(28.55)$$

$$n = 1.96$$

Total no. of bolts = 2 (1.96) = 3.92 or 4 bolts

Select 4 bolts either by shear or bearing.
8. The section has to be checked for the tensile strength and the block shear by the procedure of Chapter 9.

SLIP-CRITICAL CONNECTIONS

In a slip-critical connection, the bolts are not subjected to any stress. The resistance to slip is equal to the product of the tensile force between the connected parts and the static coefficient of friction. This is given by

$$P_u = 1.13 \phi h_{sc} \mu T_b N_s n_b \qquad (13.9)$$

where
T_b is the minimum bolt pretension given in Table 13.1
N_s is the number of slip (shear) planes
n_b is the number of bolts in connection
h_{sc} is the hole factor determined as follows: (a) For standard holes, $h_{sc} = 1.0$; (b) for oversized or short-slotted holes, $h_{sc} = 0.85$; and (c) for long-slotted holes, $h_{sc} = 0.70$
μ is the slip (friction) coefficient as given is Table 13.3
ϕ is the resistance factor has different values as follows: (a) For connections in which slip is prevented up to the service loads, $\phi = 1$, and (b) for connections in which slip is prevented at the required (factored) strength level, $\phi = 0.85$

Normally, the first condition could be considered.
Although, there is no bearing on bolts in a slip-critical connection, the AISC requires that it should also be checked as a bearing-type connection by Equation 13.3 and a relevant equation out of Equations 13.4 through 13.6.

Example 13.2

A double angle tensile member consisting of 2 L 3 × 2½ × 1/4 is connected by a gusset plate 3/4 in. thick. It is designed for a service load of 15 k and live load of 30 k. No slip is permitted. Use 5/8 in. A325 bolts and A572 steel.

TABLE 13.3
Slip (Friction) Coefficient

Class	Surface	μ
Class A	Unpainted clean mill scale or Class A coating on blast cleaned steel or hot dipped galvanized and roughened surface	0.35
Class B	Unpainted blast cleaned surface or Class B coating on blast cleaned steel	0.5

Solution

A. Factored design load

$$P_u = 1.2\,(15) + 1.6\,(30) = 66\,k$$

B. For the slip-critical limit state
   1. Slip prevented at service load, $\phi = 1$
   2. Standard hole, $h_{sc} = 1$
   3. Class A surface, $\mu = 0.35$
   4. From Table 13.1, $T_b = 19\,ksi$ for 5/8 in. bolts
   5. For double shear, $N_s = 2$
      From Equation 13.9,

$$n_b = \frac{P_u}{1.13\phi h_{sc}\mu T_b N_s}$$

$$= \frac{66}{1.13(1)(1)(0.35)(19)(2)} = 4.39\ \text{bolts}$$

C. Check for the shear limit state as a bearing-type connection
   1. $A_b = (\pi/2)(5/8)^2 = 0.613\ \text{in.}^2$
   2. For A325-X, $F_{nv} = 60\,ksi$
   3. From Equation 13.3,

$$\text{No. of bolts} = \frac{P_u}{0.75F_{nv}A_b}$$

$$= \frac{66}{0.75(60)(0.613)} = 2.39\ \text{bolts}$$

D. Check for the bearing limit state as a bearing-type connection
   1. Minimum edge distance

$$L_e = 1\tfrac{3}{4}\left(\frac{5}{8}\right) = 1.09\ \text{in., use 1.5 in.}$$

   2. Minimum spacing

$$s = 3\left(\frac{5}{8}\right) = 1.88\ \text{in., use 2 in.}$$

   3. $h = d + 1/8 = 3/4\ \text{in.}$
   4. For holes near edge

$$L_e = L_e - \frac{h}{2}$$

$$= 1.5 - \frac{6}{16} = 1.125\,\text{in.}$$

$$t = 2\,(1/4) = 0.5\ \text{in.} \leftarrow \text{thinner than the gusset plate}$$

For a standard size hole of deformation <0.25 in.

Strength/bolt $= 1.2\phi L_c t F_u$

$$= 1.2(0.75)(1.125)(0.5)(65) = 32.9k \leftarrow \text{Control}$$

Upper limit $= 2.4\phi dt F_u$

$$= 2.4(0.75)\left(\frac{5}{8}\right)(0.5)(65) = 36.56k$$

5. For interior holes

$$L_c = s - h = 2 - \left(\frac{3}{4}\right) = 1.25 \text{ in.}$$

6. Strength/bolt $= 1.2(0.75)(1.25)(0.5)(65) = 36.56k$

   Upper limit $= 2.4(0.75)\left(\frac{5}{8}\right)(0.5)(65) = 36.56k \leftarrow \text{Controls}$

7. Suppose there are n lines of holes with two bolts in each, then

$$P_u = 66 = n(32.9) + n(36.56)$$

$$n = 1$$

   Total no. of bolts $= 2$
E. The slip-critical limit controls the design
   Number of bolts selected $= 6$ for symmetry
F. Check for the tensile strength of bolt
G. Check for the block shear

TENSILE LOAD ON BOLTS

This section applies to tensile loads on bolts, both in the bearing type of connections as well as the slip-critical connections. The connections subjected to pure tensile loads (without shear) are limited. These connections exist in the hanger-type connections for bridges, the flange connections for piping systems, and wind-bracing systems in tall buildings. A hanger-type connection is shown in Figure 13.16.

A tension by the external loads acts to relieve the clamping force between the connected parts that causes a reduction in the slip resistance. This has been considered in the next section. However, as far as the tensile strength of the bolt is concerned, it is computed without giving any consideration to the initial tightening force or pretension.

In Equation 13.2, for the limit state of tension rupture, F_n is the nominal unit tensile strength of bolt, F_{nt} times the area of bolt, A_b which is taken as the unthreaded or the body area.

Thus the tensile limit state follows the standard form;

$$T_u \leq 0.75 F_{nt} A_b \cdot n_b \tag{13.10}$$

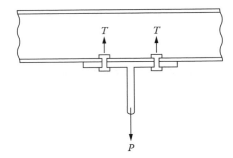

FIGURE 13.16 T-type hanger connection.

where
  T_u is the factored design tensile load
  F_{nt} is the nominal unit tensile strength as given in Table 13.4

Example 13.3

Design the hanger connection shown in Figure 13.17 for the service dead and live loads of 30 and 50k, respectively. Use A325 bolts.

Solution

1. Factored design load

$$P_u = 1.2(30) + 1.6(50) = 116\,k$$

2. Use 7/8 in. bolt

$$A_b = \frac{\pi}{4}\left(\frac{7}{8}\right) = 0.601\,in.^2$$

3. From Equation 3.10,

$$n_b = \frac{P_u}{0.75 F_{nt} A_b}$$

$$= \frac{116}{0.75(90)(0.601)} = 2.86, use\,4\,bolts, 2\,on\,each\,side$$

TABLE 13.4
Nominal Unit
Tensile Strength, F_{nt}

Bolt Type	F_{nt}, ksi
A307	45
A325	90
A490	113

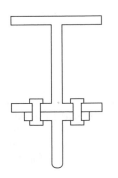

FIGURE 13.17 A tensile connection for Example 13.3.

COMBINED SHEAR AND TENSILE FORCES ON BOLTS

COMBINED SHEAR AND TENSION ON BEARING-TYPE CONNECTIONS

Many connections are subjected to a combination of shear and tension. A common case is a diagonal bracing attached to a column.

When both tension and shear are imposed, the interaction of these two forces in terms of the combined stress must be considered to determine the capacity of the bolt. A simplified approach to deal with this interaction is to reduce the unit tensile strength of bolt to F'_{nt} (from the original F_{nt}). Thus, the limiting state equation is

$$T_u \le 0.75F'_{nt}A_b \cdot n_b \tag{13.11}$$

where the adjusted nominal unit tensile strength is given as follows:

$$F'_{nt} = 1.3F_{nt} - \left(\frac{F_{nt}}{0.75F_{nv}}\right)f_v \le F_{nt} \tag{13.12}$$

where f_v is the actual shear stress given by the design shear force divided by the area of the number bolts in the connection.

To summarize, for the combined shear and tension in a bearing-type connection, the procedure comprises of the following steps:

1. Use the unmodified shear limiting state Equation 13.3
2. Use the tension limiting state Equation 13.11 for check
3. Use the relevant bearing, limiting state equation from Equations 13.4 through 13.6 for check

Example 13.4

A WT12 × 27.5 bracket of A36 steel is connected to a column, as shown in Figure 13.18, to transmit the service dead and live loads of 15 and 45 k. Design the bearing-type connection between the column and the bracket using 7/8 in. A325 bolts.

Solution

A. Design loads
　1. $P_u = 1.2\,(15) + 1.6\,(45) = 90\,k$
　2. Design shear, $V_u = P_u\,(3/5) = 54\,k$
　3. Design tension, $T_u = P_u\,(4/5) = 72\,k$
B. For the shear limiting state
　1. $A_b = (\pi/4)(7/8)^2 = 0.601\ \text{in.}^2$
　2. For A325, $F_{nv} = 60\,\text{ksi}$
　3. From Equation 13.3

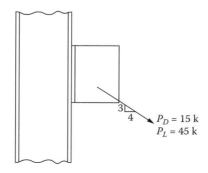

FIGURE 13.18 A column-bracket connection.

$$n_b = \frac{V_u}{0.75 F_{nv} A_b}$$

$$= \frac{54}{0.75(60)(0.601)} = 1.99 \text{ bolts}$$

Use 4 bolts, 2 on each side (minimum 2 bolts on each side)

C. For the tensile limiting state
1. $F_{nt} = 90 \text{ ksi}$
2. Actual shear stress

$$f_v = \frac{V_u}{A_b n_b} = \frac{54}{(0.601)(4)} = 22.46 \text{ ksi}$$

3. Adjusted unit tensile strength from Equation 13.12

$$F'_{nt} = 1.3 F_{nt} - \left(\frac{F_{nt}}{0.75 F_{nv}} \right) f_v \le F_{nt}$$

$$= 1.3(90) - \frac{90}{0.75(60)}(22.46) = 72.08 < 90 \text{ ksi} \quad \textbf{OK}$$

4. From Equation 13.11

$$n_b = \frac{T_u}{0.75 F'_{nt} A_b}$$

$$= \frac{72}{0.75(72.08)(0.601)} = 2.12 < 4 \text{ bolts} \quad \textbf{OK}$$

D. Check for the bearing limit state
1. Minimum edge distance

$$L_e = 1\tfrac{3}{4} \left(\frac{7}{8} \right) = 1.53 \text{ in., use 2 in.}$$

2. Minimum spacing

$$s = 3 \left(\frac{7}{8} \right) = 2.63 \text{ in., use 3 in.}$$

3. $h = d + 1/8 = 1 \text{ in.}$
4. For holes near edge

$$L_e = L_e - \left(\frac{h}{2} \right)$$

$$= 2 - \left(\frac{1}{2} \right) = 1.5 \text{ in.}$$

$t = 0.5 \text{ in.} \leftarrow$ thickness of WT flange

For a standard size hole of deformation <0.25 in.

Strength $= 1.2\phi L_c t F_u n_b$

$$= 1.2(0.75)(1.5)(0.5)(58)(4) = 156.6\,k \leftarrow Controls > 54k \quad \textbf{OK}$$

Upper limit $= 2.4\phi dt F_u n_b$

$$= 2.4(0.75)\left(\frac{7}{8}\right)(0.5)(58)(4) = 182.7\,k$$

COMBINED SHEAR AND TENSION ON SLIP-CRITICAL CONNECTIONS

As discussed in the "Tensile Load on Bolts" section, the externally applied tension tends to reduce the clamping force and the slip-resisting capacity. A reduction factor k_s is applied to the previously described slip-critical strength. Thus, for the combined shear and tension, the slip-critical limit state is

$$V_u = 1.13\phi h_{sc} \mu T_b N_s n_b k_s \tag{13.13}$$

where

$$k_s = 1 - \frac{T_u}{1.13 T_b n_b} \tag{13.14}$$

V_u is the factored shear load on the connection
T_u is the factored tension load on the connection
T_b is the minimum bolt pretension given in Table 13.1
N_s is the number of slip (shear) planes
n_b is the number of bolts in connection
h_{sc} is the hole factor determined as follows: (a) For standard holes, $h_{sc} = 1.0$; (b) for oversized or short-slotted holes, $h_{sc} = 0.85$; and (c) for long-slotted holes, $h_{sc} = 0.70$
μ is the slip (friction) coefficient as given is Table 13.3

To summarize, for the combined shear and tension in a slip-critical connection, the procedure is as follows:

1. Use the shear limiting state Equation 13.3*
2. Use the (original) tensile limit state Equation 13.10
3. Use the relevant bearing limiting state Equations 13.4 through 13.6 for check
4. Use the (modified) slip-critical limit state Equation 13.13

Example 13.5

Design Example 13.4 as a slip-critical connection

Solution

A. Design loads from Example 13.3
   1. $V_u = 54\,k$
   2. $T_u = 72\,k$

---

* The slip-critical connections also are required to be checked for bearing capacity and shear strength.

B. For the shear limiting state

$n_b = 1.99$ from Example 13.3, use 4 bolts

C. For the tensile limiting state

$$n_b = \frac{T_u}{0.75 F_{nt} A_b}$$

$$= \frac{72}{0.75(90)(0.601)} = 1.77 < 4 \text{ bolts} \quad \textbf{OK}$$

D. For the bearing limit state

Strength $= 156.6\,\text{k}$ (from Example 13.3) $> 54\,\text{k}$ \quad **OK**

E. For the slip-critical limit state
   1. Slip prevented to service loads, $\phi = 1$
   2. Standard holes, $h_{sc} = 1$
   3. Class A surface, $\mu = 0.35$
   4. From Table 13.1, $T_b = 19\,\text{ksi}$
   5. For single shear, $N_s = 1$
   6. $k_s = 1 - \dfrac{T_u}{1.13 T_b n_b}$

      $$= 1 - \frac{72}{1.13(39)(4)} = 0.59$$
   7. From Equation 13.13

      $$V_u = 1.13 \phi h_{sc} \mu T_b N_s n_b k_s$$

      $$= 1.13(1)(1)(0.35)(39)(1)(4)(0.59) = 36.4\,\text{k} < 54(V_u) \quad \textbf{NG}$$

   8. Select 6 bolts 3 on each side of web

      $$k_s = 1 - \frac{72}{1.13(39)(6)} = 0.73$$

   9. $V_u = 1.13 \phi h_{sc} \mu T_b N_s n_b k_s$

      $$= 1.13(1)(1)(0.35)(39)(1)(6)(0.73) = 67.6\,\text{k} > 54(V_u) \quad \textbf{OK}$$

WELDED CONNECTIONS

Welding is a process in which the heat of an electric arc melts the welding electrode and the adjacent material of the part being connected simultaneously. The electrode is deposited as a filler metal into the steel, which is referred to as the *base metal*. There are two types of welding processes. The *shielded metal arc welding* (SMAW), usually done manually, is the process used for

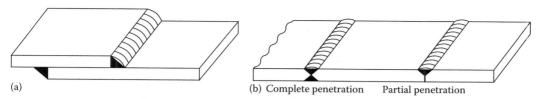

FIGURE 13.19   Types of weld: (a) fillet and (b) groove or butt welds.

field welding. The *submerged arc welding* (SAW) is an automatic or semiautomatic process used in shop welding. The strength of a weld depends on the weld metal used, which is the strength of the electrode used. An electrode is specified by the letter E followed by the tensile strength in ksi and the last two digits specifying the type of coating. Since strength is a main concern, the last two digits are specified by XX, a typical designation of E 70 XX. The electrode should be selected to have a larger tensile strength than the base metal (steel). For steel of 58 ksi strength, the electrode E 70 XX and for 65 ksi steel, the electrode E 80 XX is used. The electrodes of high strength E 120 XX are available.

The two common types of welds are the fillet welds and the grove or the butt welds, as shown in Figure 13.19. The groove welds are stronger and more expensive than the fillet welds. Most of the welded connections are made by the fillet welds because of a larger allowed tolerance.

The codes and standards for welds are prepared by the American Welding Society (AWS). These have been adopted in the *AISC Manual 2005*.

FILLET WELDS

EFFECTIVE AREA OF WELD

The cross section of a fillet weld is assumed to be a 45° right angle triangle, as shown in Figure 13.20. Any additional buildup of weld is neglected. The size of the fillet weld is denoted by the sides of the triangle, w and the throat dimension, given by the hypotenuse, t which is equal to $0.707w$. When the SAW process is used, the greater heat input produces a deeper penetration.

The effective throat size is taken as follows:
For SMAW process,

$$T_e = t = 0.707w \tag{13.15}$$

For SAW process,
(a)   When $w \le 3/8$ in.

$$T_e = t = 0.707w \tag{13.16a}$$

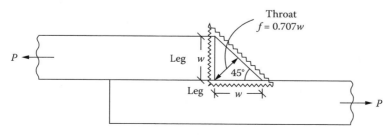

FIGURE 13.20   Fillet weld dimensions.

(b) When $w > 3/8$ in.

$$T_e = 0.707w + 0.11 \qquad (13.16b)$$

MINIMUM SIZE OF WELD

The minimum size should not be less than the dimension shown in Table 13.5.

TABLE 13.5
Minimum Size of Weld, in.

Base Material Thickness of Thinner Part, in.	w, in.
≤1/4	1/8
>1/4 to ≤1/2	3/16
>3/4 to ≤3/4	1/4
>3/4	5/16

MAXIMUM SIZE OF WELD

1. Along the edges of material less than 1/4 in. thick, the weld size should not be greater than the thickness of the material.
2. Along the edges of material 1/4 in. or more, the weld size should not be greater than the thickness of the material less 1/16 in.

LENGTH OF WELD

1. The effective length is equal to the actual length for the length up to 100 times the weld size. To the portion of length exceeding 100 times the weld size, a reduction of 20% is made.
2. The effective length should not be less than four times the weld size.
3. If only the longitudinal welds are used, the length of each side should not be less than the perpendicular distance between the welds.

STRENGTH OF WELD

COMPLETE JOINT PENETRATION (CJP) GROOVE WELDS

Since the weld metal is always stronger than the base metal (steel), the strength of CJP groove weld is taken as the strength of the base metal. The design of connection is not done in the normal sense.

For the combined shear and tension acting on CJP groove weld, there is no explicit approach. The generalized approach is to reduce the tensile strength by a factor of $(f_v/F_t)^2$ subject to a maximum reduction of 36% of the tensile strength.

PARTIAL JOINT PENETRATION (PJP) WELDS AND FILLET WELDS

Weld is weakest in shear and is always assumed to fail in the shear mode. Although, a length of weld can be loaded in shear, compression, or tension, the failure of a weld is assumed to occur in the shear rupture through the throat of the weld. Thus

$$P_u = \phi F_w A_w \qquad (13.17)$$

where
  ϕ is the resistance factor $= 0.75$
  F_w is the strength of weld $= 0.6F_{EXX}$
  F_{EXX} is the strength of electrode
  A_w is the effective area of weld $= T_e L$

However, there is a requirement that the weld shear strength cannot be larger than the base metal shear strength. For the base metal, the shear yield and shear rupture strengths are taken to be 0.6 times the tensile yield of steel and 0.6 times the ultimate strength of steel, respectively. The yield strength is applied to the gross area and the rupture strength to the net area of shear surface, but in the case of a weld both areas are the same. The resistance factor is 1 for shear yield and 0.75 for shear rupture.

Thus, the PJP groove and the fillet welds should be designed to meet the strengths of the weld and the base metal, whichever is smaller as follows:

1. Weld shear rupture limit state
   By the substitution of the terms in Equation 13.17

$$P_u = 0.45F_{EXX}T_eL \tag{13.18}$$

where
   F_{EXX} is the strength of electrode, ksi
   L is the length of weld
   T_e is the effective throat dimension from Equations 13.15 or 13.16

2. Base metal shear limit state
   a. Shear yield strength

$$P_u = 0.6F_y tL \tag{13.19}$$

   where t is the thickness of thinner connected member
   b. Shear rupture strength

$$P_u = 0.45F_u tL \tag{13.20}$$

Example 13.6

A tensile member consisting of one L $3\frac{1}{2} \times 3\frac{1}{2} \times 1/2$ section carries a service dead load of 30 k and live load of 50 k, as shown in Figure 13.21. A single 3/4 in. plate is directly welded to the column flange using the CJP groove. Fillet welds attach the angles to the plate. Design the welded connection. The longitudinal length of the weld cannot exceed 5 in. Use the return (transverse) weld, if necessary. Use E70 electrodes. Steel is A36.

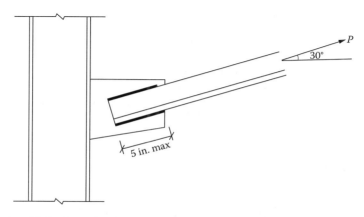

FIGURE 13.21   A welded column-bracket connection.

Solution

A. Angle plate (bracket) connection
  1. Factored load

$$P_u = 1.2(30) + 1.6(50) = 116\,\text{k}$$

  2. Maximum weld size. For thinner member, thickness of angle, $t = 1/2$ in.

$$w = t - \left(\frac{1}{16}\right) = \frac{7}{16}\,\text{in.}$$

  3. Throat dimension, SMAW process

$$T_e = 0.707\left(\frac{7}{16}\right) = 0.309\,\text{in.}$$

  4. For weld shear limit state, from Equation 13.18

$$L = \frac{P_u}{0.45 F_{EXX} T_e}$$

$$= \frac{116}{0.45(70)(0.309)} = 11.92 \approx 12\,\text{in.} \leftarrow \text{Controls}$$

  5. For steel shear yield limit state, from Equation 13.19

$$L = \frac{P_u}{0.6 F_y t}$$

$$= \frac{116}{0.6(36)(1/2)} = 10.74\,\text{in.}$$

  6. For steel rupture limit state, from Equation 13.20

$$L = \frac{P_u}{0.45 F_u t}$$

$$= \frac{116}{0.45(58)(1/2)} = 8.9\,\text{in.}$$

  7. Provide a 5 in. long weld on each side* (maximum in this problem) with 1 in. return on each side.
  8. The longitudinal length of welds (5 in.) should be at least equal to the transverse distance between the longitudinal weld (3½); it is **OK**

B. Column-bracket connection
  1. The connection is subjected to tension and shear as follows:

$$T_u = P_u \cos 30° = 116 \cos 30° = 100.5\,\text{k}$$

$$V_u = P_u \sin 30° = 116 \sin 30° = 58\,\text{k}$$

---

* Theoretically, the lengths on two sides are unequally distributed so that the centroid of the weld passes through the center of gravity of the angle member.

2. For the CJP groove, the design strengths are the same as for the base metal.
3. This is the case of the combined shear and tension in groove weld. Using a maximum reduction of 36%*, tensile strength $=0.76F_t$
4. For the base material tensile limit state

$$T_u = \phi(0.76F_t)tL, \quad \text{where } t \text{ is gusset plate thickness}$$

$$100.5=0.9(0.76)(36)\left(\frac{3}{4}\right)L$$

or

$L = 5.44$ in. $\leftarrow$ Controls

Use 6 in. weld length
5. For the base metal shear yield limit state

$$V_u = 0.6F_y tL$$

$$58 = 0.6(36)\left(\frac{3}{4}\right)L$$

or

$L = 3.6$ in.

6. For the base metal shear rupture limit state

$$V_u = 0.45F_u tL$$

$$58 = 0.45(58)\left(\frac{3}{4}\right)L$$

or

$L = 3.0$ in.

FRAME CONNECTIONS

There are three types of beam-to-column frame connections:

1. Type FR (fully restrained) or rigid frame or moment frame connection
   • It transfers the full joint moment and shear force
   • It retains the original angle between the members or the rotation is not permitted
2. Type simple or pinned frame or shear frame connection
   • It transfers shears force only
   • It permits the rotation between the members
3. Type PR (partially restraint) frame connection
   • It transfers some moment and the entire shear force
   • It permits a specified amount of rotation

The relationship between the applied moment and the rotation (variation of angle) of members for the rigid, semirigid, and simple framing is shown in Figure 13.22.

* See "Complete Joint Penetration (CJP) Groove Welds" section.

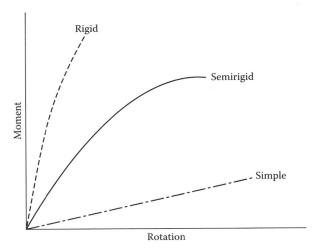

FIGURE 13.22 Moment–rotation characteristics.

A fully rigid joint will have a small change in angle with the application of moment. A simple joint will be able to support some moment (though theoretically the moment capacity should be zero). A semirigid joint is where the actual moment and rotation are accounted for.

SHEAR OR SIMPLE CONNECTION FOR FRAMES

There are a variety of beam-to-column or beam-to-girder connections that are purposely made flexible for rotation at the ends of the beam. These are designed for the end reaction (shear force). These are used for structures where the lateral forces due to wind or earthquake are resisted by the other systems like truss framing or shear walls. Following are the main categories of simple connections.

SINGLE-PLATE SHEAR CONNECTION OR SHEAR TAB

This is a simple and economical approach which is becoming very popular. The holes are pre-punched in a plate and in the web of the beam to be supported. The plate is welded (usually shop welded) to the supporting column or beam. The prepunched beam is bolted to the plate at the site. This is shown in Figure 13.23.

FRAMED-BEAM CONNECTION

The web of the beam to be supported is connected to the support-ing column through a pair of angles, as shown in Figure 13.24.

SEATED-BEAM CONNECTION

The beam to be supported sits on an angle attached to the support-ing column flange, as shown in Figure 13.25.

END-PLATE CONNECTION

A plate is welded against the end of the beam to be supported. This plate is then bolted to the supporting column or beam at the site.

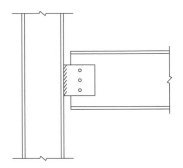

FIGURE 13.23 Single-plate or shear tab connection.

This is shown in Figure 13.26. These connections are becoming popular but not as much as the single-plate connection.

The design of the simple connections proceed on the lines of the bearing-type connections described in the "Bearing Type Connections" section. The limiting states considered are as follows: (1) the shear on bolts; (2) the bearing yield strength; (3) the shear rupture strength between the bolt and the connected part, as discussed in the "Bearing Type Connections" section; and (4) the block shear strength of the connected part.

The *AISC Manual 2005* includes a series of tables to design the different types of bolted and welded connections. The design of only a single-plate shear connection for frames is presented here.

SINGLE-PLATE SHEAR CONNECTION FOR FRAMES

The following are the conventional configurations for a single-plate shear connection:

1. A single row of bolts comprising of 2–12 bolts.
2. The distance between the bolt line to weld line not to exceed 3.5 in.
3. Provision of the standard or short-slotted holes.
4. The horizontal distance to edge $L_e \geq 2d_b$ (bolt diameter).
5. The plate and beam must satisfy $t \leq (d_b/2) + (1/16)$.
6. For welded connections, the weld shear rupture and the base metal shear limits should be satisfied.
7. For bolted connections, the bolt shear, the plate shear, and the bearing limit states should be satisfied.
8. The block shear of plate should be satisfactory.

Example 13.7

Design a single-plate shear connection for a W14 × 82 beam joining to a W12 × 96 column by a 3/8 in. plate, as shown in Figure 13.27. The factored reaction at the support of the beam is 50 k. Use 3/4 in. diameter A325 bolts, A36 steel, and E70 electrodes.

Solution

A. Design load

$$P_u = R_u = 50\,\text{k}$$

B. For W14 × 82,

$$d = 14.3\,\text{in.}, \quad t_f = 0.855\,\text{in.}, \quad t_w = 0.51\,\text{in.}, \quad b_f = 14.7\,\text{in.}$$

$$F_y = 36\,\text{ksi}, \quad F_u = 58\,\text{ksi}$$

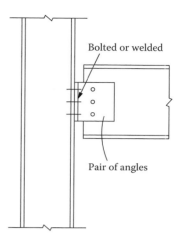

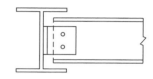

FIGURE 13.24 Framed beam connection.

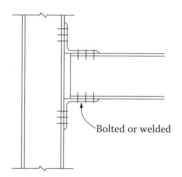

FIGURE 13.25 Seated beam connection.

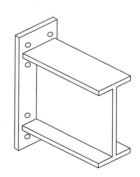

FIGURE 13.26 End-plate connection.

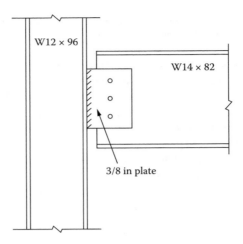

FIGURE 13.27 Example of a single-plate connection.

C. For W12 × 96,

$d = 12.7$ in., $t_f = 0.9$ in., $t_w = 0.55$ in., $b_f = 12.2$ in.

$F_y = 36$ ksi, $F_u = 58$ ksi

D. Column–plate welded connection
   1. For 3/8 in. plate,

$$\text{max size} = t - \left(\frac{1}{16}\right) = \left(\frac{3}{8}\right) - \left(\frac{1}{16}\right) = \left(\frac{5}{16}\right) \text{in.}$$

   2. $T_e = 0.707\ (5/16) = 0.22$ in.
   3. The weld shear limit state, from Equation 13.18

$$L = \frac{P_u}{0.45 F_{EXX} T_e}$$

$$= \frac{50}{0.45(70)(0.22)} = 7.21 \approx 8\,\text{in.} \leftarrow \text{Controls}$$

   4. The steel shear yield limit state, from Equation 13.19

$$L = \frac{P_u}{0.6 F_y t}$$

$$= \frac{50}{0.6(36)(3/8)} = 6.17 < 8\,\text{in.}$$

   5. The steel rupture limit state, from Equation 13.20

$$L = \frac{P_u}{0.45 F_u t}$$

$$= \frac{50}{0.45(58)(3/8)} = 5.10 < 8\,\text{in.}$$

E. Beam–plate bolted connection
  E.1 Bolt single shear limit state
    1. $A_b = (\pi/4)(3/4)^2 = 0.441$ in.^{2}
    2. For A325-X, $F_{nv} = 60$ ksi
    3. From Equation 13.3

$$\text{No. of bolts} = \frac{P_u}{0.75 F_{nv} A_b}$$

$$= \frac{50}{0.75(60)(0441)} = 2.51 \text{ or } 3 \text{ bolts}$$

  E.2 The bearing limit state
    1. Minimum edge distance

$$L_e = 1\tfrac{3}{4} d_b = 1\tfrac{3}{4}(3/4) = 1.31 \text{ in.,} \quad \text{use 1.5 in.}$$

    2. Minimum spacing

$$s = 3 d_b = 3(3/4) = 2.25 \text{ in.,} \quad \text{use 2.5 in.}$$

    3. $h = d + 1/8 = 7/8$ in.
    4. For holes near edge

$$L_c = L_e - \left(\frac{h}{2}\right)$$

$$= 1.5 - \left(\frac{7}{16}\right) = 1.063 \text{ in.}$$

$t = 3/8$ in. thinner member
For a standard size hole of deformation <0.25 in.

Strength/bolt $= 1.2 \phi L_{ct} F_u$

$$= 1.2(0.75)(1.063)\left(\frac{3}{8}\right)(58) = 20.81 \text{ k} \leftarrow \text{Controls}$$

Upper limit $= 2.4 \phi dt F_u$

$$= 2.4(0.75)\left(\frac{3}{4}\right)\left(\frac{3}{8}\right)(58) = 29.36 \text{ k}$$

    5. For other holes

$$L_c = s - h = 2.5 - \left(\frac{7}{8}\right) = 1.625 \text{ in.}$$

    6. Strength/bolt $= 1.2(0.75)(1.625)\left(\frac{3}{8}\right)(58) = 31.81 \text{ k}$

$$\text{Upper limit} = 2.4(0.75)\left(\frac{3}{4}\right)\left(\frac{3}{8}\right)(58) = 29.36\text{k} \leftarrow \text{Controls}$$

7. Total strength for 3 bolts–two near edges

$$P_u = 2(20.81) + 29.36 = 71\text{k} > 50\text{k} \quad \textbf{OK}$$

8. The section has to be checked for the block shear by the procedure given in Chapter 9.

MOMENT-RESISTING CONNECTION FOR FRAMES

The fully restraint (rigid) and partially restraint (semirigid) are two types of moment-resisting connections. It is customary to design a semirigid connection for some specific moment capacity, which is less than the full moment capacity.

Figure 13.28 shows a moment-resisting connection that has to resist a moment, M and a shear force (reaction), V.

The two components of the connection are designed separately. The moment is transmitted to the column flange as a couple by the two tees attached at the top and bottom flanges of the beam. This results in tension, T on the top flange and compression, C on the bottom flange.

From the couple expression, the two forces are given by

$$C = T = \frac{M}{h} \tag{13.21}$$

where h is taken as the depth of the beam.

The moment is taken care by designing the tee connection for the tension T. It should be noted that the magnitude of the force T can be decreased by increasing the distance between the tees (by a deeper beam).

The shear load is transmitted to the column by the beam–web connection. This is designed as a simple connection of the type discussed in the "Shear or Simple Connection for Frames" section via single plate, two angles (framed), or seat angle.

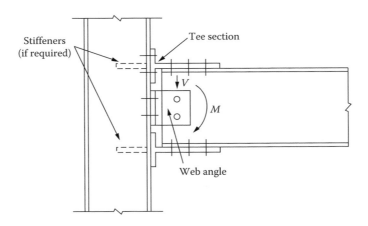

FIGURE 13.28 Moment-resisting connection.

The connecting tee element is subjected to prying action as shown in Figure 13.29. This prying action could be eliminated by connecting the beam section directly to the column via CJP groove weld, as shown in Figure 13.30.

Example 13.8

Design the connection of Example 13.7 as a moment-resisting connection subjected to a factored moment of 200 ft-k and a factored end shear force (reaction) of 50 k. The beam flanges are groove welded to the column.

Solution

A. Design for the shear force has been done in Example 13.7
B. Flanges welded to the column.

    1. $C = T = \dfrac{M}{d}$

$$= \frac{200(12)}{14.3} = 167.83\,k$$

    2.   The base material limit state

$$T_u = \phi F y t L, \quad \text{where } t = t_f$$

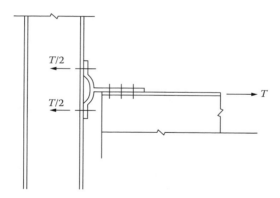

FIGURE 13.29   Prying action in connection.

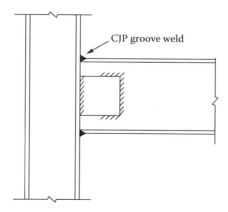

FIGURE 13.30   Welded moment-resisting connection.

or

$$L = \frac{T_u}{\phi Fyt}$$

$$= \frac{167.83}{(0.9)(36)(0.855)} = 6.06 \text{in.} < b_f$$

Provide a 6 in. long CJP weld.

PROBLEMS

13.1 Determine the strength of the bearing-type connection shown in Figure P13.1. Use A36 steel, A325 7/8 in. bolts. The threads are not excluded from shear plane.

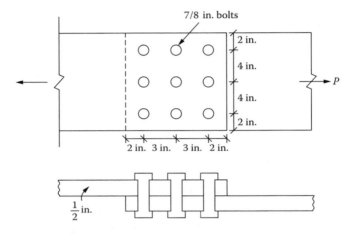

FIGURE P13.1   Connection for Problem 13.1.

13.2 Determine the strength of the bearing-type connection shown in Figure P13.2. Use A36 steel, A325 of 7/8 in. bolts. The threads are not excluded from shear plane.

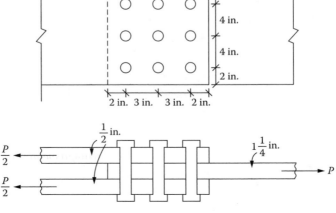

FIGURE P13.2   Connection for Problem 13.2.

13.3 Design the bearing-type connection for the bolt joint shown in Figure P13.3. Steel is A572 and bolts are A325, 3/4 in. diameter. The threads are excluded from shear plane.

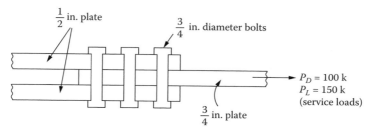

FIGURE P13.3   Connection for Problem 13.3.

13.4 A chord of a truss shown in Figure P13.4 consist of 2 C9×20 of A36 steel connected by a 1 in. gusset plate. Design the bearing-type connection by A490 bolt assuming threads are excluded from shear plane.

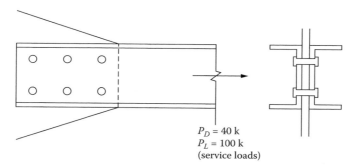

FIGURE P13.4   Truss chord connection for Problem 13.4.

13.5 Design the bearing-type connection shown in Figure P13.5 (threads excluded from shear plane) made with 7/8 in. A490 bolts. Use A572 steel.

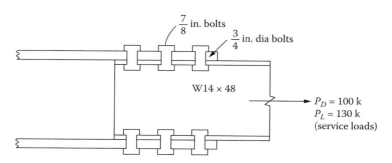

FIGURE P13.5   Connection for Problem 13.5.

13.6 Solve Problem 13.1 for the slip-critical connection of unpainted clean mill scale surface.
13.7 Solve Problem 13.2 for the slip-critical connection of unpainted blast cleaned surface.
13.8 Design a slip-critical connection for the plates shown in Figure P13.8 to resist service dead load of 30 k and live load of 50 k. Use 1 in. A325 bolts and A572 steel. Assume painted class A surface.

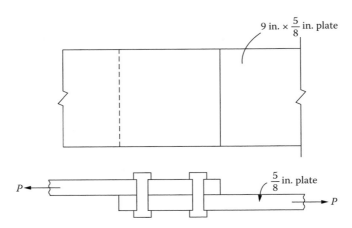

FIGURE P13.8   Connection for Problem 13.8.

13.9 A single angle 3½×3×1/4 tensile member is connected by a 3/8 in. thick gusset plate. Design a no slip (slip-critical) connection for the service dead and live loads of 8 and 24 k, respectively. Use 7/8 in. A325 bolts and A36 steel. Assume the unpainted blast cleaned surface.

13.10 A tensile member shown in Figure P13.10 consisting of two L4×3½×1/2 carries a service wind load of 110 k acting at 30°. A bracket consisting of a tee section connects this tensile member to a column flange. The connection is slip-critical. Design the bolts for the tensile member only. Use 7/8 in. A490-X bolts and A572 steel. Assume unpainted blast clean surface.

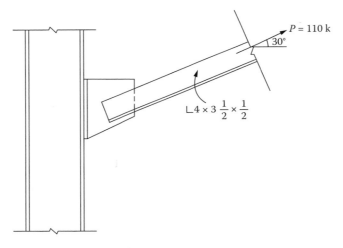

FIGURE P13.10   Connection for Problem 13.10.

13.11 Determine the strength of the bolts in hanger connection shown in Figure P13.11 (neglect the prying action).

13.12 Are the bolts in the hanger connection adequate in Figure P13.12?

13.13 A WT12×31 is attached to a 3/4 in. plate as a hanger connection, to support the service dead and live loads of 25 and 55 k. Design the connection for 7/8 in. A325 bolts and A572 steel (neglect the prying action).

In Problems 13.14 through 13.16, the threads are excluded from shear planes.

13.14 Design the column-to-bracket connection of Problem 13.10. The slip is permitted.

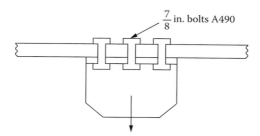

FIGURE P13.11   Hanger type connection for Problem 13.11.

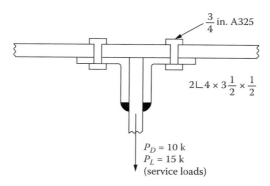

FIGURE P13.12   Hanger type connection for Problem 13.12.

13.15 In the bearing-type connection shown in Figure P13.15, determine the load capacity, P_u.

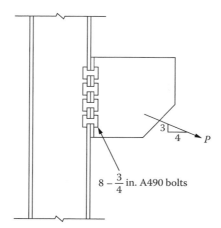

FIGURE P13.15   Combined shear-tension connection for Problem 13.15.

13.16 A tensile member is subjected to the service dead and live loads of 30 and 50 k, respectively, via 7/8 in. plate, as shown in Figure P13.16. Design the bearing-type connection. Steel is A572 and bolts are 3/4 in., A325.

In Problems 13.17 through 13.19, the connecting surface is unpainted clean mill scale.

13.17 Design connection of Problem 13.14 as the slip-critical connection.

13.18 Solve Problem 13.15 as the slip-critical connection.

13.19 Design connection of Problem 13.16 as the slip-critical connection.

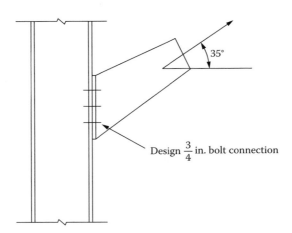

FIGURE P13.16   Combined shear-tension connection for Problem 13.16.

13.20 Determine the design strength of the connection shown in Figure P13.20. The fillet welds are 7/16 in. SMAW process. Steel is A572 and electrodes are E 70.

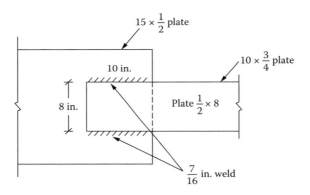

FIGURE P13.20   Welded connection for Problem 13.20.

13.21 Solve Problem 13.20 for SAW process.

13.22 A 1/4 in. thick flat plate is connected to a gusset plate of 5/16 in. thickness by a 3/16 in. weld as shown in Figure P13.22. The maximum longitudinal length is 4 in. Use the return (transverse) weld, if necessary. The connection has to resist a dead load of 10 k and live load of 20 k. What is the length of weld? Use E 70 electrodes, SMAW process. Steel is A36.

13.23 Two $1/2 \times 10$ in. A36 plates are to be connected by a lap joint. Design the connection for a factored load of 80 k. Use E 80 electrode, SAW process. Steel is A36.

13.24 Design the longitudinal fillet welds to connect a L $4 \times 3 \times 1/2$ angle tensile member shown in Figure 13.24 to resist a service dead load of 50 k and live load of 80 k. Use E 70 electrodes, SMAW process. Steel is A572.

13.25 A tensile member consists of 2 L $4 \times 3 \times 1/2$ carries a service dead load of 50 k and live load of 100 k, as shown in Figure P13.25. The angles are welded to a 3/4 in. gusset plate, which is welded to a column flange. Design the connection of the angles to the gusset plate and the gusset plate to the column. The gusset plate is connected to the column by a CJP groove and the angles are connected by a fillet weld. Use E 70 electrodes SMAW process. Steel is A572.

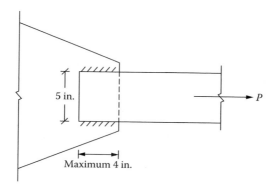

FIGURE P13.22 Welded connection for Problem 13.22.

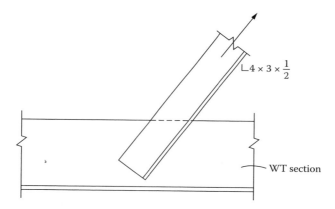

FIGURE P13.24 Welded connection design for Problem 13.24.

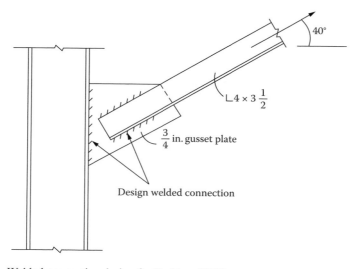

FIGURE P13.25 Welded connection design for Problem 13.25.

13.26 Design a single-plate shear connection for a W14×53 beam joining a W14×99 column by a 1/4 in. plate. The factored reaction is 60 k. Use A36 steel. Use 5/8 in. A325 bolts and E70 welds.

13.27 Design a single-plate shear connection for a W16×67 beam joining a W18×71 column by a 5/16 in. plate to support a factored beam reaction of 70 k. Use 3/4 in. A490 bolts and E 70 weld. The beam and columns have A992 steel and plate is A36 steel.

13.28 Design connection for Problem 13.26 as a moment connection to resist a factored moment of 200 ft-k in addition to the factored reaction of 60 k.

13.29 Design connection for Problem 13.27 as a moment-resisting connection to resist a factored moment of 300 ft-k and a factored shear force of 70 k.

Part IV

Reinforced Concrete Structures

14 Flexural Reinforced Concrete Members

PROPERTIES OF REINFORCED CONCRETE

Concrete is a mixture of cement, sand, gravel, crushed rock, and water. Water reacts with cement in a chemical reaction known as *hydration* that sets the cement with other ingredients into a solid mass, high in compression strength. The strength of concrete depends on the proportion of the ingredients. A most important factor for concrete strength is the water–cement ratio. More water results in a weaker concrete. However, its adequate amount is needed for concrete to be workable and easy to mix. An adequate ratio is above 0.25 by weight. The process of selecting the relative amounts of ingredients for concrete to achieve a required strength at the hardened state and to be workable in the plastic (mixed) state is known as the *concrete mix design*. The specification of concrete in terms of the proportions of cement, fine (sand) aggregate, and coarse (gravel and rocks) aggregate is called the *nominal mix*. For example a 1:2:4 nominal mix has one part of cement, two parts of sand, and four parts of gravel and rocks by volume. The nominal mixes having the same proportions could vary in strength. For this reason, another expression for specification known as the *standard mix* uses the minimum compression strength of concrete as a basis. The procedure for designing a concrete mix is a trial-and-error method. The first step is to fix the water–cement ratio for the desired concrete strength using an empirical relationship between the compressive strength and the water–cement ratio. Then, based on the characteristics of the aggregates and the proportioning desired, the quantities of the other materials comprising of cement, fine aggregate, and coarse aggregate are determined.

There are some other substances that are not regularly used in the proportioning of the mix. These, known as the *mixtures*, are usually chemicals that are added to change certain characteristics of concrete such as to accelerate or slow the setting time, to improve the workability of concrete, to decrease the water–cement ratio.

Concrete is quite strong in compression but it is very weak in tension. In a structural system, the steel bars are placed in the tension zone to compensate for this weakness. Such concrete is known as the *reinforced concrete*. At times, steel bars are used in the compression zone also to gain extra strength with a leaner concrete size as in the reinforced concrete columns and doubly reinforced beams.

COMPRESSION STRENGTH OF CONCRETE

The strength of concrete varies with time. The specified compression strength denoted as f_c' is the value that concrete attains 28 days after the placement. Beyond that stage the increase in strength is very small. The strength f_c' ranges from 2500 to 9000 psi with a common value between 3000 and 5000 psi.

The stress–strain diagram of concrete is not linear to any appreciable extent; thus concrete does not behave elastically over a major range. Moreover, the concrete of different strengths have stress–strain curves that have different slopes. Therefore, in concrete, the modulus of elasticity cannot be ascertained directly from a stress–strain diagram.

The American Concrete Institute (ACI), which is a primary agency in the United States that prepares the national standards for structural concrete, provides the empirical relations for the modulus of elasticity based on the compression strength, f_c'.

Although the stress–strain curves have different slopes for concrete of different strengths, the following two characteristics are common to all concretes:

1. The maximum compression strength, f_c' in all concrete is attained at a strain level of approximately 0.002 in./in.
2. The point of rupture of all curves lies in the strain range of 0.003–0.004 in./in. Thus, it is assumed that concrete fails at a strain level of 0.003 in./in.

DESIGN STRENGTH OF CONCRETE

To understand the development and distribution of stress in concrete, let us consider a simple rectangular beam section with steel bars at bottom (in the tensile zone), which is loaded by an increasing transverse load.

The tensile strength of concrete being small, the concrete will crack at bottom soon at a low transverse load. The stress at this level is known as the *modulus of rupture* and the bending moment is referred to as the *cracking moment*. Beyond this level, the tensile stress will be handled by the steel bars and the compression stress by the concrete section above the neutral axis. Concrete being a brittle (not a ductile) material, the distribution of stress within the compression zone could be considered linear only up to a moderate load level when the stress attained by concrete is less than 1/2 f_c', as shown in Figure 14.1. In this case, the stress and strain bear a direct proportional relationship.

As the transverse load increases further, the strain distribution will remain linear (Figure 14.2b) but the stress distribution will acquire a curvilinear shape similar to the shape of the stress–strain curve. As the steel bars reach the yield level, the distribution of strain and stress at this load will be as shown in Figure 14.2b and 14.2c.

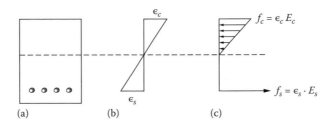

(a)   (b)   (c)

FIGURE 14.1 Stress–strain distribution at moderate loads: (a) section, (b) strain, and (c) stress.

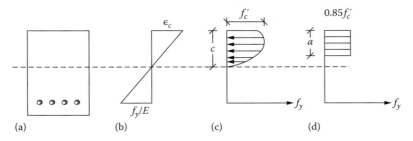

(a)   (b)   (c)   (d)

FIGURE 14.2 Stress–strain distribution at ultimate load: (a) section, (b) strain, (c) stress, and (d) equivalent stress.

For simplification, Whitney (1942) proposed a fictitious but an equivalent rectangular stress distribution of intensity $0.85 f_c'$, as shown in Figure 14.2d. This has since been adopted by the ACI. The property of this rectangular block of depth a is such that the centroid of this rectangular block is the same as the centroid of actual curved shape and that the area under the two diagrams in Figure 14.2c and d are the same. Thus, for the design purpose, the ultimate compression of concrete is taken to be $0.85 f_c'$, uniformly distributed over the depth, a.

STRENGTH OF REINFORCING STEEL

The steel bars used for reinforcing are round, deformed bars with some forms of patterned ribbed projections onto their surfaces. The bar sizes are designated from #3 through #18. For #3 to #8 sizes, the designation represents the bar diameter in one-eighths of an inch, i.e., #5 bar has a diameter of 5/8 in. The #9, #10, and #11 sizes have diameters that provide areas equal to the areas of the 1 in. × 1 in. square bar, 1⅛ in. × 1⅛ in. square bar, and 1¼ in. × 1¼ in. square bar, respectively. Sizes #14 and #18 are available only by special order. They have diameters equal to the areas of 1½ in. × 1½ in. square and 2 in. × 2 in. square bar, respectively. The diameter, area, and unit weight per foot for various sizes of bars are given in Appendix D.1.

The most useful properties of reinforcing steel are the yield stress, f_y and the modulus of elasticity, E. A large percentage of reinforcing steel bars are not made from new steel but are rolled from melted, reclaimed steel. These are available in different grades. Grade 40, Grade 50, and Grade 60 are common where Grade 40 means the steel having an yield stress of 40 ksi and so on. The modulus of elasticity of reinforcing steel of different grades varies over a very small range. It is adopted as 29,000 ksi for all grades of steel.

Concrete structures are composed of the beams, columns, or column–beam types of structures where they are subjected to flexure, compression, or the combination of flexure and compression. The theory and design of simple beams and columns have been presented in the book.

LRFD BASIS OF CONCRETE DESIGN

Until mid-1950, concrete structures were designed by the elastic or *working stress design* (WSD) method. The structures were proportioned so that the stresses in concrete and steel did not exceed a fraction of the ultimate strength, known as the *allowable* or *permissible* stresses. It was assumed that the stress within the compression portion of concrete was linearly distributed. However, beyond a moderate load when the stress level is only about one-half the compressive strength of concrete, the stress distribution in concrete section is not linear.

In 1956, the ACI introduced a more rational method wherein the members were designed for a nonlinear distribution of stress and the full strength level was to be explored. This method was called the *ultimate strength design* (USD) method. Since then, the name has been changed to the *strength design* method.

The same approach is known as the *load resistance factor design* (LRFD) method in steel and wood structures. Thus, the concrete structures were the first ones to adopt the LFRD method of design in the United States.

The ACI Publication #318, revised numerous times, contains the codes and standards for concrete buildings. The ACI codes 318-56 of 1956 for the first time included the codes and standards for the ultimate strength design in an appendix to the code. The ACI 318-63 code provided an equal status to the WSD and ultimate strength methods bringing both of them within the main body of the code. The ACI 318-02 code made the ultimate strength design, with a changed name of the strength design as the mandatory method of design. The ACI 318-08 code contains the latest design provisions.

In the strength design method, the service loads are amplified using the load factors. The member's strength at failure known as the theoretical or the nominal capacity is somewhat reduced by a strength reduction factor to represent the usable strength of the member. The amplified loads must not exceed the usable strength of member, namely,

$$\text{Amplified loads on member} \leq \text{usable strength of member} \qquad (14.1)$$

Depending upon the types of structure, the loads are the compression forces, shear forces, or bending moments.

REINFORCED CONCRETE BEAMS

A concrete beam is a composite structure where a group of steel bars are embedded into the tension zone of the section to support the tensile component of the flexural stress. The areas of the group of bars are given in Appendix D.2. The minimum widths of beam that can accommodate a specified number of bars in a single layer are indicated in Appendix D.3. These tables are very helpful in designs.

Equation 14.1 in the case of beams takes the following form similar to wood and steel structures:

$$M_u \leq \phi M_n \qquad (14.2)$$

where
  M_u is the maximum moment due to application of the factored loads
  M_n is the nominal or theoretical capacity of the member
  ϕ is the strength reduction (resistance) factor for flexure

According to the flexure theory, $M_n = F_b S$ where F_b is the ultimate bending stress and S is the section modulus of the section. The application of this formula is straightforward for a homogeneous section for which the section modulus or the moment of inertia could be directly found. However, for a composite concrete–steel section and a nonlinear stress distribution, the flexure formula presents a problem. A different approach termed as the *internal couple method* is followed for concrete beams.

In the internal couple method, two forces act on the beam cross section represented by a compressive force, C acting on one side of the neutral axis (above the neutral axis in a simply supported beam) and a tensile force, T acting on the other side. Since the forces acting on any cross section of the beam must be in equilibrium, C must be equal and opposite of T, thus representing a couple. The magnitude of this internal couple is the force (C or T) times the distance Z between the two forces called the *moment arm*. This internal couple must be equal and opposite to the bending moment acting at the section due to the external loads. This is a very general and a convenient method for determining the nominal moment, M_n in concrete structures.

DERIVATION OF THE BEAM RELATIONS

The stress distribution across a beam cross section at the ultimate load is shown in Figure 14.3 representing the concrete stress by a rectangular block according to the "Design Strength of Concrete" section.

The ratio of stress block and depth to the neutral axis is defined by a factor β_1 as follows:

$$\beta_1 = \frac{a}{c} \qquad (14.3)$$

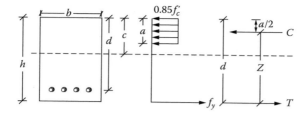

FIGURE 14.3   Internal forces and couple on a section.

Sufficient test data are available to evaluate β_1. According to the ACI

1. For $f_c' \leq 4000\,\text{psi}$ $\beta_1 = 0.85$ (14.4a)
2. For $f_c' > 4000\,\text{psi}$ but $\leq 8000\,\text{psi}$

$$\beta_1 = 0.85 - \left(\frac{f_c' - 4000}{1000}\right)(0.05)$$ (14.4b)

3. For $f' > 8000\,\text{psi}$ $\beta_1 = 0.65$ (14.4c)

With reference to Figure 14.3, since force = (stress) (area)

$$C = (0.85 f_c')(ab)$$ (a)

$$T = f_y A_s$$ (b)

Since, $C = T$,

$$(0.85 f_c')(ab) = f_y A_s$$ (c)

or

$$a = \frac{A_s f_y}{0.85 f_c' b}$$ (d)

or

$$a = \frac{\rho f_y d}{0.85 f_c'}$$ (14.5)

where

$$\rho = \text{steel ratio} = \frac{A_s}{bd}$$ (14.6)

Since moment = (force) (moment arm)

$$M_n = T\left(d - \frac{a}{2}\right) = f_y A_s \left(d - \frac{a}{2}\right) \tag{e}$$

Substituting a from Equation 14.5 and A_s from Equation 14.6 into (e)

$$M_n = \rho f_y b d^2 \left(1 - \frac{\rho f_y}{1.7 f_c'}\right) \tag{f}$$

Substituting (f) into Equation 14.2

$$\frac{M_u}{\phi b d^2} = \rho f_y \left(1 - \frac{\rho f_y}{1.7 f_c'}\right) \tag{14.7}$$

Equation 14.7 is a very useful relation to analyze and design a beam.

If we arbitrarily define the expression to the right side of Equation 14.7 by $\bar{K}$ called the *coefficient of resistance*, then Equation 14.7 becomes

$$M_u = \phi b d^2 \bar{K} \tag{14.8}$$

where

$$\bar{K} = \rho f_y \left(1 - \frac{\rho f_y}{1.7 f_c'}\right) \tag{14.9}$$

The coefficient $\bar{K}$ depends on (1) ρ, (2) f_y, and (3) f_c'. The values of $\bar{K}$ for different combinations of ρ, f_y, and f_c' are listed in Appendices D.4 through D.10.

In place of Equation 14.7, these tables can be directly used in beam computations.

THE STRAIN DIAGRAM AND MODES OF FAILURE

The strain diagrams in Figures 14.1 and 14.2 show a straight line variation of the concrete compression strain ε_C to the steel tensile strain, ε_s; the line passes through the neutral axis. Concrete can have a maximum strain of 0.003 and the strain at which steel yields is $\varepsilon_y = f_y / E$. When the strain diagram is such that the maximum concrete strain of 0.003 and the steel yield strain of ε_y are attained at the same time, it is said to be a balanced section, as shown by the solid line labeled I in Figure 14.4.

In this case, the amount of steel and the amount of concrete balance each other out and both of these will reach the failing level (will attain the maximum strains) simultaneously. If a beam has more steel than the balanced condition, then the concrete will reach a strain level of 0.003 before the steel attains the yield strain of ε_y. This is shown by condition II in Figure 14.4. The neutral axis moves down in this case.

The failure will be initiated by crushing of concrete which will be sudden since concrete is brittle. This, mode of failure in compression, is undesirable because a structure will fail suddenly without any warning.

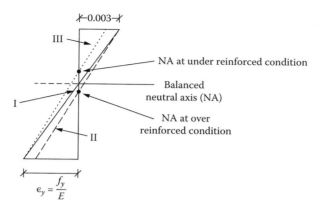

FIGURE 14.4 Strain stages in beam.

If a beam has lesser steel than the balanced condition, then steel will attain its yield strain before the concrete could reach the maximum strain level of 0.003. This is shown by condition III in Figure 14.4. The neutral axis moves up in this case. The failure will be initiated by the yielding of the steel, which will be gradual because of the ductility of steel. This is a tensile mode of failure, which is more desirable because at least there is an adequate warning of an impending failure. The ACI recommends the tensile mode of failure or the under-reinforcement design for a concrete structure.

BALANCED AND RECOMMENDED STEEL PERCENTAGES

To ensure the under-reinforcement conditions, the percent of steel should be less than the balanced steel percentage, ρ_b, which is the percentage of steel required for the balanced condition.

From Figure 14.4, for the balanced condition,

$$\frac{0.003}{c} = \frac{f_y/E}{d-c} \tag{a}$$

By substituting $c = a/\beta_1$ from Equation 14.3 and $a = \rho f_y d / 0.85 f_c'$ from Equation 14.5 and $E = 29 \times 10^6$ psi in Equation (a), the following expression for the balanced steel is obtained:

$$\rho_b = \left(\frac{0.85\beta_1 f_c'}{f_y} \right) \left(\frac{870,000}{87,000 + f_y} \right) \tag{14.10}$$

The values for the balanced steel ratio, ρ_b, calculated for different values of f_c' and f_y are tabulated in Appendix D.11. Although a tensile mode of failure ensues when the percent of steel is less than the balanced steel, the ACI code defines a section as *tension controlled* only when the tensile strain in steel ε_t is equal to or greater than 0.005 as the concrete reaches to its strain limit of 0.003. The strain range between $\varepsilon_y = (f_y/E)$ and 0.005 is regarded as the *transition zone*.

The values of the percentage of steel for which ε_t is equal to 0.005 are also listed in Appendix D.11 for different grades of steel and concrete. It is recommended to design beams with the percentage of steel, which is not larger than these listed values for ε_t of 0.005.

If a larger percentage of steel is used than for $\varepsilon_t = 0.005$, to be in the transition region, the strength reduction factor ϕ should be adjusted, as discussed in the "Strength Reduction Factor for Concrete" section.

MINIMUM PERCENTAGE OF STEEL

Just as the maximum amount of steel is prescribed to ensure the tensile mode of failure, a minimum limit is also set to ensure that the steel is not too small so as to cause the failure by rupture (cracking) of the concrete in the tension zone. The ACI recommends the higher of the following two values for the minimum steel in flexure members.

$$(A_s)\min = \frac{3\sqrt{f_c'}}{f_y}bd \tag{14.11}$$

or

$$(A_s)_{\min} = \frac{200}{f_y}bd \tag{14.12}$$

where
    b is the width of beam
    d is the effective depth of beam

The values of $\rho_{\min}$ which is $(A_s)_{\min}/bd$, are also listed in Appendix D.11, where higher of the values from Equations 14.10 and 14.11 have been tabulated.

The minimum amount steel for slabs is controlled by shrinkage and temperature requirements, as discussed in the "Specifications for Slab" section.

STRENGTH REDUCTION FACTOR FOR CONCRETE

In Equations 14.2 and 14.7, a strength reduction factor ϕ is applied to account for all kinds of uncertainties involved in strength of materials, design and analysis, and workmanship. The values of the factor recommended by the ACI are listed in Table 14.1.

For the transition region between the compression-controlled and the tension-controlled stages when ε_t is between ε_y (assumed to be 0.002) and 0.005 as discussed above, the value of ϕ is interpolated between 0.65 and 0.9 by the following relation

$$\phi = 0.65 + (\varepsilon_t - 0.002)\left(\frac{250}{3}\right)^* \tag{14.13}$$

The values[†] of ε_t for different percentages of steel are also indicated in Appendices D.4 through D.10. When it is not listed in these tables, it means that ε_t is larger than 0.005.

TABLE 14.1
Strength Reduction Factors

Structural System	ϕ
1. Tension-controlled beams and slabs	0.9
2. Compression-controlled columns	
Spiral	0.70
Tied	0.65
3. Shear and torsion	0.75
4. Bearing on concrete	0.65

SPECIFICATIONS FOR BEAMS

The ACI specifications for beams are as follows:

1. *Width-to-depth ratio*: There is no code requirement for b/d ratio. From experience, the desirable b/d ratio lies between 1/2 and 2/3.

---

* For spiral reinforcement this is $\phi = 0.70 + (\varepsilon_t - 0.002)(250/3)$.
† ε_t is calculated by the formula $\varepsilon_t = \left(0.00255 f_c'\beta_1/\rho f_y\right) - 0.003$.

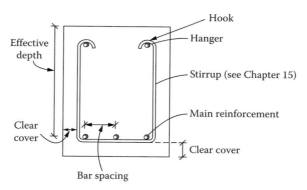

FIGURE 14.5 Specifications sketch of beam.

2. *Selection of steel*: After a required reinforcement area is computed, Appendix D.2 is used to select the number of bars that provide for the necessary area.
3. Minimum beam width required to accommodate multiples of various size bars are given in Appendix D.3. This is an useful design aid as demonstrated in the example.
4. The reinforcement is located at certain distance from the surface of the concrete called the *cover*. The cover requirements in the ACI code are extensive. For beams, girders, and columns that are not exposed to weather or are not in contact with the ground, the minimum clear distance from the bottom of the steel to the concrete surface is 1½ in. There is a minimum cover requirement of 1½ in. from the outermost longitudinal bars to the edge toward the width of the beam, as shown in Figure 14.5.

TABLE 14.2
First Estimate of Beam Weight

Design Moment, M_u, ft-k	Estimated Weight, lbs/ft
≤200	300
>200 but ≤300	350
>300 but ≤400	400
>400 but ≤500	450
>500	500

5. *Bar spacing*: The clear spacing between the bars in a single layer should not be less than any of the following:
   • 1 in.
   • The bar diameter
   • 1⅓×maximum aggregate size
6. *Bars placement*: If the bars are placed in more than one layer, those in the upper layers are required to be placed directly over the bars in the lower layers and the clear distance between the layers must not be less than 1 in.
7. *Concrete weight*: Concrete is a heavy material. The weight of the beam is significant. An estimated weight should be included. If it is found to be appreciably less than the weight of the section designed, then the design should to be revised. For a good estimation of concrete weight, Table 14.2 could be used as a guide.

ANALYSIS OF BEAMS

Analysis relates to determining of the factored or service moment or the load capacity of a beam of known dimensions and known reinforcement.

The procedure of analysis follows:

1. Calculate the steel ratio from Equation 14.6

$$\rho = \frac{A_s}{bd}$$

2. Calculate $(A_s)_{min}$ from Equations 14.11 and 14.12 or use Appendix D.11.
   Compare this to the A_s of beam to ensure that it is more than the minimum.
3. For known ρ, read ε_t from Appendices D.4 through D.10. If no value is given, then $\varepsilon_t = 0.005$.
   If $\varepsilon_t < 0.005$, determine ϕ from Equation 14.13.
4. For known ρ, compute $\overline{K}$ from Equation 14.9 or read the value from Appendices D.4 through D.10.
5. Calculate M_u from Equation 14.7

$$M_u = \phi b d^2 \overline{K}$$

6. Break down into the loads if required.

Example 14.1

The loads on a beam section are shown in Figure 14.6. Whether the beam is adequate to support the loads. $f_c' = 4{,}000\,\text{psi}$ and $f_y = 60{,}000\,\text{psi}$.

Solution

A. Design loads and moments
   1. Weight of beam/ft $= (12/12) \times (20/12) \times 1 \times 150 = 250\,\text{lb/ft}$ or $0.25\,\text{k/ft}$
   2. Factored dead load, $w_u = 1.2\,(1.25) = 1.5\,\text{k/ft}$
   3. Factored live load, $P_u = 1.6\,(15) = 24\,\text{k}$
   4. Design moment due to dead load $= w_u L^2/8 = 1.5(20)^2/8 = 75\,\text{ft-k}$
   5. Design moment due to live load $= P_u L/4 = 24(20)/4 = 120\,\text{ft-k}$
   6. Total design moment, $M_u = 195\,\text{ft-k}$
   7. $A_s^{\bullet} = 3.16\,\text{in.}^2$ (from Appendix D.2 for 4 bars of #8)
   8. $\rho = A_s/bd = 3.16/12 \times 17 = 0.0155$
   9. $\rho_{min} = 0.0033$ (from Appendix D.11) < 0.0155 OK
   10. $\varepsilon_t \geq 0.005$ (value not listed in Appendix D.9), $\phi = 0.9$
   11. $\overline{K} = 0.8029\,\text{ksi}$ (for $\rho = 0.0155$ from Appendix D.9)
   12. $Mu = \phi b d^2 \overline{K}$
       $$= (0.9)(12)(17)^2(0.8029) = 2506\,\text{in.-k}$$ or $209\,\text{ft-k} > 195\,\text{ft-k}$ OK

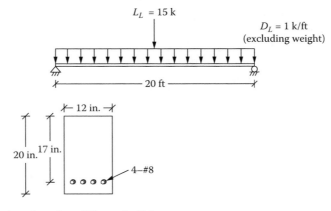

FIGURE 14.6 Loads and section of Example 14.1.

DESIGN OF BEAMS

In wood beam design in Chapter 7 and steel beam design in Chapter 11, beams were designed for bending moment capacity and checked for shear and deflection.

In concrete beams, shear is handled independently, as discussed in Chapter 15. For deflection, the ACI stipulates that when certain depth requirements are met, deflection will not interfere with the use or cause a damage to the structure. These limiting values are given in Table 14.3 for normal weight (120–150 lb/ft³) concrete and Grade 60 steel. For other grade concrete and steel, the adjustments are made as indicated in the footnotes to the Table 14.3.

When the minimum depth requirement is met, deflection needs not be computed. For members of lesser thickness than those listed in Table 14.3, the deflections should be computed to check for safe limits. This book assumes that the minimum depth requirement is satisfied.

The beam design falls into the two categories discussed below.

DESIGN FOR REINFORCEMENT ONLY

When a beam section has been fixed from the architectural or any other consideration, only the amount steel has to be selected. The procedure is as follows:

1. Determine the design moment, M_u including the beam weight for various critical load combinations.
2. Using $d = h - 3$, and $\phi = 0.9$, calculate the required $\bar{K}$ from Equation 14.8 expressed as

$$\bar{K} = \frac{M_u}{\phi bd^2}$$

3. From Appendices D.4 through D.10, find the value of ρ corresponding to $\bar{K}$ of step 2. From the same appendices, confirm that $\varepsilon_t \geq 0.005$. If $\varepsilon_t < 0.005$, reduce ϕ by Equation 14.13 and recompute $\bar{K}$ and find the corresponding ρ.
4. Compute the required steel area A_s from Equation 14.6

$$A_s = \rho bd$$

TABLE 14.3
Minimum Thickness of Beams and Slabs, for Normal Weight Concrete and Grade 60 Steel

	Minimum Thickness, h, in.			
Member	Simply Supported	Cantilever	One End Continuous	Both Ends Continuous
Beam	$L/16$	$L/18.5$	$L/21$	$L/8$
Slab (one-way)	$L/20$	$L/24$	$L/28$	$L/10$

Notes: L is the span in inches.

For lightweight concrete of unit weight 90–120 lb/ft³, the table values should be multiplied by $(1.65 - 0.005 W_c)$ but not less than 1.09, where W_c is the unit weight in lb/ft³.

For other than Grade 60 steel, the table value should be multiplied by $(0.4 + f_y/100)$, where f_y in ksi.

---

5. Check for minimum steel $(\rho)_{min}$ from Appendix D.11.
6. Select the bar size and the number of bars from Appendix D.2. From Appendix D.3, check whether the selected steel (size and number) can fit into width of the beam, preferably in a single layer. They can be arranged in two layers. Check to confirm that the actual depth is at least equal to $h-3$.
7. Sketch the design.

Example 14.2

Design a rectangular reinforced beam to carry a service dead of 1.6 k/ft and a live load of 1.5 k/ft on a span of 20 ft. The architectural consideration requires the width to be 10 in. and depth to be 24 in. Use $f'_c = 3,000$ psi and $f_y = 60,000$ psi.

Solution

1. Weight of beam/ft = $(10/12) \times (24/12) \times 1 \times 150 = 250$ lb/ft or 0.25 k/ft
2. $w_u = 1.2 (1.6 + 0.25) + 1.6 (1.5) = 4.62$ k/ft
3. $M_u = w_u L^2/8 = 4.62(20)^2/8 = 231$ ft-k or 2772 in.-k
4. $d = 24 - 3 = 21$ in.
5. $\bar{K} = 2772/(0.9)(10)(21)^2 = 0.698$ ksi
6. $\rho = 0.0139$ $\varepsilon_t = 0.0048$ (from Appendix D.6)
7. From Equation 14.13, $\phi = 0.65 + (0.0048 - 0.002)(250/3) = 0.88$
8. Revised $\bar{K} = 2772/(0.88)(10)(21)^2 = 0.714$ ksi
9. Revised $\rho = 0.0143$ (from Appendix D.6)
10. $A_s = \rho bd = (0.0143)(10)(21) = 3$ in.^{2}
11. $\rho_{(min)} = 0.0033$ (from Appendix D.11) < 0.0143 OK
12. Selection of steel

Bar Size	No. of Bars	A_s, from Appendix D.2	Minimum Width in One Layer from Appendix D.3
#6	7	3.08	15 NG
#7	5	3.0	12.5 NG
#9	3	3.0	9.5 OK

Select 3 bars of #9.

13. Beam section is shown in Figure 14.7.

DESIGN OF BEAM SECTION AND REINFORCEMENT

The design comprises of determining the beam dimensions and selecting the amount of steel. The procedure is as follows:

1. Determine the design moment, M_u including the beam weight for various critical load combinations.
2. Select steel ratio ρ corresponding to $\varepsilon_t = 0.005$ from Appendix D.11.
3. From Appendices D.4 through D.10, find $\bar{K}$ for the steel ratio of step 2.
4. For b/d ratio of 1/2 and 2/3, find two values of d from the following expression

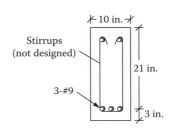

FIGURE 14.7 Beam section of Example 14.2.

$$d = \left[\frac{M_u}{\phi (b/d) \overline{K}} \right]^{1/3}$$

(14.14)*

5. Select the effective depth to be between the two values of step 4.
6. If the depth from Table 14.3 is larger, use that value.
7. Determine the corresponding width b from

$$b = \frac{M_u}{\phi d^2 \overline{K}}$$

(14.15)

8. Estimate h and compute the weight of beam. If this is excessive as compared to the assumed value of step 1, repeat steps 1 through 7.
9. From now on, follow the steps 4 through 7 of the design procedure of the "Design for Reinforcement Only" section for the selection of steel.

Example 14.3

Design a rectangular reinforce beam for the service loads shown in Figure 14.8. Use $f_c' = 3,000\,\text{psi}$ and $f_y = 60,000\,\text{psi}$.

Solution

1. Assume a weight of beam/ft = 0.5 k/ft
2. Factored dead load, $w_u = 1.2\ (1.5 + 0.5) = 2.4\,\text{k/ft}$
3. Factored live load, $P_u = 1.6\ (20) = 32\,\text{k}$
4. Design moment due to dead load $= w_u L^2/8 = 2.4(30)^2/8 = 270\,\text{ft-k}$
5. Design moment due to live load $= P_u L/3 = 32(30)/3 = 320\,\text{ft-k}$
6. Total design moment, $M_u = 590\,\text{ft-k}$ or $7080\,\text{in.-k}$
7. $\rho = 0.0136$ (from Appendix D.11 for $\varepsilon_t = 0.005$)
8. $\overline{K} = 0.684\,\text{ksi}$ (from Appendix D.6)
9.

	Calculate *d* from
Select *b/d* ratio	Equation 14.14
1/2	28.3[a]
2/3	25.8

  [a] $[7080/0.9(1/2)(0.684)]^{1/3}$.

10. Depth for deflection (from Table 14.3)

$$h = \frac{L}{16} = \frac{30 \times 12}{16} = 22.5\,\text{in.}$$

or $d = h - 3 = 22.5 - 3 = 19.5\,\text{in.}$
Use $d = 27\,\text{in.}$

11. From Equation 14.15

$$b = \frac{7080}{(0.9)(27)^2(0.684)} = 15.75\,\text{in. use 16 in.}$$

---

* This relation is the same as $M_u\, bd^2\ \overline{K}$ or $M_u = \phi\ (b/d)d^3\ \overline{K}$.

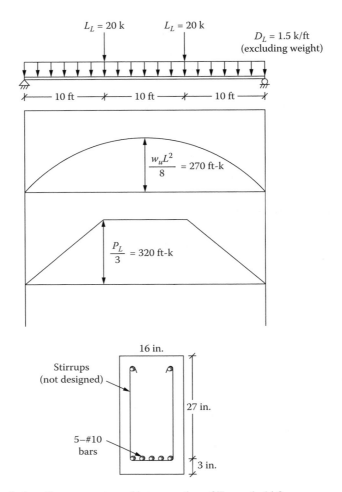

FIGURE 14.8 Loads, bending moments, and beam section of Example 14.3.

12. $h = d + 3 = 30$ in.
Weight of beam/ft $= 16/12 \times 30/12 \times 1 \times 150 = 500$ lb/ft or 0.50 k/ft OK
13. $A_s = \rho bd = (0.0136)(16)(27) = 5.88$ in.^{2}
14. Selection of steel

Bar Size	No. of Bars	A_s, from Appendix D.2	Minimum Width in One Layer from Appendix D.3
#9	6	6.00	16.5 NG
#10	5	6.35	15.5

Select 5 bars of #10.

ONE-WAY SLAB

Slabs are the concrete floor systems supported by reinforced concrete beams, steel beams, concrete columns, steel columns, concrete walls, or masonry walls. If they are supported on two opposite sides only, they are referred to as *one-way slabs* because the bending is in one direction

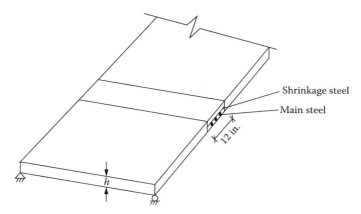

FIGURE 14.9 Simply supported one-way slab.

only; perpendicular to the supported edge. When slabs are supported on all four edges, they are called *two-way slabs* because the bending is in both the directions. A rectangular floor plan has slab supported on all four sides. However, if the long side is two or more times of the short side, the slab could be considered as a one-way slab spanning in the short direction.

A one-way slab is analyzed and designed as 12 in. wide beam segments placed side by side having a total depth equal to the slab thickness, as shown in Figure 14.9.

The amount of steel computed is considered to exist in 12 in. width on an average. Appendix D.12 is used for this purpose that indicates for the different bar sizes the center-to-center spacing of the bars for a specified area of steel. The relationship is as follows:

$$\text{Bar spacing center to center} = \frac{\text{required steel area}}{\text{area of 1 bar}} \times 12 \qquad (14.16)$$

SPECIFICATIONS FOR SLABS

The ACI specifications for one-way slab follow:

1. *Thickness*: Table 14.3 indicates the minimum thickness for one-way slabs where deflections are not to be calculated. The slab thickness is rounded off to the nearest 1/4 in. on the higher side for slabs up to 6 in. and to the nearest 1/2 in. for slabs thicker than 6 in.
2. *Cover*: (1) For slabs that are not exposed to the weather or are not in contact with the ground the minimum cover is 3/4 in for #11 and smaller bars, and (2) for slabs exposed to the weather or are in contact with the ground, the minimum cover is 3 in.
3. *Spacing of bars*: The main reinforcement should not be spaced on center-to-center more than (1) three times the slab thickness or (2) 18 in., whichever is smaller.
4. *Shrinkage steel*: Some steel is placed in the direction perpendicular to the main steel to resist shrinkage and temperature stresses. The minimum area of such steel is
   1. For Grade 40 or 50 steel, shrinkage $A_s = 0.002bh$
   2. For Grade 60 steel, shrinkage $A_s = 0.0018bh$, where $b = 12$ in.
      The shrinkage and temperature steel should not be spaced farther apart than (1) five times the slab thickness or (2) 18 in., whichever is smaller.
5. *Minimum main reinforcement*: Minimum amount of main steel should not be less than the shrinkage and temperature steel.

ANALYSIS OF ONE-WAY SLAB

The procedure of analysis is as follows:

1. For the given bar size and spacing, read A_s from Appendix D.12
2. Find the steel ratio

$$\rho = \frac{A_s}{bd} \text{ where } b = 12 \text{ in}, d = h - 0.75 \text{ in.} - 1/2(\text{bar diameter})^*$$

3. Check for the minimum shrinkage steel and also that the main reinforcement A_s is more than $A_{s(\min)}$.

$$A_{s(\min)} = 0.002bh \text{ or } 0.0018bh$$

4. For ρ of step 2, read $\bar{K}$ and ε_t (if given in the same appendices) from Appendices D.4 through D.10.
5. Correct ϕ from Equation 14.13 if $\varepsilon_t < 0.005$
6. Find out M_u as follows and convert to loads if necessary

$$M_u = \phi b d^2 \bar{K}$$

Example 14.4

The slab of an interior floor system has a cross section as shown in Figure 14.10. Determine the service live load that the slab can support in addition to its own weight on a span of 10 ft. $f_c' = 3{,}000$ psi, $f_y = 40{,}000$ psi.

Solution

1. $A_s = 0.75$ in.^{2} (from Appendix D.12)
2. $d = 6 - 1/2(0.625) = 5.688$ in. and $\rho = A_s/bd = 0.75/(12)(5.688) = 0.011$
3. $A_{s(\min)} = 0.002bh = 0.002\ (12)\ (6.75) = 0.162 < 0.75$ in.^{2} **OK**
4. $\bar{K} = 0.402$ (from Appendix D.4) $\varepsilon_t > 0.005$ for $\rho = 0.011$
5. $M_u = \phi b d^2\ \bar{K} = (0.9)(12)(5.688)^2(0.402) = 140.46$ in.-k or 11.71 ft-k
6. $M_u = w_u L^2/8$ or $w_u = 8M_u/L^2 = 8(11.71)/10^2 = 0.94$ k/ft
7. Weight of a slab/ft = $12/12 \times 6.75/12 \times 1 \times 150/1000 = 0.08$ k/ft
8. $w_u = 1.2(w_D) + 1.6(w_L)$ or $0.94 = 1.2(0.084) + 1.6\ w_L$ or $w_L = 0.52$ k/ft
   Since the slab width is 12 in., live load is 0.52 k/ft^2.

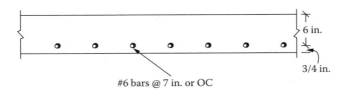

#6 bars @ 7 in. or OC

FIGURE 14.10 Slab cross section of a floor system (perpendicular to span).

---

* For slabs laid on ground $d = h - 3 - 1/2$ (bar diameter).

DESIGN OF ONE-WAY SLAB

1. Determine the minimum h from Table 14.3. Compute the slab weight/ft for $b = 12$ in.
2. Compute the design moment M_u. The unit load per square feet automatically becomes load/ft since the slab width = 12 in.
3. Calculate an effective depth, d from

$$d = h - 1 - \tfrac{1}{2} \times \text{assumed bar diameter}$$

4. Compute $\bar{K}$ assuming $\phi = 0.90$,

$$\bar{K} = \frac{M_u}{\phi b d^2}$$

5. From Appendices D.4 through D.10, find the steel ratio ρ and note the value of ε_t (if ε_t is not listed then $\varepsilon_t > 0.005$).
6. If $\varepsilon_t < 0.005$, correct ϕ from Equation 14.13 and repeat steps 4 and 5.
7. Compute the required A_s

$$A_s = \rho b d$$

8. From the table in Appendix D.12, select the main steel satisfying the condition that the bar spacing is $\leq 3h$ or 18 in.
9. Select shrinkage and temperature of steel

$$\text{Shrinkage } A_s = 0.002bh \text{ (Grade 40 or 50 steel)}$$

or

$$0.0018bh \text{ (Grade 60 steel)}$$

10. From Appendix D.12, select size and spacing of shrinkage steel with a maximum spacing of $5h$ or 18 in., whichever is smaller.
11. Check that the main steel area of step 7 is not less than the shrinkage steel area of step 9.
12. Sketch the design.

Example 14.5

Design an exterior one-way slab exposed to the weather to span 12 ft and to carry a service dead load of 100 psf and live load of 300 psf in addition to the slab weight. Use $f_c' = 3,000$ psi and $f_y = 40,000$ psi.

Solution

1. Minimum thickness for deflection from Table 14.3

$$h = \frac{L}{20} = \frac{12(12)}{20} = 7.2 \text{ in.}^2$$

For exterior slab use $h = 10$ in.

2. Weight of slab = $12/12 \times 10 \times 12 \times 1 \times 150/1000 = 0.125$ k/ft

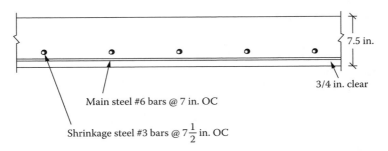

7.5 in.

3/4 in. clear

Main steel #6 bars @ 7 in. OC

Shrinkage steel #3 bars @ $7\frac{1}{2}$ in. OC

FIGURE 14.11   Design section of Example 14.5.

3. $w_u = 1.2(0.1 + 0.125) + 1.6(0.3) = 0.75$ k/ft
4. $M_u = w_u L^2/8 = 13.5$ ft-k or 163 in.-k
5. Assuming #8 size bar (diameter = 1 in.)

$$d = h - \text{cover} - \tfrac{1}{2}(\text{bar diameter}) = 10 - 3\text{-}\tfrac{1}{2} = 6.5\,\text{in.}$$

6. $K = M_u/\phi bd^2 = 162/(0.9)(12)(6.5)^2 = 0.355$
7. $\rho = 0.0096$, $\varepsilon_t > 0.005$ (from Appendix D.4)
8. $A_s = \rho bd = (0.0096)(12)(6.5) = 0.75$ in.^{2}/ft
   Provide #8 bars @ 12 in. on center (from Appendix D.12), $A_s = 0.79$ in.^{2}
9. Check for maximum spacing
   a. $3h = 3$ (7.5) = 22.5 in.
   b.   18 in. > 7 in. **OK**
10. Shrinkage and temperature steel

$$A_s = 0.002bh$$
$$= 0.002(12)(7.5) = 0.18 \ \text{in.}^2/\text{ft}$$

   Provide #3 bars @ 7½ in. on center (from Table D.12) $A_s = 0.18$ in.^{2}
11. Check for maximum spacing of shrinkage steel
   a. $5h = 5$ (7.5) = 37.5 in.
   b.   18 in. > 7 ½ in. **OK**
12. Main steel > shrinkage steel **OK**
13. A designed section is shown in Figure 14.11

PROBLEMS

14.1 A beam cross section is shown in Figure P14.1. Determine the service dead load and live load/ft for a span of 20 ft. The service dead load is one-half of the live load. $f_c' = 4,000$ psi, $f_y = 60,000$ psi.

14.2 Calculate the design moment for a rectangular reinforced concrete beam having a width of 16 in. and an effective depth of 24 in. The tensile reinforcement is 5 bars of #8. $f_c' = 4,000$ psi, $f_y = 40,000$ psi.

14.3 A reinforced concrete beam has a cross section shown in Figure P14.3 for a simple span of 25 ft. It supports a dead load of 2 k/ft (excluding beam weight) and live load of 3 k/ft. Is the beam adequate? $f_c' = 4,000$ psi, $f_y = 60,000$ psi.

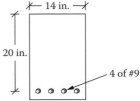

├─ 14 in. ─┤

20 in.

4 of #9

FIGURE P14.1   Beam section for Problem 14.1.

14.4 Determine the dead load (excluding the beam weight) for the beam section shown in Figure P14.4 of a span of 30 ft. The service dead load and live load are equal. $f_c' = 5,000$ psi, $f_y = 60,000$ psi.

14.5 The loads on a beam and its cross section are shown in Figure P14.5. Is this beam adequate? $f_c' = 4,000$ psi, $f_y = 50,000$ psi.

14.6 Design a reinforced concrete beam to resist a factored design moment of 150 ft-k. It is required that the beam width be 12 in. and the overall depth to be 24 in. use $f_c' = 3,000$ psi, $f_y = 60,000$ psi.

14.7 Design a reinforced concrete beam of a span of 30 ft. The service dead load is 0.85 k/ft (excluding weight) and live load is 1 k/ft. The beam has to be 12 in. wide and 26 in. deep. $f_c' = 4,000$ psi, $f_y = 60,000$ psi.

14.8 Design a reinforced beam for a simple span of 30 ft. There is no dead load except the weight of beam and the service live load is 1.5 k/ft. The beam can be 12 in. wide and 28 in. overall depth. $f_c' = 5,000$ psi, $f_y = 60,000$ psi.

14.9 A beam carries the service loads shown in Figure P14.9. From architectural consideration, the beam width is 12 in. and the overall depth is 20 in. Design the beam reinforcement. $f_c' = 4,000$ psi, $f_y = 60,000$ psi.

14.10 In Problem 14.9, the point dead load has a magnitude of 6.5 k (instead of 4 k). Design the reinforcement for the beam of the same size of Problem 14.9. $f_c' = 4,000$ psi, $f_y = 60,000$ psi.

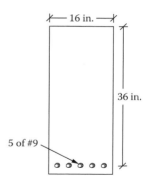

FIGURE P14.3 Beam section for Problem 14.3.

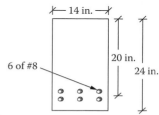

FIGURE P14.4 Beam section for Problem 14.4.

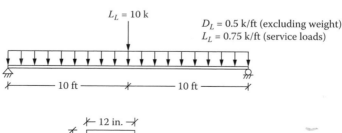

$L_L = 10$ k

$D_L = 0.5$ k/ft (excluding weight)
$L_L = 0.75$ k/ft (service loads)

10 ft   10 ft

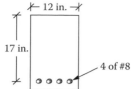

12 in.

17 in.

4 of #8

FIGURE P14.5 Beam loads and section for Problem 14.5.

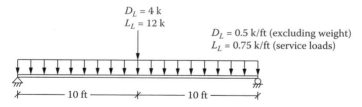

$D_L = 4$ k
$L_L = 12$ k

$D_L = 0.5$ k/ft (excluding weight)
$L_L = 0.75$ k/ft (service loads)

10 ft   10 ft

FIGURE P14.9 Beam loads for Problem 14.9.

14.11 Design a rectangular reinforced beam for a simple span of 30 ft. The uniform service loads are dead load of 1.5 k/ft (excluding beam weight) and live load of 2 k/ft. $f_c' = 4,000$ psi, $f_y = 60,000$ psi.

14.12 Design a simply supported rectangular reinforced beam for the service loads shown in Figure P14.12. Provide the reinforcement in a single layer. Sketch the design. $f_c' = 4,000$ psi, $f_y = 60,000$ psi.

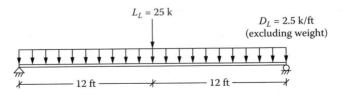

FIGURE P14.12   Beam loads for Problem 14.12.

14.13 Design a simply supported rectangular reinforced beam for the service loads shown in Figure P14.13. Provide the reinforcement in a single layer. Sketch the design. $f_c' = 3,000$ psi, $f_y = 40,000$ psi.

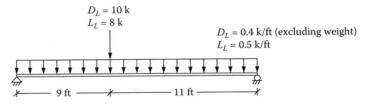

FIGURE P14.13   Beam loads for Problem 14.13.

14.14 Design the cantilever rectangular reinforced beam shown in Figure P14.14. Provide the maximum of #8 size bars, in two rows if necessary. Sketch the design. $f_c' = 3,000$ psi, $f_y = 50,000$ psi.
[*Hint*: Reinforcement will be at the top. Design as usual.]

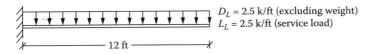

FIGURE P14.14   Cantilever beam for Problem 14.14.

14.15 Design the beam for the floor shown in Figure P14.15. The service dead load (excluding beam weight) is 100 psf and live load is 300 psf. $f_c' = 3,000$ psi, $f_y = 40,000$ psi.

14.16 A 9 in. thick one-way interior slab supports a service live load of 500 psf on a simple span of 15 ft. The main reinforcement consists of #7 bars at 7 in. on center. Check whether the slab can support the load in addition to its own weight. Use $f_c' = 3,000$ psi, $f_y = 60,000$ psi.

14.17 A one-way interior slab shown in Figure P14.17 spans 12 ft. Determine the service load that the slab can carry in addition to its own weight. $f_c' = 3,000$ psi, $f_y = 40,000$ psi.

14.18 A one-way slab, exposed to the weather, has a thickness of 9 in. The main reinforcement consists of #8 bars at 7 in. on center. The slab carries a dead load of 500 psf in addition to its own weight on a span of 10 ft. What is the service live load that the slab can carry. $f_c' = 4,000$ psi, $f_y = 60,000$ psi.

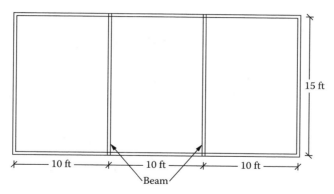

FIGURE P14.15 Floor system for Problem 14.15.

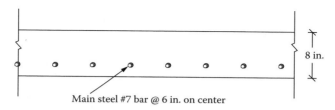

FIGURE P14.17 Cross section of slab for Problem 14.17.

14.19 A 8-½ in. thick one-way slab interior spans 10 ft. It was designed with the reinforcement of #6 bars at 6.5 in. on center, to be placed with a cover of 0.75 in. However, the same steel was misplaced at a clear distance of 2 in. from the bottom. How much is the reduction in the capacity of the slab to carry the superimposed service live load in addition to its own weight. $f_c' = 4,000$ psi, $f_y = 60,000$ psi.

14.20 Design a simply supported one-way interior slab to span 15 ft and to support the service dead and live loads of 150 and 250 psf in addition to its own weight. Sketch the design. $f_c' = 4,000$ psi, $f_y = 50,000$ psi.

14.21 Design the concrete floor slab as shown in Figure P14.21. Sketch the design. $f_c' = 3,000$ psi, $f_y = 40,000$ psi.

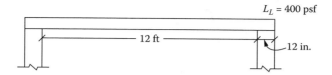

FIGURE P14.21 One-way slab span for Problem 14.21.

14.22 Design the slab of the floor system of Problem 14.15. $f_c' = 3,000$ psi, $f_y = 40,000$ psi.
  [*Hint*: The slab weight is included in the service dead load.]

14.23 For Problem 14.15, design the thinnest slab so that the strain in steel is not less 0.005. $f_c' = 3,000$ psi, $f_y = 40,000$ psi.

14.24 Design a balcony slab exposed to the weather. The cantilevered span is 8 ft and the service live load is 100 psf. Sketch the design. $f_c' = 4,000$ psi, $f_y = 60,000$ psi.
  [*Hint*: Reinforcement is placed on top. For thickness of slab, in addition to the provision of main steel and shrinkage steel, at least 3 in. of depth (cover) should exist over and below the steel.]

15 Shear and Torsion in Reinforced Concrete

STRESS DISTRIBUTION IN BEAM

The transverse loads on a beam segment cause a bending moment and a shear force that vary across the beam cross section and along the beam length. At point (1) in a beam shown in Figure 15.1, these contribute to the bending (flexure) stress and the shear stress, respectively, expressed as follows:

$$f_b = \frac{My}{I} \tag{15.1a}$$

and

$$f_v = \frac{VQ}{Ib} \tag{15.1b}$$

where
- M is the bending moment at a horizontal distance X
- y is the vertical distance of point (1) from the neutral axis
- I is the moment of inertia of section
- V is the shear force at X
- Q is the moment taken at the neutral axis of the cross-sectional area of beam above point (1)
- b is the width of section at (1)

The distribution of these stresses is shown in Figure 15.1. At any point (2) on the neutral axis, the bending stress is zero and the shear stress is maximum (for a rectangular section). On a small element at point (2), the vertical shear stresses act on the two faces balancing each other, as shown in Figure 15.2. According to the laws of mechanics, the complementary shear stresses of equal magnitude and opposite sign act on the horizontal faces as shown, so as not to cause any rotation to the element.

If we consider a free-body diagram along the diagonal a–b, as shown in Figure 15.3, and resolve the forces (shear stress times area) parallel and perpendicular to the plane a–b, the parallel force will cancel and the total perpendicular force acting in tension will be $1.414 f_v A$. Dividing by the area $1.41A$ along a–b, the tensile stress acting on plane a–b will be f_v. Similarly, if we consider a free-body diagram along the diagonal c–d, as shown in Figure 15.4, the total compression stress on the plane c–d will be f_v. Thus, the planes a–b and c–d are subjected to tensile stress and compression stress, respectively, that has a magnitude equal to the shear stress on horizontal and vertical faces. These stresses on the planes a–b and c–d are the principal stresses (since they are not accompanied any shear stress). The concrete is strong in compression but weak in tension. Thus, the stress on plane a–b, known as the *diagonal tension*, is of great significance. It is not the direct shear strength of concrete but the shear-induced diagonal tension that is considered in the analysis and design of concrete beams.

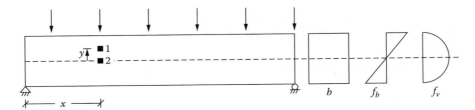

FIGURE 15.1   Transverse loaded beam.

DIAGONAL CRACKING OF CONCRETE

There is a tendency for concrete to crack along the plane sub-jected to tension when the level of stress exceeds a certain value. The cracks will form near the mid-depth where the shear stress (including the diagonal tension) is maximum and will move in a diagonal path to the tensile surface, as shown in Figure 15.5. These are known as the *web-shear cracks*. These are nearer to the support where shear is high. In a region where the moment is higher than the cracking moment capacity, the vertical flexure cracks will appear first and the diagonal shear cracks will develop as an extension to the flexure cracks. Such cracks are known as the *flexure-shear cracks*. These are more frequent in beams. The longitudinal (tensile) reinforcement does not prevent shear cracks but it restrains the cracks from widening up.

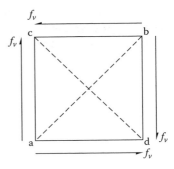

FIGURE 15.2   Shear stress on an element at neutral axis.

After a crack develops, the shear resistance along the cracked plane is provided by the following factors:

1. Shear resistance provided by the uncracked section above the crack, V_{cz}. This is about 20%–40% of the total shear resistance of cracked section.
2. Friction developed due to interlocking of the aggregates on opposite sides of the crack, V_a. This is about 30%–50% of the total.
3. Frictional resistance between concrete and longitudinal (main) reinforcement called the *dowel action*, V_d. This is about 15%–25% of the total.

In a deep beam, some tie–arch action is achieved by the longitudinal bars acting as a tie and the uncracked concrete above and to the sides of the crack acting as an arch.

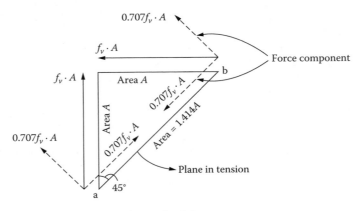

FIGURE 15.3   Free body diagram along plane a–b of element.

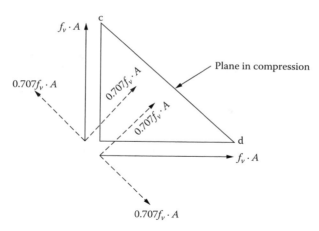

FIGURE 15.4 Free body diagram along plane c–d of element.

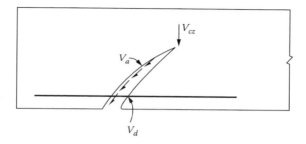

FIGURE 15.5 Shear resistance of cracked concrete.

Once the applied shear force exceeds the shear resistance offered by the above three factors in a cracked section, the beam will fail suddenly unless a reinforcement known as the *web* or *shear reinforcement* is provided to prevent the further opening up of the crack. It should be understood that the web reinforcement does not prevent the diagonal cracks that will happen at almost at the same loads with or without a web reinforcement. It is only after a crack develops that the tension which was previously held by the concrete is transferred to the web reinforcement.

STRENGTH OF WEB (SHEAR) REINFORCED BEAM

As stated above, the web reinforcement handles the tension that cannot be sustained by a diagonally cracked section. The actual behavior of web reinforcement is not clearly understood in spite of many theories presented. The truss analogy is the classic theory, which is very simple and widely used. The theory assumes that a reinforced concrete beam with web reinforcement behaves like a truss. A concrete beam with the vertical web reinforcement in a diagonally cracked section is shown in Figure 15.6. The truss members shown by dotted lines are superimposed in the Figure 15.6. The analogy between the beam and the truss members is as shown in Table 15.1.

According to the above concept, the web reinforcement represents the tensile member. Thus, the entire applied shear force that induces the diagonal tension is resisted by the web reinforcement only. But the observations have shown that the tensile stress in the web reinforcement was much smaller than the tension produced by the entire shear force. Accordingly, the truss analogy theory was modified to consider that the applied shear force is resisted by the two components: the web reinforcement and the cracked concrete section. Thus,

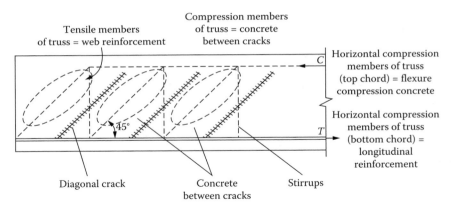

FIGURE 15.6   Truss analogy of beam.

TABLE 15.1
Beam–Truss Analogy

Truss	Beam
Horizontal tensile member (bottom chord)	Longitudinal reinforcement
Horizontal compression member (top chord)	Flexure compression concrete
Vertical tensile members	Web reinforcement
Diagonal compression members	Web concrete between the cracks in the compression zone

$$V_n = V_c + V_s \qquad (15.2)$$

Including a capacity reduction factor, ϕ

$$V_u \le \phi V_n \qquad (15.3)$$

For the limiting condition

$$V_u = \phi V_c + \phi V_s \qquad (15.4)$$

where
   V_n is the nominal shear strength
   V_u is the factored design shear force
   V_c is the shear contribution of concrete
   V_s is the shear contribution of web reinforcement
   ϕ is the capacity reduction factor for shear = 0.75 (Table 14.1)

Equation 15.4 serves as a design basis for web (shear) reinforcement design.

SHEAR CONTRIBUTION OF CONCRETE

Concrete (with flexure reinforcement but without web reinforcement) does not contribute in resisting the diagonal tension once the diagonal crack is formed. Therefore, the shear stress in concrete at the time of diagonal cracking can be assumed to be the ultimate strength of concrete in shear.

Many empirical relations have been suggested for the shear strength. The ACI has suggested the following relation:

$$V_c = 2\lambda\sqrt{f_c'}bd \tag{15.5}$$

The expression λ has been introduced in the ACI 2008 code, to account for lightweight concrete. An alternative much complicated expression has been proposed by the ACI, which is a function of the longitudinal reinforcement, bending moment, and shear force at various points of beam.

SHEAR CONTRIBUTION OF WEB REINFORCEMENT

The web reinforcement takes a form of stirrups that run along the face of a beam. The stirrups enclose the longitudinal reinforcement. The common types of stirrups, as shown in Figure 15.7, are ⊔ shaped or ⊔⊔ shaped that are arranged vertically or diagonally. When a significant amount of torsion is present, the closed stirrups are used, as shown in Figure 15.7c.

The strength of a stirrup of area A_v is $F_y A_v$. If n number of stirrups cross a diagonal crack, then the shear strength by stirrups across a diagonal will be

$$V_s = f_y A_v n \tag{15.6}$$

In a 45° diagonal crack, the horizontal length of crack equals the effective depth d, as shown in Figure 15.8. For stirrups spaced s on center, $n = d/s$. Substituting this in Equation 15.6:

$$V_s = f_y A_v \frac{d}{s} \tag{15.7}$$

where
  A_v is the area of stirrups
  s is the spacing of stirrups

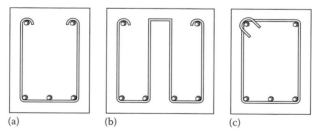

(a)   (b)   (c)

FIGURE 15.7 Types of stirrups. (a) Open stirrup, (b) double stirrup, and (c) closed stirrup.

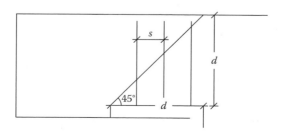

FIGURE 15.8 Vertical stirrups in a diagonal crack.

For a ⌴ shaped stirrup, A_v is twice the area of the bar and for a ⌴⌴ stirrup, A_v is four times the bar area.

When the stirrups are inclined at 45°, the shear force component along the diagonal will match the stirrups (web reinforcement) strength, or

$$V_s = 1.414 f_y A_v \frac{d}{s} \qquad (15.8)$$

Equations 15.7 and 15.8 can be expressed as a single relation:

$$V_s = \alpha f_y A_v \frac{d}{s} \qquad (15.9)$$

where, $\alpha = 1$ for the vertical stirrups, and 1.414 for the inclined stirrups.

SPECIFICATIONS FOR WEB (SHEAR) REINFORCEMENT

ACI 318-08 requirements for web reinforcement are summarized below:

1. According to Equation 15.4, when $V_u \leq \phi V_c$, no web reinforcement is necessary. However, the ACI code requires that a minimum web reinforcement should be provided when V_u exceeds $1/2\phi V_c$, except for slabs, shallow beams ($\leq$10 in.), and footing.
2. *Minimum steel*: When a web reinforcement is provided, its amount should fall between the specified lower and upper limits. The reinforcing should not be so low as to make the web reinforcement steel yield as soon as a diagonal crack develops. The minimum web reinforcement area should be *higher* of the following two values:

   a.
   $$(A_v)_{\min} = \frac{0.75\sqrt{f_c'} bs}{f_y} \qquad (15.10)$$

   or

   b.
   $$(A_v)_{\min} = \frac{50bs}{f_y} \qquad (15.11)$$

3. *Maximum steel*: The maximum limit of web reinforcement is set because the concrete will eventually disintegrate no matter how much steel is added. The upper limit is

   $$(A_v)_{\max} = \frac{8\sqrt{f_c'} bs}{f_y} \qquad (15.12)$$

4. *Stirrup size*: The most common stirrup size is #3 bar. Where the value of shear force is large, #4 size bar might be used. The use of larger than #4 size is unusual. For a beam width of $\leq$24 in., a single loop stirrup, ⌴ is satisfactory. Up to a width of 48 in. a double loop ⌴⌴ is satisfactory.
5. *Stirrup spacing*
   a. *Minimum spacing*: The vertical stirrups are generally not closer than 4 in. on center.
   b. *Maximum spacing* when $V_s \leq 4\sqrt{f_c'} bd \cdot$

The maximum spacing is smaller of the following:

1. $s_{max} = \dfrac{d}{2}$

2. $s_{max} = 24\,\text{in.}$

3. $s_{max} = \dfrac{A_v f_y}{0.75\sqrt{f_c'}\,b}$ (based on Equation 15.10)

4. $s_{max} = \dfrac{A_v f_y}{50b}$ (based on Equation 15.11)

c. *Maximum spacing* when $V_s > 4\sqrt{f_c'}\,bd$
   The maximum spacing is smaller of the following:

1. $s_{max} = \dfrac{d}{4}$

2. $s_{max} = 12\,\text{in.}$

3. $s_{max} = \dfrac{A_v f_y}{0.75\sqrt{f_c'}\,b}$ (based on Equation 15.10)

4. $s_{max} = \dfrac{A_v f_y}{50b}$ (based on Equation 15.11)

6. *Stirrups pattern*: The size of stirrups is held constant while the spacing of stirrups is varied. Generally the shear force decreases from the support toward the middle of the span indicating that the stirrups spacing can continually increase from the end toward the center. From a practical point of view, the stirrups are placed in groups; each group has the same spacing. Only two to three such groups of the incremental spacing are used within a pattern. The increment of spacing shall be in the multiple of whole inches perhaps in the multiple of 3 in. or 4 in.

7. *Critical section*: For a normal kind of loading where a beam is loaded at the top and there is no concentrated load applied within a distance d (effective depth) from the support, the section located at a distance d from the face of the support is called the *critical section*. The shear force at the critical section is taken as the design shear value V_u, and the shear force from the face of the support to the critical section is assumed to be the same as at the critical section.

Some designers place their first stirrup at a distance d from the face of the support while others place the first stirrup at one-half of the spacing calculated at the end.

When the support reaction is in tension at the end region of a beam or the loads are applied at bottom (to the tension flange), no design shear force reduction is permitted and the critical section is taken at the face of the support itself.

ANALYSIS FOR SHEAR CAPACITY

The process involves the following steps to check for the shear strength of an existing member and to verify the other code requirements:

1. Compute the concrete shear capacity by Equation 15.5.
2. Compute the web reinforcement shear capacity by Equation 15.9.

3. Determine the total shear capacity by Equation 15.4. This should be more than the applied factored shear force on the beam.
4. Check for the spacing of the stirrups from the "Specifications for Web (Shear) Reinforcement" section, step 5.

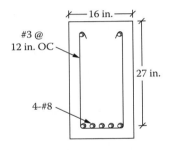

FIGURE 15.9 Concrete beam section of Example 15.1.

Example 15.1

Determine the factored shear force permitted on a reinforced concrete beam shown in Figure 15.9. Check for the web reinforcement spacing. Use $f'_c = 4,000\,\text{psi}$, $f_y = 60,000\,\text{psi}$.

Solution

A. Concrete shear capacity from Equation 15.5

$$V_c = 2\lambda\sqrt{f'_c}bd$$

$$= 2(1)\sqrt{4000}(16)(27) = 5464\,\text{lb or } 54.64\,\text{k}$$

B. Web shear capacity from Equation 15.9

$$A_v = 2(0.11) = 0.22\,\text{in.}^2$$

$$V_s = \alpha f_y A_v \frac{d}{s}$$

$$= 1(600,000(0.22)\left(\frac{27}{12}\right) = 29,700\,\text{lb or } 29.7\,\text{k}$$

C. Design shear force from Equation 15.4

$$V_u = \phi V_c + \phi V_s$$

$$= 0.75(54.64) + 0.75(29.7) = 63.26\,\text{k}$$

D. For maximum spacing
1. $4\sqrt{f'_c}bd/1000 = 4\sqrt{4000}(16)(27)/1000 = 109.3\,\text{k}$
2. Since V_s of $29.7\,\text{k} < 109.3\,\text{k}$
   spacing is smaller of

   a. $\dfrac{d}{2} = \dfrac{27}{2} = 13.2\,\text{in} \leftarrow \text{controls} >$

   b. 24 in.

   c. $s_{max} = \dfrac{A_v f_y}{0.75\sqrt{f'_c}b}$

   $$= \frac{(0.22)(60,000)}{(0.75)\sqrt{4,000}(16)} = 17.4$$

d. $s_{max} = \dfrac{A_v f_y}{50b}$

$$= \frac{(0.22)(60,000)}{50(16)} = 16.5 \text{ in.}$$

DESIGN FOR SHEAR CAPACITY

A summary of the steps to design for web reinforcement is presented below:

1. Based on the factored loads and clear span, draw a shear force, V_u diagram.
2. Calculate the critical V_u at a distance d from the support and show this on the V_u diagram as the critical section.
3. Calculate $\phi V_c = (0.75)2\sqrt{f_c'}bd$ and draw a horizontal line at ϕV_c level on the V_u diagram. The portion of the V_u diagram above this line represents ϕV_s that has to be provided with stirrups.
4. Calculate $1/2\phi V_c$ and show it by a point on the V_u diagram. The stirrups are needed from the support to this point. Below the $1/2\phi V_c$ point on the diagram toward the center, no stirrups are needed.
5. Make the tabular computations indicated in steps 5, 6, and 7 for the theoretical stirrups spacing.

   Starting at the critical section, divide the span into a number of segments. Determine V_u at the beginning of each segment from the slope of the V_u diagram. At each segment, calculate V_s from the following rearranged Equation 15.4:

$$V_s = \frac{(V_u - \phi V_c)}{\phi}.$$

6. Calculate the stirrup spacing for a selected stirrup size at each segment from the following rearranged Equation 15.9:

$$s = \alpha f_y A_v \frac{d}{V_s} \quad (\alpha \text{ being 1 for vertical stirrup}).$$

7. Compute the maximum stirrup spacing from the equations at the "Specifications for Web (Shear) Reinforcement" section, step 5.
8. Draw a spacing vs. distance diagram from step 6. On this diagram, draw a horizontal line at the maximum spacing of step 7 and a vertical line from step 4 for the cut of limit of stirrup.
9. From the diagram, select a few groups of different spacing and sketch the design.

Example 15.2

The service loads on a reinforced beam are shown in Figure 15.10 along with the designed beam section. Design the web reinforcement. Use $f_c' = 4{,}000 \text{ psi}$ and $f_y = 60{,}000 \text{ psi}$.

Solution

A: V_u diagram
1. Weight of beam $= (15/12) \times (21/12) \times 1 \times (150/1000) = 0.33 \text{ k/ft}$.
2. $w_u = 1.2(3 + 0.33) = 4 \text{ k/ft}$

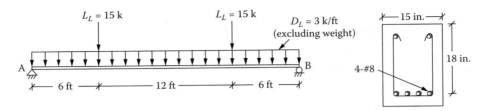

FIGURE 15.10   Loads and section of beam of Example 15.2.

3. $P_u = 1.6(15) = 24\,k$
4. $M @ B = 0$

   $R_A(24) - 4(24)(12) - 24(18) - 24(6) = 0$

   $R_A = 72\,k$
5. Shear force diagram is shown in Figure 15.11
6. V_u diagram for one-half span is shown in Figure 15.12

B. Concrete and steel strengths
1. Critical V_u at distance $= 72 - (18/12)(4) = 66\,k$
2. $\phi V_c = 0.75(2)\sqrt{f_c'}bd = 0.75(2)\sqrt{4000}(15)(18)/1000 = 25.61k$

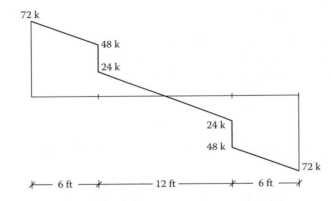

FIGURE 15.11   Shear force diagram of Example 15.2.

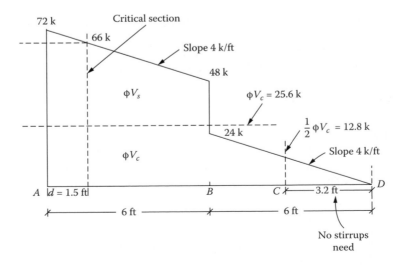

FIGURE 15.12 V_u diagram of Example 15.2.

3. $1/2\phi V_c = 12.81$ k
4. Distance from the beam center line to $(1/2)(\phi V_c/\text{slope}) = 12.81/4 = 3.2$ ft
C. Stirrups design: Use #3 stirrups

Distance from Support, x, ft	V_u, k	$V_s = \dfrac{V_u - \phi V_c}{\phi}$, k	$s = f_y A_v \dfrac{d}{V_s}$, in.
1	2	3[c]	4[d]
$D = 1.5$	66[a]	53.85	4.41
2	64	51.19	4.64
4	56	40.52	5.86
6⁻	48	29.85	7.96
6⁺	24[b]	0	∞

[a] $V_u = V_u@\text{end} - (\text{slope})(\text{distance}) = 72 - 4$ (Col. 1).
[b] $V_u = V_u@\text{B in Figure 15.12} - (\text{slope})(\text{distance} - 6) = 24 - 4$ (Col. 1 − 6).
[c] (Col. 2 − 25.61)/0.75.
[d] (0.22)(60,000)(18)/Col. 3 (1,000).

Distance vs. spacing from the above table are plotted on Figure 15.13.
D. Maximum spacing
1. $4\sqrt{f_c'}bd/1000 = 4\sqrt{4000}(15)(18)/1000 = 68.3$ k
2. $V_{s\ critical}$ of 66 k < 68.3 k
3. Maximum spacing is the smaller of

    a. $\dfrac{d}{2} = \dfrac{18}{2} = 9$ in. ← Controls

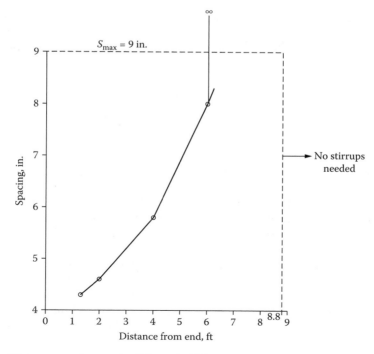

FIGURE 15.13 Distance–spacing graph of Example 15.2.

b. 24 in.

c. $S_{max} = \dfrac{A_v f_y}{0.75\sqrt{f'_c}\,b}$

$= \dfrac{(0.22)(60,000)}{(0.75)\sqrt{4,000}\,(15)} = 18.55$

d. $S_{max} = \dfrac{A_v f_y}{50b}$

$= \dfrac{(0.22)(60,000)}{50(15)} = 17.6\,\text{in.}$

The s_{max} line is shown on Figure 15.13.

E. Selected spacings

Distance Covered, ft	Spacing, in.	No. of Stirrups
0–5	4	15
5–6	6	2
6–8.8	9	4

TORSION IN CONCRETE

Torsion occurs when a member is subjected to a twist about its longitudinal axis due to a load acting off center of the longitudinal axis. Such a situation can be seen in a spandrel girder shown in Figure 15.14.

The moment developed at the end of the beam will produce a torsion in the spandrel girder. A similar situation develops when a beam supports a member that overhangs across the beam. An earthquake can cause substantial torsion to the members. The magnitude of torsion can be given by

$$T = Fr \qquad\qquad (15.13)$$

where

F is the force or reaction
r is the perpendicular distance of the force from the longitudinal axis

A load factor is applied to the torsion to convert T to T_u similar to the moment. A torsion produces torsional shear on all faces of a member. The torsional shear leads to diagonal tensile stress very similar to that caused the flexure shear. The concrete will crack along the spiral lines that will run at 45° from the faces of a member when this diagonal tension exceeds the strength of concrete. After the cracks develop, any additional torsion will make concrete to fail suddenly unless the torsional reinforcement is provided. Similar to the shear reinforcement, providing of a torsional reinforcement will not change the magnitude of the torsion at which the cracks will form. However, once the cracks are formed the torsional tension will be taken over by the torsional reinforcement to provide the additional strength against the torsional tension.

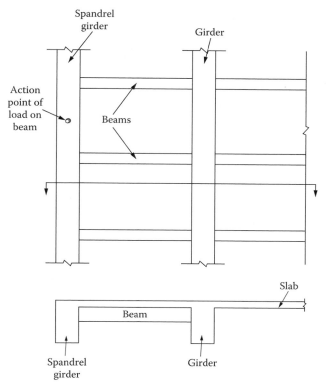

FIGURE 15.14   Beam subjected to torsion.

PROVISION FOR TORSIONAL REINFORCEMENT

The ACI 318-08 provides that as long as the factored applied torsion, T_u is less than one-fourth of the cracking torque T_r, the torsional reinforcement is not required. Equating T_u to one-fourth of cracking torque T_r, the threshold limit is expressed as

$$[T_u]_{\text{limit}} = \phi\sqrt{f_c'}\,\frac{A_{cp}^2}{P_{cp}} \tag{15.14}$$

where
  T_u is the factored design torsion
  A_{cp} is the area enclosed by the outside parameter of the concrete section = width × height
  P_{cp} is the outside parameter of concrete = 2 $(b+h)$
  $\phi = 0.75$ for torsion

  When T_u exceeds the above threshold limit, torsional reinforcement has to be designed. The process consists of performing the following computations:

1. To verify from Equation 15.14 that the cross-sectional dimensions of the member are sufficiently large to support the torsion acting on the beam.
2. If required, to design the closed loop stirrups to support the torsional tension ($T_u = \phi T_n$) as well as the shear-induced tension ($V_u = \phi V_n$).

3. To compute the additional longitudinal reinforcement to resist the horizontal component of the torsional tension. There must be a longitudinal bar in each corner of the stirrups.

When an appreciable torsion is present that exceeds the threshold value, it might be more expedient and economical to select a larger section than would normally be chosen, to satisfy Equation 15.14, so that the torsional reinforcement has not to be provided. The book uses this approach.

Example 15.3

The concentrated service loads, as shown in Figure 15.15, are located at the end of a balcony cantilever section, 6 in. to one side of the centerline. Is the section adequate without any torsional reinforcement? If not, redesign the section so that no torsional reinforcement has to be provided. Use $f'_c = 4,000\,\text{psi}$ and $f_y = 60,000\,\text{psi}$.

Solution
The beam is subjected to moment, shear force, and torsion. It is being analyzed for torsion only.

A. Checking the existing section:
1. Design load

$$P_u = 1.2(10) + 1.6(15) = 36\,\text{k}$$

2. Design torsion

$$T_u = 36\left(\frac{6}{12}\right) = 18\,\text{ft k}$$

3. Area enclosed by the outside parameter

$$A_{cp} = bh = 18 \times 24 = 432\,\text{in.}^2$$

4. Outside parameter

$$P_{cp} = 2(b + h) = 2(18 + 24) = 84\,\text{in.}$$

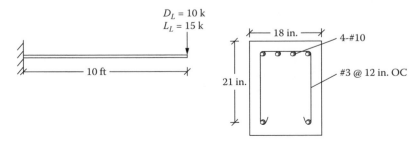

FIGURE 15.15   Cantilever beam and section of Example 15.3.

5. Torsional capacity of concrete

$$= \phi\sqrt{f_c'} \frac{A_{cp}^2}{P_{cp}}$$

$$= (0.75)\sqrt{4000} \frac{(432)^2}{84}$$

$$= 105{,}385.2 \text{ in.-lb or } 8.78 \text{ ft-k} < 18 \text{ k } \textbf{NG}$$

B. Redesigning the section
   1. Assume a width of 24 in.
   2. Area enclosed by the outside parameter $A_{cp} = (24h)$
   3. Parameter enclosed $P_{cp} = 2(24 + h)$
   4. Torsional capacity $= \phi\sqrt{f_c'} \frac{A_{cp}^2}{P_{cp}}$

$$= (0.75)\sqrt{4000} \frac{(24h)^2}{2(24+h)}$$

$$= 13{,}661\frac{h^2}{(24+h)} \text{ in.-lb or } 1.138 \frac{h^2}{(24+h)} \text{ ft-k}$$

5. For no torsional reinforcement

$$T_u = \phi\sqrt{f_c'} \frac{A_{cp}^2}{P_{cp}}$$

or

$$18 = 1.138\frac{h^2}{(24+h)}$$

or

$$h = 29 \text{ in.}$$

A section 24×29 will be adequate.

PROBLEMS

15.1–15.3 Determine the concrete shear capacity, web reinforcement shear capacity, and design shear force permitted on the beam sections shown in Figures P15.1 through P15.3. Check for web reinforcement. Use $f_c' = 3{,}000 \text{ psi}$ and $f_y = 40{,}000 \text{ psi}$.

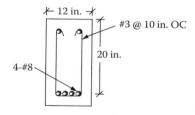

FIGURE P15.1 Beam section.

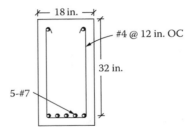

FIGURE P15.2 Beam section.

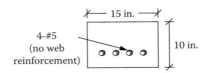

FIGURE P15.3 Beam section.

15.4 A reinforced beam of span 20 ft shown in Figure P15.4 is subjected to a dead load of 1 k/ft (excluding beam weight) and live load of 2 k/ft. Is the beam satisfactory to resist the maximum shear force? Use $f_c' = 3,000$ psi and $f_y = 60,000$ psi.

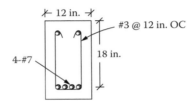

FIGURE P15.4 Beam section.

15.5 If the service dead load (excluding beam) is one-half of the service live load on the beam of span 25 ft shown in Figure P15.5. What are the magnitude of these loads from shear consideration? Use $f_c' = 4,000$ psi and $f_y = 60,000$ psi.

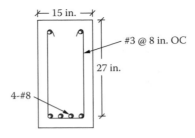

FIGURE P15.5 Load capacity of beam.

15.6 A simply supported beam is 15 in. wide and has an effective depth of 24 in. It supports a total factored load of 10 k/ft (including the beam weight) on a clear span of 22 ft. Design the web reinforcement. Use $f_c' = 4,000$ psi and $f_y = 60,000$ psi.

15.7 Design the web reinforcement for the service loads shown in Figure P15.7. Use $f_c' = 4,000$ psi and $f_y = 60,000$ psi.

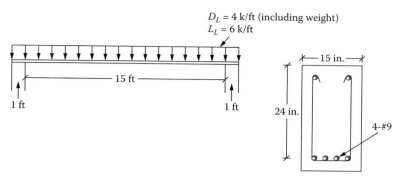

FIGURE P15.7   Loads and section of beam.

15.8   For the beam and service loads shown in Figure P15.8, design the web reinforcement use #4 stirrups. Use $f'_c = 5,000\,\text{psi}$ and $f_y = 60,000\,\text{psi}$.

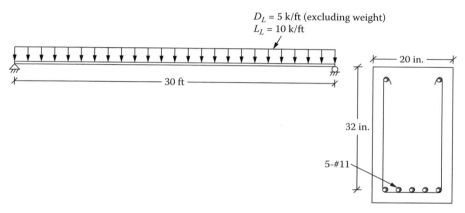

FIGURE P15.8   Loads and section of beam.

15.9   For the service loads on a beam (excluding beam weight) shown in Figure P15.9, design the web reinforcement. Use $f'_c = 4,000\,\text{psi}$ and $f_y = 50,000\,\text{psi}$.

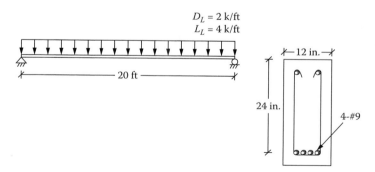

FIGURE P15.9   Loads and section of beam.

15.10   Design the web reinforcement for the service loads on the beam shown in Figure P15.10. Use $f'_c = 3,000\,\text{psi}$ and $f_y = 40,000\,\text{psi}$.

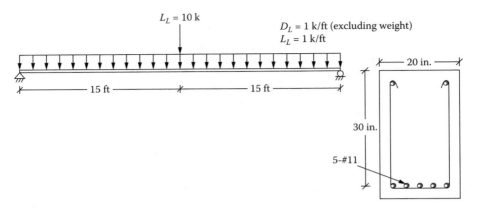

FIGURE P15.10   Loads and section of beam.

15.11 A simply supported beam carries the service loads (excluding the beam weight), as shown in Figure P15.11. Design the web reinforcement. Use $f'_c = 4,000\,\text{psi}$ and $f_y = 60,000\,\text{psi}$.

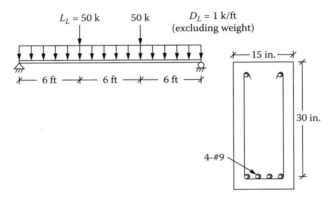

FIGURE P15.11   Loads and section of beam.

15.12 A simply supported beam carries the service loads (excluding the beam weight), as shown in Figure P15.12. Design the web reinforcement. Use #4 size stirrups. Use $f'_c = 4,000\,\text{psi}$ and $f_y = 60,000\,\text{psi}$.

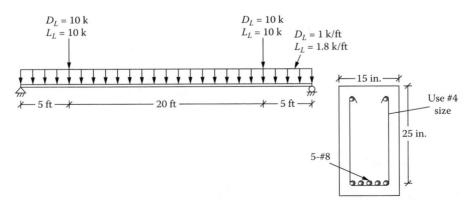

FIGURE P15.12   Loads and section of beam.

15.13 A cantilever beam carries the service loads including the beam weight, as shown in Figure P15.13. Design the web reinforcement. Use $f_c' = 4,000\,\text{psi}$ and $f_y = 60,000\,\text{psi}$.

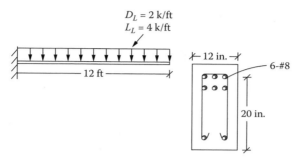

FIGURE P15.13   Loads and section of cantilever beam.

15.14 A beam carries the factored loads (including beam weight), as shown in Figure P15.14. Design the #3 size web reinforcement. Use $f_c' = 3,000\,\text{psi}$ and $f_y = 40,000\,\text{psi}$.

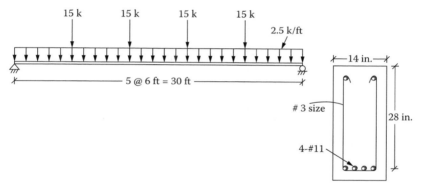

FIGURE P15.14   Loads and section of beam.

15.15 A beam supported on the walls carries the uniform distributed loads and the concentrated loads from the upper floor, as shown in Figure P15.15. The loads are service loads

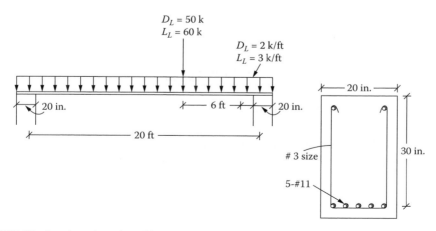

FIGURE P15.15   Loads and section of beam.

including the weight of beam. Design the #3 size web reinforcement. Use $f'_c = 4,000$ psi and $f_y = 50,000$ psi.

15.16 Determine the torsional capacity of the beam section in Figure P15.16 without torsional reinforcement. Use $f'_c = 4,000$ psi and $f_y = 60,000$ psi.

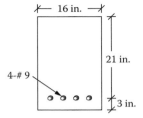

FIGURE P15.16   Beam section under torsion.

15.17 Determine the torsional capacity of the cantilever beam section shown in Figure P15.17 without torsional reinforcement. Use $f'_c = 3,000$ psi and $f_y = 40,000$ psi.

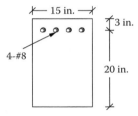

FIGURE P15.17   Cantilever section under torsion.

15.18 A spandrel beam shown in Figure P15.18 is subjected to a factored torsion of 8 ft-k. Is this beam adequate if no torsional reinforcement is used? If not redesign the section. The width cannot exceed 16 in. Use $f'_c = 4,000$ psi and $f_y = 50,000$ psi.

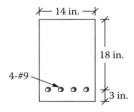

FIGURE P15.18   Beam section under torsion.

15.19 Determine the total depth of a 24 in. wide beam if no torsional reinforcement is used. The service loads, as shown in Figure P15.19, act 5 in. to one side of the center line. Use $f'_c = 4,000$ psi and $f_y = 60,000$ psi.

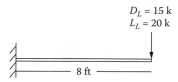

$D_L = 15\,\text{k}$
$L_L = 20\,\text{k}$

8 ft

FIGURE P15.19   Torsion loads on cantilever.

15.20 A spandrel beam is exposed to a service dead load of 8 k and live load of 14 k acting 8 in. off center of the beam. The beam section 20 in. wide and 25 in. deep. Is the section adequate without torsional reinforcement? If not, redesign the section of the same width. Use $f_c' = 5{,}000\,\text{psi}$ and $f_y = 60{,}000\,\text{psi}$.

16 Compression and Combined Forces Reinforced Concrete Members

TYPES OF COLUMNS

Concrete columns are divided into four categories.

PEDESTALS

The column height is less than three times of the least lateral dimension. A pedestal is designed with plain concrete (without reinforcement) for a maximum compression strength of 0.85 $\phi f_c' A_g$, where ϕ is 0.65 and A_g is the cross-sectional area of the column.

COLUMNS WITH AXIAL LOADS

The compressive load acts coinciding with the longitudinal axis of the column or at a small eccentricity so that there is no induced moment or there is a moment of little significance. This is a basic case although not quite common in practice.

SHORT COLUMNS WITH COMBINED LOADS

The columns are subjected to an axial force and a bending moment. However, the buckling effect is not present and the failure is initiated by crushing of the material.

LARGE OR SLENDER COLUMNS WITH COMBINED LOADS

Due to the axial load P buckling a column axis by an amount Δ, the column is subjected to the secondary moment or the P–Δ moment.

Since concrete and steel both can share compression loads, steel bars directly add to the strength of a concrete column. The compression strain is equally distributed between concrete and steel that are bonded together. It causes a lengthwise shortening and a lateral expansion of the column due to Poisson's effect. The column capacity can be enhanced by providing a lateral restraint. The column is known as a *tied* or a *spiral* column depending upon whether the lateral restraint is in the form of the closely spaced ties or the helical spirals wrapped around the longitudinal bars, as shown in Figure 16.1a and b.

Tied columns are ordinarily square or rectangular and spiral columns are round but they could be otherwise too. The spiral columns are more effective in terms of the column strength because of their hoop stress capacity. But they are more expensive. As such tied columns are more common and spiral columns are used only for heavy loads.

The *composite columns* are reinforced by steel shapes that are contained within the concrete sections or by concrete being filled in within the steel section or tubing as shown in Figure 16.1c and d. The latter are commonly called the *lally* columns.

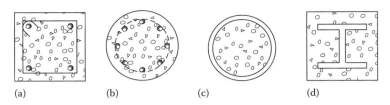

FIGURE 16.1 Types of column. (a) Tied column, (b) spiral column, (c) and (d) composite column.

AXIALLY LOADED COLUMNS

This category includes the columns with a small eccentricity. The small eccentricity is defined when the compression load acts at a distance, e, from the longitudinal axis controlled by the following conditions:

$$\text{For spiral columns:} \quad e \leq 0.05h \tag{16.1}$$

$$\text{For tied columns:} \quad e \leq 0.1h \tag{16.2}$$

where h is the column dimension along distance, e.

In the case of columns, unlike beams, it does not matter whether the concrete or steel reaches ultimate strength first because both of them deform/strain together that distributes the matching stresses between them.

Also, the high strength is more effective in columns because the entire concrete area contributes to the strength unlike the contribution from the concrete in compression zone only in beams, which is about 30%–40% of the total area.

The basis of design is the same as for wood or steel columns, i.e.,

$$P_u \leq \phi P_n \tag{16.3}$$

where
P_u is the factored axial load on column
P_n is the nominal axial strength
ϕ = the strength reduction factor
  = 0.70 for spiral column
  = 0.65 for tied column

The nominal strength is the sum of the strength of concrete and the strength of steel. The concrete strength is the ultimate (uniform) stress $0.85f'_c$ times the concrete area $(A_g - A_{st})$ and the steel strength is the yield stress, f_y times the steel area, A_{st}. However, to account for the small eccentricity, a factor (0.85 for spiral and 0.8 for tied) is applied.

Thus,

$$P_n = 0.85[0.85f'_c(A_g - A_{st}) + f_y A_{st}] \text{ for spiral columns} \tag{16.4}$$

$$P_n = 0.80[0.85f'_c(A_g - A_{st}) + f_y A_{st}] \text{ for tied columns} \tag{16.5}$$

Including a strength reduction factor of 0.7 for spiral and 0.65 for tied columns in the above equations, Equation 16.3 for column design is as follows:

For spiral columns with $e \leq 0.05h$

$$P_u = 0.60[0.85 f_c'(A_g - A_{st}) + f_y A_{st}] \tag{16.6}$$

For tied columns with $e \leq 0.1\,h$

$$P_u = 0.52[0.85 f_c'(A_g - A_{st}) + f_y A_{st}] \tag{16.7}$$

STRENGTH OF SPIRALS

It could be noticed that a higher factor is used for spiral columns than tied columns. The reason is that in a tied column, as soon as the shell of a column spalls off, the longitudinal bars will buckle immediately with the lateral support gone. But a spiral column will continue to stand and resist more load with the spiral and longitudinal bars forming a cage to confine the concrete.

Because the utility of a column is lost once its shell spalls off, the ACI assigns only a slightly more strength to the spiral as compared to strength of the shell that gets spalled off.

With reference to Figure 16.2,

$$\text{Strength of shell} = 0.85 f_c'(A_g - A_c) \tag{a}$$

$$\text{Hoop tension in spiral} = 2 f_y A_{sp} = 2 f_y \rho_s A_c \tag{b}$$

where ρ_s is the spiral steel ratio $= A_{sp}/A_c$.

Equating the two expressions (a) and (b) and solving for ρ_s

$$\rho_s = 0.425 \frac{f_c'}{f_y}\left(\frac{A_g}{A_c} - 1\right) \tag{c}$$

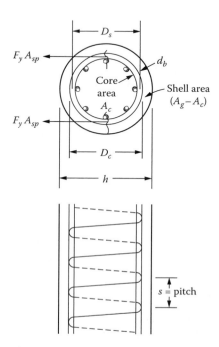

FIGURE 16.2   Spiral column section and profile.

Making the spiral a little stronger

$$\rho_s = 0.45 \frac{f'_c}{f_y} \left(\frac{A_g}{A_c} - 1 \right)$$ (16.8)

Once the spiral steel is determined, the following expression derived from the definition of ρ_s is used to set the spacing or pitch of spiral.

By definition, from Figure 16.2,

$$\rho_s = \frac{\text{volume of spiral in one loop}}{\text{volume of concrete in pitch}, s}$$ (d)

$$= \frac{\pi(D_c - d_b) A_{sp}}{(\pi D_c^2/4) s}$$ (e)

If the diameter difference, i.e., d_b is neglected,

$$\rho_s = \frac{4 A_{sp}}{D_c s}$$ (f)

or

$$s = \frac{4 A_{sp}}{D_c \rho_s}$$ (16.9)

Appendix D.13, based on Equations 16.8 and 16.9, can be used to select the size and pitch of spirals for a given diameter of a column.

SPECIFICATIONS FOR COLUMNS

1. *Main steel ratio*: The steel ratio, ρ_g should not be less than 0.01 (1%) and not more than 0.08. Usually a ratio of 0.03 is adopted.
2. *Minimum number of bars*: A minimum of four bars are used within the rectangular or circular ties and six within the spirals.
3. *Cover*: A minimum cover over the ties or spiral shall be 1-½ in.
4. *Spacing*: The clear distance between the longitudinal bars should neither be less than 1.5 times the bar diameter nor 1-½ in. To meet these requirements, Appendix D.14 can be used to determine the maximum number of bars that can be accommodated in one row for a given size of a column.
5. *Ties requirements*:
   a. The minimum size of the tie bars is #3 when the size of longitudinal bars is #10 or smaller or when the column diameter is 18 in. or less. The minimum size is #4 for the longitudinal bars larger than #10 or the column larger than 18 in. Usually, #5 is a maximum size.
   b. The center-to-center spacing of ties should be smaller of the following
      i. 16 times the diameter of longitudinal bars
      ii. 48 times the diameter of ties
      iii. Least column dimension
   c. The ties shall be so arranged that every corner and alternate longitudinal bar will have the lateral support provided by the corner of a tie having an included angle of not more than 135°. Figure 16.3 shows the tie arrangements for several columns.

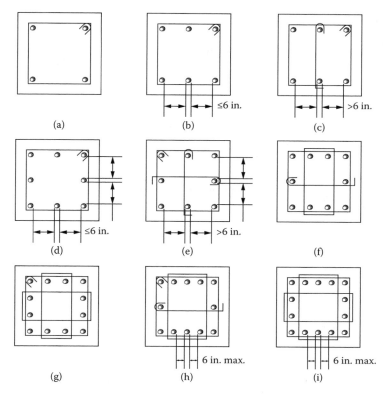

FIGURE 16.3   Tie arrangements.

   d.   No longitudinal bar shall have more than 6 in. clear distance on either side of a tie. If it
      is more than 6 in., a tie is provided as shown in Figure 16.3c and e.
 6. *Spiral requirements*:
   a.   The minimum spiral size is 3/8 in. (#3). Usually the maximum size is 5/8 in. (#5).
   b.   The clear space between spirals should not be less than 1 in. or not more than 3 in.

ANALYSIS OF AXIALLY LOADED COLUMNS

The analysis of columns of small eccentricity comprises of determining the maximum design load
capacity and verifying the amount and details of the reinforcement according to the code. The procedure is summarized below:

 1. Check that the column meets the eccentricity requirement ($\leq 0.05h$ for spiral and $\leq 0.1h$ for
    tied column)
 2. Check that the steel ratio, ρ_g is within 0.01–0.08
 3. Check that there are at least four bars for a tied column and six bars for a spiral column
    and that the clear spacing between bars is according to the "Specifications for Columns—
    Spacing" section
 4. Calculate the design column capacity by Equations 16.6 or 16.7
 5. For ties, check for the size, spacing, and arrangement according to the "Specifications for
    Columns—Ties Requirements" section. For spiral, check for the size and spacing, accord-
    ing to the "Specifications for Columns—Spiral Requirements" section.

Example 16.1

Determine the design axial load on a 16 in. square axially loaded column reinforced with eight #8 size bars. Ties are #3 at 12 in. on center. Use $f'_c = 4,000$ psi and $f_y = 60,000$ psi.

Solution

1. $A_{st} = 6.32$ in.^{2} (From Appendix D.2)
2. $A_g = 16 \times 16 = 256$ in.^{2}
3. $\rho_g = \dfrac{A_{st}}{A_g} = \dfrac{6.32}{256} = 0.0247$

   This is >0.01 and <0.08   **OK**
4. $h = 2(\text{cover}) + 2(\text{ties diameter}) + 3(\text{bar diameter}) + 2(\text{spacing})$
   or $16 = 2(1.5) + 2(0.375) + 3(1) + 2(s)$
   or $s = 4.625$ in.
   $s_{min} = 1.5(1) = 1.5$ in.   **OK**
   $s_{max} = 6$ in.   **OK**
5. From Equation 16.7
   $P_u = 0.52[0.85(4,000)(256 - 6.32) + (60,000)(6.32)]/1,000$
   $\qquad = 638.6\,k$
6. Check the ties
   a. #3 size   **OK**
   b. The spacing should be smaller of the following:
       i. $16 \times 1 = 16$ in. ← Controls
       ii. $48 \times 0.375 = 18$ in.
       iii. 16 in.
   c. Clear distance from the tie = 4.625 in. (Step 4) < 6 in.   **OK**

Example 16.2

A service dead load of 150 k and live load of 220 k is axially applied on a 15 in. diameter circular spiral column reinforced with 6-#9 bars. The lateral reinforcement consists of 3/8 in. spiral at 2 in. on center. Is the column adequate? Use $f'_c = 4,000$ psi and $f_y = 60,000$ psi.

Solution

1. $A_{st} = 6$ in.^{2} (From Appendix D.2)
2. $A_g = \dfrac{\pi}{4}(15)^2 = 176.63$ in.^{2}
3. $\rho_g = \dfrac{A_{st}}{A_g} = \dfrac{6}{176.63} = 0.034$

   This is >0.01 and <0.08:   **OK**
4. $(D_c - d_b) = h - 2\,(\text{cover}) - 2\,(\text{spiral diameter})$
   $\qquad = 15 - 2(1.5) - 2(0.375) = 11.25$ in.
5. Perimeter, $p = \pi(D_c - d_b) = \pi\,(11.25) = 35.33$ in.
   $p = 6(\text{bar diameter}) + 6(\text{spacing})$
   or $35.33 = 6(1.128) + 6(s)$
   or $s = 4.76$ in.
   $s_{min} = 1.5(1) = 1.5$ in.   **OK**
   $s_{max} = 6$ in.   **OK**
6. $\phi\, P_n = 0.60[0.85(4,000)(176.63 - 6) + (60,000)(6)]/1,000 = 564\,k$
7. $P_u = 1.2(150) + 1.6(220) = 532\,k < 564\,k$ **OK**
8. Check for spiral
   a. 3/8 in. diameter   **OK**
       $D_c = 15 - 3 = 12$ in.

b. $A_c = \dfrac{\pi}{4}(12)^2 = 113.04\,\text{in.}^2$

$A_{sp} = 0.11\,\text{in.}^2$

From Equation 16.8

$$\rho_s = 0.45 \dfrac{(4)}{(60)}\left(\dfrac{176.63}{113.04} - 1\right) = 0.017$$

From Equation 16.9

$$s = \dfrac{4(0.11)}{(12)(0.017)} = 2.16\text{ in.} > 2\text{ in. (given) } \textbf{OK}$$

c. Clear distance $= 2 - 3/8 = 1.625$ in. > 1 in. **OK**

DESIGN OF AXIALLY LOADED COLUMNS

Design involves fixing of the column dimensions, selecting of reinforcement, and deciding the size and spacing of ties and spirals. For a direct application, Equations 16.6 and 16.7 are rearranged as follows by substituting $A_{st} = \rho_g A_g$.
For spiral columns:

$$P_u = 0.60 A_g[0.85 f_c'(1-\rho_g) + f_y \rho_g] \tag{16.10}$$

For tied columns:

$$P_u = 0.52 A_g[0.85 f_c'(1-\rho_g) + f_y \rho_g] \tag{16.11}$$

The design procedure comprises of the following:

1. Determine the factored design load for various load combinations.
2. Assume $\rho_g = 0.03$. A lower or higher value could be taken depending upon the bigger or smaller size of column being acceptable.
3. Determine the gross area, A_g from Equations 16.10 or 16.11. Select the column dimensions to a full-inch increment.
4. For the actual gross area, calculate the adjusted steel area from Equations 16.6 or 16.7. Make the selection of steel using Appendix D.2 and check from Appendix D.14 that the number of bars can fit in a single row of the column.
5. (For spirals) select the spiral size and pitch from Appendix D.13. (For ties) select the size of tie, decide the spacing, and arrange ties by the "Specifications for Columns" section.
6. Sketch the design.

Example 16.3

Design a tied column for an axial service dead load of 200 k and service live load of 280 k. Use $f_c' = 4,000$ psi and $f_y = 60,000$ psi.

Solution

1. $P_u = 1.2(200) + 1.6(280) = 688\,k$
2. For $\rho_g = 0.03$, from Equation 16.11

$$A_g = \frac{P_u}{0.52[0.85f'_c(1-\rho_g) + f_y\rho_g]}$$

$$= \frac{P_u}{0.52[0.85(4)(1-0.03) + 60(0.03)]}$$

$$= 259.5\ \text{in.}^2$$

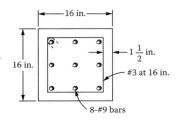

FIGURE 16.4 Tied column section of Example 16.3.

For a square column, $h = \sqrt{259.5} = 16.1\ \text{in.}$, use 16 in. × 16 in., $A_g = 256\ \text{in.}^2$

3. From Equation 16.7
   $688 = 0.52\,[0.85(4)(256 - A_{st}) + 60(A_{st})]$
   or $688 = 0.52(870.4 + 56.6A_{st})$
   or $A_{st} = 8\ \text{in.}^2$
   Select 8 bars of #9 size, A_{st} (provided) $= 8\ \text{in.}^2$
   From Appendix D.14, for a core size of $16 - 3 = 13$ in., 8 bars of #9 can be arranged in a row.
4. Design of ties:
   a. Select #3 size
   b. Spacing should be smaller of the following:
       i. $16(1.128) = 18$ in.
       ii. $48(0.375) = 18$ in.
       iii. 16 in. ← Controls
   c. Clear distance
       $16 = 2(\text{cover}) + 2(\text{ties diameter}) + 3(\text{bar diameter}) + 2(\text{spacing})$
       $16 = 2(1.5) + 2\,(0.375) + 3(1.128) + 2s$
       or $s = 4.43$ in. < 6 in.   **OK**
5. Sketch shown in Figure 16.4.

Example 16.4

For Example 16.3, design a circular spiral column.

Solution

1. $P_u = 1.2(200) + 1.6(280) = 688\,k$
2. For $\rho_g = 0.03$, from Equation 16.10

$$A_g = \frac{P_u}{0.60[0.85f'_c(1-\rho_g) + f_y\rho_g]}$$

$$= \frac{P_u}{0.60[0.85(4)(1-0.03) + 60(0.03)]}$$

$$= 225\ \text{in.}^2$$

For a circular column, $\dfrac{\pi h^2}{4} = 225$, $h = 16.93\ \text{in.}$, use 17 in., $A_g = 227\ \text{in.}^2$

3. From Equation 16.6
   $688 = 0.60[0.85(4)(227 - A_{st}) + 60(A_{st})]$
   or $A_{st} = 8\ \text{in.}^2$
   Select 8 bars of #9 size, A_{st} (provided) $= 8\ \text{in.}^2$

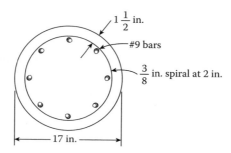

FIGURE 16.5　Spiral column section of Example 16.5.

From Appendix D.14, for a core size of $17 - 3 = 14$ in., 9 bars of #9 can be arranged in a single row. **OK**

4. Design of spiral:
   a. From Table D.13, for 17 in. diameter column,
      spiral size = 3/8 in.
      pitch = 2 in.
   b. Clear distance
      $2 - 0.375 = 1.625$ in. > 1 in. **OK**
5. Sketch shown in Figure 16.5.

SHORT COLUMNS WITH COMBINED LOADS

Most of the reinforced concrete columns belong to this category. The condition of an axial loading or a small eccentricity is rare. The rigidity of the connection between beam and column makes the column to rotate with the beam resulting in a moment at the end. Even an interior column of equally spanned beams will receive unequal loads due to variations in the applied loads, producing a moment on the column.

Consider that a load, P_u acts at an eccentricity, e, as shown in Figure 16.6a. Apply a pair of loads P_u, one acting up and one acting down through the column axis, as shown in Figure 16.5b. The applied loads cancel each other and, as such, have no technical significance. When we combine the

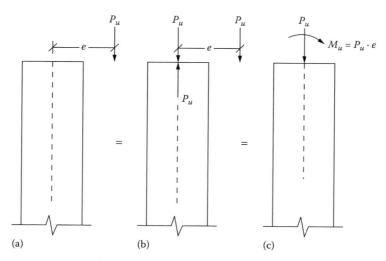

FIGURE 16.6　Equivalent force systems on a column. (a) Eccentric load on a column, (b) equivalent loaded column, and (c) column with load and equivalent moment.

load P_u acting down at an eccentricity, e with the load, P_u acting upward through the axis, a couple, $M_u = P_u e$ is produced. In addition, the downward load P_u acts through the axis. Thus, a system of force acting at an eccentricity is equivalent to a force and a moment acting through the axis, as shown in Figure 16.6c. Inverse to this, a force and a moment when acting together are equivalent to a force acting with an eccentricity.

As discussed with wood and steel structures, buckling is a common phenomenon associated with columns. However, concrete columns are stocky and a great number of columns are not affected by buckling. These are classified as the *short columns*. It is the slenderness ratio that determines whether a column could be considered a short or a slender (long) column. The ACI sets the following limits when the slenderness effects could be ignored:

a. For members not braced against sidesway:

$$\frac{Kl}{r} \leq 22 \tag{16.12}$$

b. For members braced against sidesway:

$$\frac{Kl}{r} \leq 34 - 12\left(\frac{M_1}{M_2}\right) \leq 0 \tag{16.13}$$

where
M_1 and M_2 are the small and large end moments. The ratio, M_1/M_2 is positive if a column bends in a single curvature, i.e., the end moments have the opposite signs. It is negative for a double curvature when the end moments have the same sign. (This is opposite of the sign convention in steel in the "Magnification Factor, B_1" section in Chapter 12.)
l is the length of column
K is the effective length factor given in Figure 7.6 and the alignment charts in Figures 10.5 and 10.6
r = the radius of gyration = $\sqrt{I/A}$
$\quad$ = $0.3h$ for rectangular column
$\quad$ = $0.25h$ for circular column

If a clear bracing system in not visible, the ACI provides certain rules to decide whether a frame is braced or unbraced. However, conservatively it can be assumed to be unbraced.

The effective length factor has been discussed in detailed in the "Effective Length Factor for Slenderness Ratio" section in Chapter 10. For columns braced against sidesway, the effective length factor is one or less; conservatively it can be used as 1. For members subjected to sidesway, the effective length factor is greater than 1. It is 1.2 for a column fixed at one end and the other end has the rotation fixed but is free to translate (sway).

EFFECTS OF MOMENT ON SHORT COLUMNS

To consider the effect of an increasing moment (eccentricity) together with an axial force on a column, the following successive cases have been presented accompanied with respective stress/strain diagrams.

ONLY AXIAL LOAD ACTING (CASE 1)

The entire section will be subjected to a uniform compression stress, $\sigma_c = P_u/A_g$ and a uniform strain of $\varepsilon = \sigma_c/E_c$, as shown in Figure 16.7. The column will fail by the crushing of concrete. By another

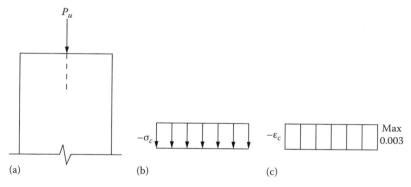

FIGURE 16.7   Axial load only on column. (a) Load on column, (b) stress, and (c) strain.

measure the column will fail when the compressive concrete strain reaches 0.003. In other cases, the strain measure will be considered because the strain diagrams are linear. The stress variations in concrete are nonlinear.

LARGE AXIAL LOAD AND SMALL MOMENT (SMALL ECCENTRICITY) (CASE 2)

Due to axial load there is a uniform strain, $-\varepsilon_c$ and due to moment, there is a bending strain of compression on one side and tension on the other side. The sum of these strains is shown in the last diagram of Figure 16.8d. Since the maximum strain due to the axial load and moment together cannot exceed 0.003, the strain due to the load will be smaller than 0.003 because a part of the contribution is made by the moment and correspondingly, the axial load P_u will be smaller than the previous case.

LARGE AXIAL LOAD AND MOMENT LARGER THAN CASE 2 SECTION (CASE 3)

This is a case when the strain is zero at one face. To attain the maximum crushing strain of 0.003 on the compression side, the strain contribution each by the axial load and moment will be 0.0015 (Figure 16.9).

LARGE AXIAL LOAD AND MOMENT LARGER THAN CASE 3 SECTION (CASE 4)

When the moment (eccentricity) increases somewhat from the previous case, the tension will develop on one side of the column as the bending strain will exceed the axial strain. The entire tensile strain

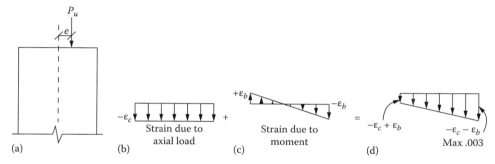

FIGURE 16.8   Axial load and small moment on column. (a) Load on column, (b) axial strain, (c) bending strain, and (d) combined strain.

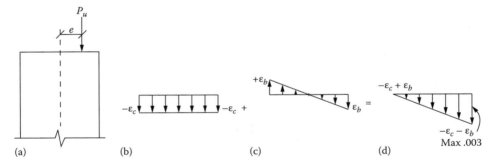

FIGURE 16.9 Axial load and moment (Case 3). (a) Load on column, (b) axial strain, (c) bending strain, and (d) combined strain.

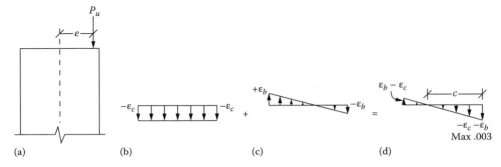

FIGURE 16.10 Axial load and moment (Case 4). (a) Load on column, (b) axial strain, (c) bending strain, and (d) combined strain.

contribution will come from steel.* The concrete on the compression side will contribute to compression strain. The strain diagram will be as shown in Figure 16.10d. The neutral axis (the point of zero strain) will be at a distance c from the compression face. Since the strain in steel is less than yielding, the failure will occur by crushing of concrete on the compression side.

BALANCED AXIAL LOAD AND MOMENT (CASE 5)

As the moment (eccentricity) continues to increase, the tensile strain steadily rises. A condition will be reached when the steel on the tension side will attain the yield strain, $\varepsilon_y = f_y/E$ (for Grade 60 steel, this strain is 0.002), simultaneously as the compression strain in concrete reaches to the crushing strain of 0.003. The failure of concrete will occur at the same time as steel yields. This is known as the *balanced condition*. The strain diagrams in this case is shown in Figure 16.11. The value of c in Figure 16.11d is less as compared to the previous case, i.e., the neutral axis moves up toward the compression side.

SMALL AXIAL LOAD AND LARGE MOMENT (CASE 6)

As the moment (eccentricity) is further increased, steel will reach to the yield strain, $\varepsilon_y = f_y/E$ before concrete attains the crushing strain of 0.003. In other words, when compared to the concrete strain of 0.003, the steel strain had already exceeded its yield limit, ε_y, as shown in Figure 16.12d. The failure will occur by yielding of steel. This is called the *tension-controlled condition*.

---

* The concrete being weak in tension, its contribution is neglected.

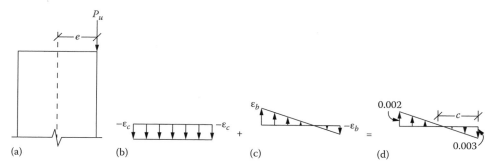

FIGURE 16.11 Balanced axial load and moment on column (Case 5). (a) Load on column, (b) axial strain, (c) bending strain, and (d) combined strain.

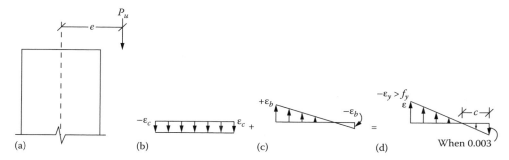

FIGURE 16.12 Small axial load and large moment on column (Case 6). (a) Load on column, (b) axial strain, (c) bending strain, and (d) combined strain.

No Appreciable Axial Load and Large Moment (Case 7)

This is the case when column acts as a beam. The eccentricity is assumed to be at infinity. The steel has long before yielded prior to concrete reaching a level of 0.003. In other words, when compared to a concrete strain of 0.003, the steel strain is 0.005 or more. This is shown in Figure 16.13b.

As discussed in the "Axially Loaded Column" section, when a member acts as a column, the strength (capacity) reduction factor, ϕ is 0.7 for the spiral columns and 0.65 for the tied columns. This is the situation for Cases 1 through 5. For beams, as in Case 7, the factor is 0.9. For Case 6, between the column and the beam condition, the magnitude of ϕ is adjusted by Equation 14.13, based on the value of strain in steel, ε_t.

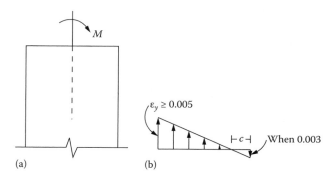

FIGURE 16.13 Moment only on column (Case 7). (a) Load on column and (b) combined strain.

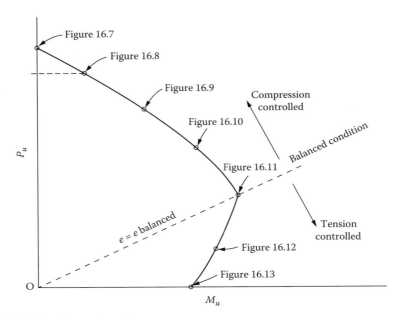

FIGURE 16.14 Column interaction diagram.

If the magnitudes of the axial loads and the moments for all seven cases are plotted, it will appear like a shape shown in Figure 16.14. This is known as the *interaction diagram.*

CHARACTERISTICS OF THE INTERACTION DIAGRAM

The interaction diagram presents the capacity of a column for various proportions of the loads and the moments. Any combination of loading that falls inside the diagram is satisfactory whereas any combination falling outside represents a failure condition.

From Cases 1 through 5 where compression control exists, as the axial load decreases the moment capacity increases. Below this stage, the position is different. First of all, for the same moment, the axial capacity is higher in the compression control zone than in the tensile control zone. Further in the tensile control zone, as the axial load increases the moment capacity also increases. This is for the fact that any axial compression load tends to reduce the tensile strain (and stress) which results in raising of the moment-resisting capacity.

Any radial line drawn from origin O to any point on the diagram represents a constant eccentricity, i.e., a constant ratio of the moment to the axial load. From O to a point on the diagram for the "Balanced Axial Load" section, represents the $e_{balanced}$ eccentricity.

For the same column, for different amounts of steel within the column, the shape of the curves will be similar (parallel).

The interaction diagram serves as a very useful tool in the analysis and design of columns for the combined loads.

APPLICATION OF THE INTERACTION DIAGRAM

The ACI has prepared the interaction diagrams in the dimensionless units for rectangular and circular columns with the different arrangements of bars for various grades of steel and various strengths of concrete. The abscissa has been represented as $R_n = M_u/\phi f_c' A_g h$ and the ordinate as $K_n = P_u/\phi f_c' A_g$. Several of these diagrams for concrete strength of 4,000 psi and steel strength of 60,000 psi are included in Appendices D.15 through D.22.

On these diagrams, the radial strain line of value=1 represents the balanced condition. Any point on or above this line represents the compression control and ϕ=0.7 (spiral) or 0.65 (tied). Similarly line of ε_t=0.005 represents the "steel has yielded" or beam behavior. Any point on or below this line will have ϕ=0.9. In between these two lines, is the transition zone for which ϕ has to be corrected by Equation 14.13.

The line labeled K_{max} indicates the maximum axial load with the limiting small eccentricity of 0.05h for spiral and 0.1h for tied columns.

The other terms in these diagrams are

1. $\rho_g = \dfrac{A_{st}}{A_g}$

2. h=column dimension in line with eccentricity (perpendicular to the plane of bending)

3. $\gamma = \dfrac{\text{center-to-center distance of outer row of steel}}{h}$ (16.14)

4. Slope of radial line from origin=h/e

ANALYSIS OF SHORT COLUMNS FOR COMBINED LOADING

This involves determining the axial load strength and the moment capacity of a known column. The steps comprise the following:

1. From Equations 16.12 or 16.13, confirm that that it is a short column (there is no slenderness effect).
2. Calculate the steel ratio, $\rho_g = A_{st}/A_g$ and check for the value to be between 0.01 and 0.08.
3. Calculate, γ from Equation 16.14
4. Select the right interaction diagram to be used based on γ, type of cross section, f_c' and f_y.
5. Calculate the slope of radial line=h/e
6. Locate a point for coordinates, K_n=1 and R_n=1/slope, or R_n=e/h (or for any value of K_n, R_n=$K_n e/h$). Draw a radial line connecting the coordinate point to the origin. Extend the line to intersect with ρ_g of step (2). If necessary, interpolate the interaction curve.
7. At the intersection point, read K_n and R_n.
8. If the intersection point is on or above the strain line=1, ϕ=0.7 or 0.65. If it is on or below ε_t=0.005, ϕ=0.9. In between, correct ϕ by Equation 14.13. This correction is rarely applied.
9. Compute $P_u=K_n\phi f_c'A_g$ and $M_u=R_n\phi f_c'A_g h$.

Example 16.5

A 10 ft long braced column with a cross section is shown in Figure 16.15. Find the axial design load and the moment capacity for an eccentricity of 6 in. The end moments are equal and have the same sign. Use f_c' =4,000 psi and f_y=60,000 psi.

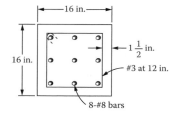

FIGURE 16.15 Eccentrically loaded column of Example 16.5.

Solution

1. For same sign (double curvature), $\dfrac{M_1}{M_2} = -1$.
2. $K=1$ (braced), $\ell=10\times 12=120$ in., $r=0.3h=0.3(16)=4.8$ in.

3. $\dfrac{K\ell}{r} = \dfrac{1(120)}{4.8} = 25$

4. From Equation 16.13

$$\frac{K\ell}{r} = 34 - 12\left(\frac{M_1}{M_2}\right)$$

$$= 34 - 12(-1) = 46 > 40, \quad \text{use } 40$$

since step (3) < step (4) short column

5. $A_g = 16 \times 16 = 256\ \text{in.}^2$
   $A_{st} = 6.32\ \text{in.}^2$

$$\rho_g = \frac{6.32}{256} = 0.025$$

6. Center to center of steel = 16 − 2(cover) − 2(ties diameter) − 1(bar diameter)
   = 16 − 2(1.5) − 2(0.375) − 1(1) = 11.25 in.

$$\gamma = \frac{11.25}{16} = 0.70$$

7. Use the interaction diagram Appendix D.17
8. $\text{slope} = \dfrac{h}{e} = \dfrac{16}{6} = 2.67$
9. $K_n = 1,\ R_n = \dfrac{1}{\text{slope}} = \dfrac{1}{2.67} = 0.375$

   Draw a radial line connecting the above coordinates to origin
10. At $\rho_g = 0.025$, $K_n = 0.48$ and $R_n = 0.18$
11. The point is above the strain line = 1, hence $\phi = 0.65$
12. $P_u = K_n \phi f'_c A_g = 0.48(0.65)(4)(256) = 319.5\ \text{k}$
    $M_u = R_n \phi f'_c A_g h = 0.18(0.65)(4)(256)(16) = 1917\ \text{in.-k or } 159.74\ \text{ft-k.}$

DESIGN OF SHORT COLUMNS FOR COMBINED LOADING

This involves determining the size, selecting steel, and fixing ties or spirals for a column. The steps are as follows:

1. Determine the design-factored axial load and moment.
2. Based on $\rho_g = 1\%$ and axial load only, estimate the column size by Equations 16.10 or 16.11, rounding on the lower side.
3. For a selected size of bars, estimate γ for the column size of step 2.
4. Select the right interaction diagram based on f'_c, f_y, the type of cross section, and γ of step 3.
5. Calculate $K_n = P_u/\phi f'_c A_g$ and $R_n = M_u/\phi f'_c A_g h$, assuming $\phi = 0.7$ (spiral) or 0.65 (ties).
6. Entering an appropriate diagram at Appendices D.15 through D.22, read ρ_g at the intersection point of K_n and R_n. This should be less than 0.05. If not, change the dimension and repeat steps 3–6.
7. Check that the interaction point of step 6 is above the strain line = 1. If not adjust ϕ and repeat steps 5 and 6.
8. Calculate required steel area, $A_{st} = \rho_g A_g$ and select reinforcement from Appendix D.2 and check that it fits in one row from Appendix D.14.

9. Design ties or spiral from steps 5 and 6 of the "Specification for Columns" section.
10. Confirm from Equations 16.12 or 16.13 that the column is short (no slenderness effect).

Example 16.6

Design a 10 ft long circular spiral column for a braced system to support the service dead and live loads of 300 k and 460 k, respectively, and the service dead and live moments of 100 ft-k each. The moment at one end is zero. Use $f_c' = 4,000$ psi and $f_y = 60,000$ psi.

Solution

1. $P_u = 1.2(300) + 1.6(460) = 1096$ k
   $M_u = 1.2(100) + 1.6(100) = 280$ ft-k
2. Assume $\rho_g = 0.01$, from Equation 16.10:

$$A_g = \frac{P_u}{0.60[0.85f_c'(1-\rho_g) + f_y\rho_g]}$$

$$= \frac{1096}{0.60[0.85(4)(1-0.01) + 60(0.01)]}$$

$$= 460.58 \text{ in.}^2$$

$$\frac{\pi h^2}{4} = 460.58$$

   or $h = 24.22$ in.
   Use $h = 24$ in., $A_g = 452$ in.^{2}
3. Assume #9 size of bar and 3/8 in. spiral
   center-to-center distance
   $= 24 - 2(\text{cover}) - 2(\text{spiral diameter}) - 1$ (bar diameter)
   $= 24 - 2(1.5) - 2(3/8) - 1.128 = 19.12$ in.

$$\gamma = \frac{19.12}{24} = 0.8$$

   Use the interaction diagram Appendix D.21
4. $K_n = \dfrac{P_u}{\phi f_c' A_g} = \dfrac{1096}{(0.7)(4)(452)} = 0.866$

$$R_n = \frac{M_u}{\phi f_c' A_g h} = \frac{3360}{(0.7)(4)(452)(24)} = 0.11$$

5. At the intersection point of K_n and R_n, $\rho_g = 0.025$
6. The point is above the strain line $= 1$, hence $\phi = 0.7$ **OK**
7. $A_{st} = (0.025)(452) = 11.3$ in.^{2}
   From Appendix D.2, select 12 bars of #9, $A_{st} = 12$ in.^{2}
   From Appendix D.14 for a core diameter of $24 - 3 = 21$ in., 15 bars of #9 can be arranged in a row
8. Selection of spirals
   From Appendix D.13, size $= 3/8$ in.
   pitch $= 2\frac{1}{4}$ in.
   Clear distance $= 2.25 - 3/8 = 1.875 > 1$ in.   **OK**

9. $K = 1$, $l = 10 \times 12 = 120$ in., $r = 0.25(24) = 6$ in.

$$\frac{Kl}{r} = \frac{1(120)}{6} = 20$$

$$\left(\frac{M_1}{M_2}\right) = 0$$

$$34 - 12\left(\frac{M_1}{M_2}\right) = 34$$

since $(Kl/r) < 34$, short column.

LONG OR SLENDER COLUMNS

When the slenderness ratio of a column exceeds the limits given by Equations 16.12 or 16.13, it is classified as a *long* or *slender* column. In physical sense, when a column bends laterally by an amount, Δ, the axial load, P introduces an additional moment equal to $P \cdot \Delta$. When this $P \cdot \Delta$ moment cannot be ignored, the column is a long or slender column.

There are two approaches to deal with this additional or secondary moment. The nonlinear second-order analysis is based on a theoretical analysis of the structure under application of an axial load, a moment, and the deflection. As an alternative approach, the ACI provides a first-order method that magnifies the moment acting on the column to account for the P–Δ effect. The magnification expressions for the braced (nonsway) and unbraced (sway) frames are similar to the steel magnification factors discussed in the "Magnification Factor, B_1" section in Chapter 12 and the "Magnification Factor for Sway, B_2" section in Chapter 12. After the moments are magnified, the procedure of short column of the "Analysis of Short Columns for Combined Loading" and "Design of Short Columns for Combined Loading" sections can be applied for analysis and design of the column using the interaction diagrams.

The computation of the magnification factors is appreciably complicated for concrete because of the involvement of the modulus of elasticity of concrete and the moment of inertia with creep and cracks in concrete.

A large percent of columns do not belong to the slender category. It is advisable to avoid the slender columns whenever possible by increasing the column dimensions, if necessary. As a rule of thumb, a column dimension of one-tenth of the column length in braced frames will meet the short column requirement. For a 10 ft length, a column of 1 ft or 12 in. or more will be a short braced column. For unbraced frames, a column dimension one-fifth of the length will satisfy the short column requirement. A 10 ft long unbraced column of 2 ft or 24 in. dimension will avoid the slenderness effect.

PROBLEMS

16.1 Determine the design axial load capacity and check whether the reinforcements meet the specifications for the column shown in Figure P16.1. Use $f_c' = 4,000$ psi and $f_y = 60,000$ psi.

16.2 Determine the design axial load capacity and check whether the reinforcements meet specifications for the column shown in Figure P16.2. Use $f_c' = 4,000$ psi and $f_y = 60,000$ psi.

16.3 Determine the design axial load capacity of the column in Figure P16.3 and check whether the reinforcement is adequate. Use $f_c' = 5,000$ psi and $f_y = 60,000$ psi.

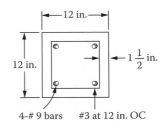

4-# 9 bars   #3 at 12 in. OC

FIGURE P16.1 Column section for Problem 16.1.

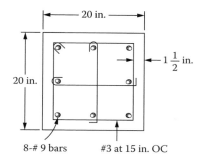

FIGURE P16.2 Column section for Problem 16.2.

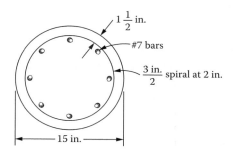

FIGURE P16.3 Column section for Problem 16.3.

16.4 Determine whether the maximum service dead load and live load carried by the column shown in Figure P16.4, if they are equal. Check for spiral steel. Use $f'_c = 3,000$ psi and $f_y = 40,000$ psi.

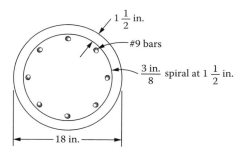

FIGURE P16.4 Column section for Problem 16.4.

16.5 Compute the maximum service live load that may be axially placed on the column shown in Figure P16.5. The service dead load is 150 k. Check for ties specifications. Use $f'_c = 3,000$ psi and $f_y = 40,000$ psi.

16.6 A service dead load of 100 k and service live load of 450 k is axially applied on a 20 in. diameter circular column reinforced with six #8 bars. The cover is 1½ in. and the spiral size is ½ in. at a 2 in. pitch. Is the column adequate? Use $f'_c = 4,000$ psi and $f_y = 60,000$ psi.

16.7 Design a tied column to carry a factored axial design load of 900 k. Use $f'_c = 5,000$ psi and $f_y = 60,000$ psi.

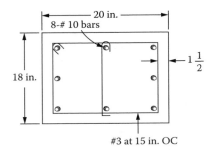

FIGURE P16.5 Column section for Problem 16.5.

16.8 For Problem 16.7, design a circular spiral column.

16.9 Design a tied column to support a service dead axial load of 300 k and live load of 480 k. Use $f'_c = 4,000$ psi and $f_y = 60,000$ psi.

16.10 Redesign a circular spiral column for Problem 16.9.

16.11 Design a rectangular tied column to support an axial service dead load of 400 k and live load of 590 k. The longer dimension of the column is approximately twice of the shorter dimension. Use $f'_c = 5,000$ psi and $f_y = 60,000$ psi.

16.12 Design a smallest circular spiral column to carry an axial service dead load of 200 k and live load of 300 k. Use $f'_c = 3,000$ psi and $f_y = 60,000$ psi. [*Hint*: It is desirable to use #11 steel to reduce the number of bars to be accommodated in a single row.]

16.13 For a 8 ft long braced column shown in Figure P16.13, determine the axial load strength and the moment capacity at an eccentricity of 5 in. Use $f'_c = 4,000$ psi and $f_y = 60,000$ psi.

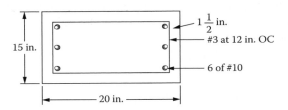

FIGURE P16.13 Column section for Problem 16.13.

16.14 An unbraced column shown in Figure P16.14 has a length of 8 ft and the cross section as shown. The factored moment-to-load ratio of the column is 0.5 ft. Determine the strength of the column. $K = 1.2$. Use $f'_c = 4,000$ psi and $f_y = 60,000$ psi.

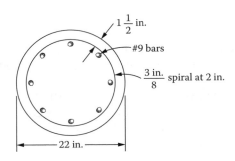

FIGURE P16.14 Column section for Problem 16.14.

16.15 On a 10 ft long column of an unbraced frame system, the load acts at an eccentricity of 5 in. The column section is shown in Figure P16.15. What are the axial load capacity and moment strength of the column? Use $f_c' = 4,000$ psi and $f_y = 60,000$ psi.

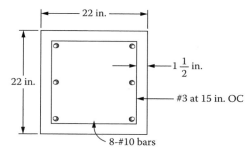

22 in.

22 in.

$1\frac{1}{2}$ in.

#3 at 15 in. OC

8-#10 bars

FIGURE P16.15   Column section for Problem 16.15.

16.16 Design a 8 ft long circular spiral column of a braced system to support the factored axial load of 1200 k and the factored moment of 300 ft-k. The end moments are equal and have the same signs. Use $f_c' = 4,000$ psi and $f_y = 60,000$ psi.

16.17 Design a tied column for Problem 16.16. Arrange the reinforcement on all faces.

16.18 For an unbraced frame, design a circular column of 10 ft length that supports the service dead and live loads of 400 k and 600 k, respectively, and the service dead and live moments of 120 ft-k and 150 ft-k respectively. The end moments are equal and have opposite signs. $K = 1.2$. Use $f_c' = 4,000$ psi and $f_y = 60,000$ psi.

16.19 Design a tied column for Problem 16.8 having reinforcement on two faces.

16.20 An unbraced frame has a 10 ft long column. Design a tied column with reinforcing bars on two end faces only to support the following service loads and moments. If necessary, adjust the column dimensions to qualify it as a short column. The column has equal end moments and a single curvature. Use $f_c' = 4,000$ psi and $f_y = 60,000$ psi.
$P_D = 150$ k, $P_L = 200$ k
$M_D = 50$ ft-k, $M_L = 70$ ft-k.

Appendix A: General

TABLE A.1
Useful Conversion Factors

Multiply	By	To Obtain
Pounds (m)	0.4356	Kilogram
Kilogram	2.205	Pounds (m)
Mass in slug	32.2	Weight in pound
Mass in kilogram	9.81	Weight in Newton (N)
Pound (f)	4.448	Newton
Newton	0.225	Pounds
U.S. or short ton	2000	Pounds
Metric ton	1000	Kilogram
U.S. ton	0.907	Metric ton
Foot	0.3048	Meter
Meter	3.281	Feet
Mile	5280	Feet
Mile	1609	Meter
	1.609	Kilometer
Square feet	0.0929	Square meter
Square mile	2.59	Square kilometer
Square kilometer	100	Hectare (ha)
Liter	1000	Cubic centimeter
Pounds per ft^2	47.88	N/m^2 or pascal
Standard atmosphere	101.325	Kilopascal (kPa)
Horsepower	550	Foot-pound/second
	745.7	Newton-meter/second or Watt
°F	5/9(°F − 32)	°C
°C	9/5(°C + 32)	°F
Log to base e (i.e., $\log_e$, where e = 2.718)	0.434	Log to base 10 (i.e., $\log_{10}$)

TABLE A.2
Geometric Properties of Common Shapes

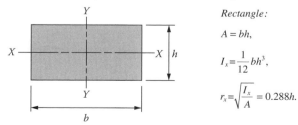

Rectangle:

$A = bh,$

$I_x = \dfrac{1}{12} bh^3,$

$r_x = \sqrt{\dfrac{I_x}{A}} = 0.288h.$

(continued)

TABLE A.2 (continued)
Geometric Properties of Common Shapes

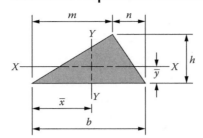

Triangle:

$$A = \frac{1}{2}bh,$$

$$\bar{y} = \frac{h}{3},$$

$$\bar{x} = \frac{b+m}{3},$$

$$I_x = \frac{1}{36}bh^3.$$

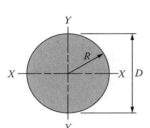

Circle:

$$A = \frac{1}{4}\pi D^2 = \pi R^2,$$

$$I_x = \frac{\pi D^4}{64} = \frac{\pi R^4}{4},$$

$$r_x = \sqrt{\frac{I_x}{A}} = \frac{D}{4} = \frac{R}{2},$$

$$J = I_x + I_y = \frac{\pi D^4}{32} = \frac{\pi R^4}{2}.$$

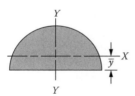

Semicircle:

$$A = \frac{1}{8}\pi D^2 = \frac{1}{2}\pi R^2,$$

$$\bar{y} = \frac{4r}{3\pi},$$

$$I_x = 0.00682D^4 = 0.11R^4,$$

$$I_y = \frac{\pi D^4}{128} = \frac{\pi R^4}{8},$$

$$r_x = 0.264R.$$

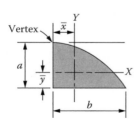

Parabola:

$$A = \frac{2}{3}ab,$$

$$\bar{x} = \frac{3}{8}b,$$

$$\bar{y} = \frac{2}{5}a.$$

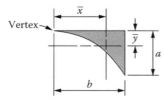

Spandrel of parabola:

$$A = \frac{1}{3}ab,$$

$$\bar{x} = \frac{3}{4}b,$$

$$\bar{y} = \frac{3}{10}a.$$

TABLE A.3.1
Shears, Moments, and Deflections

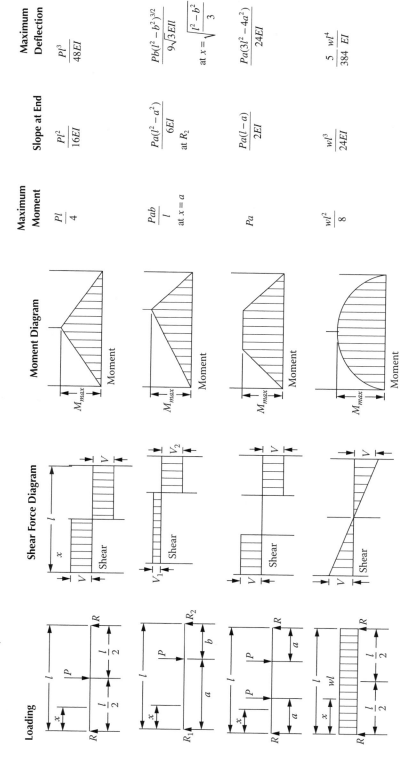

Loading	Shear Force Diagram	Moment Diagram	Maximum Moment	Slope at End	Maximum Deflection
			$\dfrac{Pl}{4}$	$\dfrac{Pl^2}{16EI}$	$\dfrac{Pl^3}{48EI}$
			$\dfrac{Pab}{l}$ at $x = a$	$\dfrac{Pa(l^2 - a^2)}{6EI}$ at R_2	$\dfrac{Pb(l^2 - b^2)^{3/2}}{9\sqrt{3}EI}$ at $x = \sqrt{\dfrac{l^2 - b^2}{3}}$
			Pa	$\dfrac{Pa(l - a)}{2EI}$	$\dfrac{Pa(3l^2 - 4a^2)}{24EI}$
			$\dfrac{wl^2}{8}$	$\dfrac{wl^3}{24EI}$	$\dfrac{5}{384}\dfrac{wl^4}{EI}$

Note: w: load per unit length; W: total load.

TABLE A.3.2
Shears, Moments, and Deflections

Loading	Shear Diagram	Moment Diagram	Maximum Moment	Slope at End	Maximum Deflection
			$\dfrac{wl^2}{9\sqrt{3}}$	$\dfrac{8wl^3}{360EI}$ at R_2	$\dfrac{2.5wl^4}{384EI}$ at $x = 0.519l$
			$\dfrac{wl^2}{12}$	$\dfrac{5wl^3}{192EI}$	$\dfrac{wl^4}{120EI}$
			$-Pl$	$\dfrac{Pl^2}{2EI}$	$\dfrac{Pl^3}{3EI}$
			$-Pb$	$\dfrac{Pb^2}{2EI}$	$\dfrac{Pb^2(3l-b)}{6EI}$ at free end

TABLE A.3.3
Shears, Moments, and Deflections

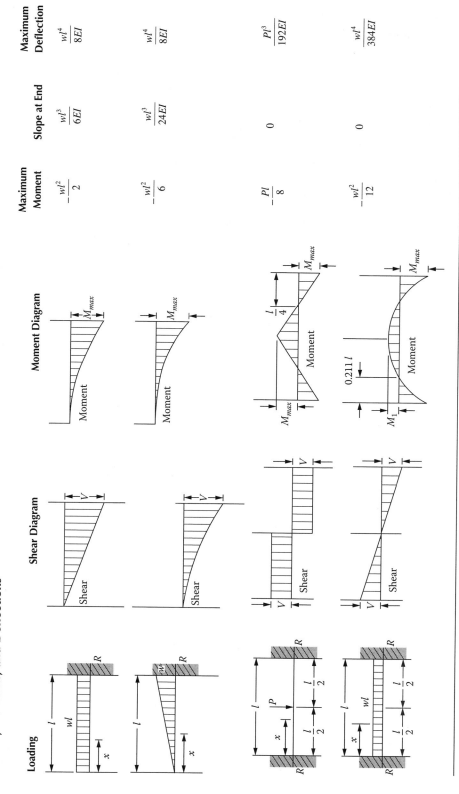

Loading	Shear Diagram	Moment Diagram	Maximum Moment	Slope at End	Maximum Deflection
			$-\dfrac{wl^2}{2}$	$\dfrac{wl^3}{6EI}$	$\dfrac{wl^4}{8EI}$
			$-\dfrac{wl^2}{6}$	$\dfrac{wl^3}{24EI}$	$\dfrac{wl^4}{8EI}$
			$-\dfrac{Pl}{8}$	0	$\dfrac{Pl^3}{192EI}$
			$-\dfrac{wl^2}{12}$	0	$\dfrac{wl^4}{384EI}$

TABLE A.4
Typical Properties of Engineering Materials

Material	Strength (psi) (Yield Values except Where Noted)			Modulus of Elasticity (E) (ksi)	Coefficient of Thermal Expansion (F^{-1}) (10^{-6})
	Tension	Compression	Shear		
Wood (dry)					
Douglas fir	6,000	3,500[a]	500	1,500	2
Redwood	6,500	4,500[a]	450	1,300	2
Southern Pine	8,500	5,000[a]	600	1,500	3
Steel	50,000	50,000	30,000	29,000	6.5
Concrete					
Structural, lightweight	150[b]	3,500[b]	130[b]	2,100	5.5
Brick masonry	300[b]	4,500[b]	300[b]	4,500	3.4
Aluminum, structural	30,000	30,000	18,000	10,000	12.8
Iron, cast	20,000[b]	85,000[b]	25,000[b]	25,000	6
Glass, plate	10,000[b]	36,000[b]	—	10,000	4.5
Polyester, glass-reinforced	10,000[b]	25,000[b]	25,000[b]	1,000	35

[a] For the parallel-to-grain direction.
[b] Denotes ultimate strength for brittle materials.

Appendix B: Wood

TABLE B.1
Section Properties of Standard Dressed (S4S) Sawn Lumber

Nominal Size, $b \times d$	Standard Dressed Size $b \times d$ (in. × in.)	Area of Section, A (in.²)	X–X Axis		Y–Y Axis	
			Section Modulus, S_{xx} (in.³)	Moment of Inertia, I_{xx} (in.⁴)	Section Modulus, S_{yy} (in.³)	Moment of Inertia, I_{yy} (in.⁴)
1×3	3/4×2-1/2	1.875	0.781	0.977	0.234	0.088
1×4	3/4×3-1/2	2.625	1.531	2.680	0.328	0.123
1×6	3/4×5-1/2	4.125	3.781	10.40	0.516	0.193
1×8	3/4×7-1/4	5.438	6.570	23.82	0.680	0.255
1×10	3/4×9-1/4	6.938	10.70	49.47	0.867	0.325
1×12	3/4×11-1/4	8.438	15.82	88.99	1.055	0.396
2×3	1-1/2×2-1/2	3.750	1.563	1.953	0.938	0.703
2×4	1-1/2×3-1/2	5.250	3.063	5.359	1.313	0.984
2×5	1-1/2×4-1/2	6.750	5.063	11.39	1.688	1.266
2×6	1-1/2×5-1/2	8.250	7.563	20.80	2.063	1.547
2×8	1-1/2×7-1/4	10.88	13.14	47.63	2.719	2.039
2×10	1-1/2×9-1/4	13.88	21.39	98.93	3.469	2.602
2×12	1-1/2×11-1/4	16.88	31.64	178.0	4.219	3.164
2×14	1-1/2×13-1/4	19.88	43.89	290.8	4.969	3.727
3×4	2-1/2×3-1/2	8.750	5.104	8.932	3.646	4.557
3×5	2-1/2×4-1/2	11.25	8.438	18.98	4.688	5.859
3×6	2-1/2×5-1/2	13.75	12.60	34.66	5.729	7.161
3×8	2-1/2×7-1/4	18.13	21.90	79.39	7.552	9.440
3×10	2-1/2×9-1/4	23.13	35.65	164.9	9.635	12.04
3×12	2-1/2×11-1/4	28.13	52.73	296.6	11.72	14.65
3×14	2-1/2×13-1/4	33.13	73.15	484.6	13.80	17.25
3×16	2-1/2×15-1/4	38.13	96.90	738.9	15.89	19.86
4×4	3-1/2×3-1/2	12.25	7.146	12.51	7.146	12.51
4×5	3-1/2×4-1/2	15.75	11.81	26.58	9.188	16.08
4×6	3-1/2×5-1/2	19.25	17.65	48.53	11.23	19.65
4×8	3-1/2×7-1/4	25.38	30.66	111.1	14.80	25.90
4×10	3-1/2×9-1/4	32.38	49.91	230.8	18.89	33.05
4×12	3-1/2×11-1/4	39.38	73.83	415.3	22.97	40.20
4×14	3-1/2×13-1/4	46.38	102.4	678.5	27.05	47.34
4×16	3-1/2×15-1/4	53.38	135.7	1,034	31.14	54.49
5×5	4-1/2×4-1/2	20.25	15.19	34.17	15.19	34.17
6×6	5-1/2×5-1/2	30.25	27.73	76.26	27.73	76.26
6×8	5-1/2×7-1/2	41.25	51.56	193.4	37.81	104.0
6×10	5-1/2×9-1/2	52.25	82.73	393.0	47.90	131.7
6×12	5-1/2×11-1/2	63.25	121.2	697.1	57.98	159.4
6×14	5-1/2×13-1/2	74.25	167.1	1,128	68.06	187.2
6×16	5-1/2×15-1/2	85.25	220.2	1,707	78.15	214.9
6×18	5-1/2×17-1/2	96.25	280.7	2,456	88.23	242.6
6×20	5-1/2×19-1/2	107.3	348.6	3,398	98.31	270.4
6×22	5-1/2×21-1/2	118.3	423.7	4,555	108.4	298.1
6×24	5-1/2×23-1/2	129.3	506.2	5,948	118.5	325.8
8×8	7-1/2×7-1/2	56.25	70.31	263.7	70.31	263.7
8×10	7-1/2×9-1/2	71.25	112.8	535.9	89.06	334.0
8×12	7-1/2×11-1/2	86.25	165.3	950.5	107.8	404.3
8×14	7-1/2×13-1/2	101.3	227.8	1,538	126.6	474.6

TABLE B.1 (continued)
Section Properties of Standard Dressed (S4S) Sawn Lumber

Nominal Size, $b \times d$	Standard Dressed Size $b \times d$ (in. $\times$ in.)	Area of Section, A (in.^{2})	X–X Axis		Y–Y Axis	
			Section Modulus, S_{xx} (in.^{3})	Moment of Inertia, I_{xx} (in.^{4})	Section Modulus, S_{yy} (in.^{3})	Moment of Inertia, I_{yy} (in.^{4})
8 × 16	7-1/2 × 15-1/2	116.3	300.3	2,327	145.3	544.9
8 × 18	7-1/2 × 17-1/2	131.3	382.8	3,350	164.1	615.2
8 × 20	7-1/2 × 19-1/2	146.3	475.3	4,634	182.8	685.5
8 × 22	7-1/2 × 21-1/2	161.3	577.8	6,211	201.6	755.9
8 × 24	7-1/2 × 23-1/2	176.3	690.3	8,111	220.3	826.2
10 × 10	9-1/2 × 9-1/2	90.25	142.9	678.8	142.9	678.8
10 × 12	9-1/2 × 11-1/2	109.3	209.4	1,204	173.0	821.7
10 × 14	9-1/2 × 13-1/2	128.3	288.6	1,948	203.1	964.5
10 × 16	9-1/2 × 15-1/2	147.3	380.4	2,948	233.1	1,107
10 × 18	9-1/2 × 17-1/2	166.3	484.9	4,243	263.2	1,250
10 × 20	9-1/2 × 19-1/2	185.3	602.1	5,870	293.3	1393
10 × 22	9-1/2 × 21-1/2	204.3	731.9	7,868	323.4	1,536
10 × 24	9-1/2 × 23-1/2	223.3	874.4	10,270	353.5	1,679
12 × 12	11-1/2 × 11-1/2	132.3	253.5	1,458	253.5	1,458
12 × 14	11-1/2 × 13-1/2	155.3	349.3	2,358	297.6	1,711
12 × 16	11-1/2 × 15-1/2	178.3	460.5	3,569	341.6	1,964
12 × 18	11-1/2 × 17-1/2	201.3	587.0	5,136	385.7	2,218
12 × 20	11-1/2 × 19-1/2	224.3	728.8	7,106	429.8	2,471
12 × 22	11-1/2 × 21-1/2	247.3	886.0	9,524	473.9	2,725
12 × 24	11-1/2 × 23-1/2	270.3	1,058	12,440	518.0	2,978
14 × 14	13-1/2 × 13-1/2	182.3	410.1	2,768	410.1	2,768
14 × 16	13-1/2 × 15-1/2	209.3	540.6	4,189	470.8	3,178
14 × 18	13-1/2 × 17-1/2	236.3	689.1	6,029	531.6	3,588
14 × 20	13-1/2 × 19-1/2	263.3	855.6	8,342	592.3	3,998
14 × 22	13-1/2 × 21-1/2	290.3	1040	11,180	653.1	4,408
14 × 24	13-1/2 × 23-1/2	317.3	1243	14,600	713.8	4,818
16 × 16	15-1/2 × 15-1/2	240.3	620.6	4,810	620.6	4,810
16 × 18	15-1/2 × 17-1/2	271.3	791.1	6,923	700.7	5,431
16 × 20	15-1/2 × 19-1/2	302.3	982.3	9,578	780.8	6,051
16 × 22	15-1/2 × 21-1/2	333.3	1194	12,840	860.9	6,672
16 × 24	15-1/2 × 23-1/2	364.3	1427	16,760	941.0	7,293
18 × 18	17-1/2 × 17-1/2	306.3	893.2	7,816	893.2	7,816
18 × 20	17-1/2 × 19-1/2	341.3	1109	10,810	995.3	8,709
18 × 22	17-1/2 × 21-1/2	376.3	1348	14,490	1097	9,602
18 × 24	17-1/2 × 23-1/2	411.3	1611	18,930	1199	10,500
20 × 20	19-1/2 × 19-1/2	380.3	1236	12,050	1236	12,050
20 × 22	19-1/2 × 21-1/2	419.3	1502	16,150	1363	13,280
20 × 24	19-1/2 × 23-1/2	458.3	1795	21,090	1489	14,520
22 × 22	21-1/2 × 21-1/2	462.3	1656	17,810	1656	17,810
22 × 24	21-1/2 × 23-1/2	505.3	1979	23,250	1810	19,460
24 × 24	23-1/2 × 23-1/2	552.3	2163	25,420	2163	25,420

Source: Courtesy of the American Wood Council, Washington, DC.

TABLE B.2 SIZE FACTOR, C_F (ALL SPECIES EXCEPT SOUTHERN PINE)

Tabulated bending, tension, and compression parallel to grain design values for dimension lumber 2–4 in. breadth shall be multiplied by the following size factors:

| Grades | Width (Depth) | F_b Thickness (Breadth) | | F_t | F_c |
		2 and 3 in.	4 in.		
Select structural, No. 1 and Btr, No. 1, No. 2, No. 3	2, 3, and 4 in.	1.5	1.5	1.5	1.15
	5 in.	1.4	1.4	1.4	1.1
	6 in.	1.3	1.3	1.3	1.1
	8 in.	1.2	1.3	1.2	1.05
	10 in.	1.1	1.2	1.1	1.0
	12 in.	1.0	1.1	1.0	1.0
	14 in. and wider	0.9	1.0	0.9	0.9
Stud	2, 3, and 4 in.	1.1	1.1	1.1	1.05
	5 and 6 in.	1.0	1.0	1.0	1.0
	8 in. and wider	Use No. 3 grade tabulated design values and size factors			
Construction standard	2, 3, and 4 in.	1.0	1.0	1.0	1.0
Utility	4 in.	1.0	1.0	1.0	1.0
	2 and 3 in.	0.4	—	0.4	0.6

TABLE B.2
Reference Design Values for Visually Graded Dimension Lumber (2–4 in. Breadth) (All Species Except Southern Pine)

Design Values in Pounds per Square Inch (psi)

Species and Commercial Grade	Size Classification	Bending, F_b	Tension Parallel to Grain, F_t	Shear Parallel to Grain, F_v	Compression Perpendicular to Grain, $F_{c\perp}$	Compression Parallel to Grain, F_c	Modulus of Elasticity E	E_{min}	Grading Rules Agency
Beech-Birch-Hickory									
Select structural	2 in. and wider	1450	850	195	715	1200	1,700,000	620,000	NELMA
No.1		1050	600	195	715	950	1,600,000	580,000	
No.2		1000	600	195	715	750	1,500,000	550,000	
No.3		575	350	195	715	425	1,300,000	470,000	
Stud	2 in. and wider	775	450	195	715	475	1,300,000	470,000	
Construction	2–4 in. wide	1150	675	195	715	1000	1,400,000	510,000	
Standard		650	375	195	715	775	1,300,000	470,000	
Utility		300	175	195	715	500	1,200,000	440,000	
Cottonwood									
Select structural	2 in. and wider	875	525	125	320	775	1,200,000	440,000	NSLB
No.1		625	375	125	320	625	1,200,000	440,000	
No.2		625	350	125	320	475	1,100,000	400,000	
No.3		350	200	125	320	275	1,000,000	370,000	
Stud	2 in. and wider	475	275	125	320	300	1,000,000	370,000	
Construction	2–4 in. wide	700	400	125	320	650	1,000,000	370,000	
Standard		400	225	125	320	500	900,000	330,000	
Utility		175	100	125	320	325	900,000	330,000	
Douglas Fir-Larch									
Select structural	2 in. and wider	1500	1000	180	625	1700	1,900,000	690,000	WCLIB WWPA
No.1 and Btr		1200	800	180	625	1550	1,800,000	660,000	
No.1		1000	675	180	625	1500	1,700,000	620,000	
No.2		900	575	180	625	1350	1,600,000	580,000	
No.3		525	325	180	625	775	1,400,000	510,000	

(continued)

TABLE B.2 (continued)
Reference Design Values for Visually Graded Dimension Lumber (2–4 in. Breadth) (All Species Except Southern Pine)

Design Values in Pounds per Square Inch (psi)

Species and Commercial Grade	Size Classification	Bending, F_b	Tension Parallel to Grain, F_t	Shear Parallel to Grain, F_v	Compression Perpendicular to Grain, $F_{c\perp}$	Compression Parallel to Grain, F_c	Modulus of Elasticity E	E_{min}	Grading Rules Agency
Stud	2 in. and wider	700	450	180	625	850	1,400,000	510,000	NLGA
Construction	2–4 in. wide	1000	650	180	625	1650	1,500,000	550,000	
Standard		575	375	180	625	1400	1,400,000	510,000	
Utility		275	175	180	625	900	1,300,000	470,000	
Douglas Fir-Larch (North)									
Select structural	2 in. and wider	1350	825	180	625	1900	1,900,000	690,000	
No.1 and Btr		1150	750	180	625	1800	1,800,000	660,000	
No.1/No.2		850	500	180	625	1400	1,600,000	580,000	
No.3		475	300	180	625	825	1,400,000	510,000	
Stud	2 in. and wider	650	400	180	625	900	1,400,000	510,000	
Construction	2–4 in. wide	950	575	180	625	1800	1,500,000	550,000	
Standard		525	325	180	625	1450	1,400,000	510,000	
Utility		250	150	180	625	950	1,300,000	470,000	
Douglas Fir-South									
Select structural	2 in. and wider	1,350	900	180	520	1600	1,400,000	510,000	WWPA
No.1		925	600	180	520	1450	1,300,000	470,000	
No.2		850	525	180	520	1350	1,200,000	440,000	
No.3		500	300	180	520	775	1,100,000	400,000	
Stud	2 in. and wider	675	425	180	520	850	1,100,000	400,000	
Construction	2–4 in. wide	975	600	180	520	1650	1,200,000	440,000	
Standard		550	350	180	520	1400	1,100,000	400,000	
Utility		250	150	180	520	900	1,000,000	370,000	
Eastern Hemlock-Balsam Fir									
Select structural	2 in. and wider	1250	575	140	335	1200	1,200,000	440,000	NELM
No.1		775	350	140	335	1000	1,100,000	400,000	NSLB

Grade	Size							Agency
No.2		575	275	140	335	825	1,100,000	400,000	
No.3		350	150	140	335	475	900,000	330,000	
Stud	2 in. and wider	450	200	140	335	525	900,000	330,000	
Construction	2–4 in. wide	675	300	140	335	1050	1,000,000	370,000	
Standard		375	175	140	335	850	900,000	330,000	
Utility		175	75	140	335	550	800,000	290,000	
Eastern Hemlock-Tamarack									
Select structural	2 in. and wider	1250	575	170	555	1200	1,200,000	440,000	NELMA
No.1		775	350	170	555	1000	1,100,000	400,000	NSLB
No.2		575	275	170	555	825	1,100,000	400,000	
No.3		350	150	170	555	475	900,000	330,000	
Stud	2 in. and wider	450	200	170	555	525	900,000	330,000	
Construction	2–4 in. wide	675	300	170	555	1050	1,000,000	370,000	
Standard		375	175	170	555	850	900,000	330,000	
Utility		175	75	170	555	550	800,000	290,000	
Eastern Softwoods									
Select structural	2 in. and wider	1250	575	140	335	1200	1,200,000	440,000	NELMA
No.1		775	350	140	335	1000	1,100,000	400,000	NSLB
No.2		575	275	140	335	825	1,100,000	400,000	
No.3		350	150	140	335	475	900,000	330,000	
Stud	2 in. and wider	450	200	140	335	525	900,000	330,000	
Construction	2–4 in. wide	675	300	140	335	1050	1,000,000	370,000	
Standard		375	175	140	335	850	900,000	330,000	
Utility		175	75	140	335	550	800,000	290,000	
Eastern White Pine									
Select structural	2 in. and wider	1250	575	135	350	1200	1,200,000	440,000	NELMA
No.1		775	350	135	350	1000	1,100,000	400,000	NSLB
No.2		575	275	135	350	825	1,100,000	400,000	
No.3		350	150	135	350	475	900,000	330,000	
Stud	2 in. and wider	450	200	135	350	525	900,000	330,000	
Construction	2–4 in. wide	675	300	135	350	1050	1,000,000	370,000	
Standard		375	175	135	350	850	900,000	330,000	
Utility		175	75	135	350	550	800,000	290,000	

(continued)

TABLE B.2 (continued)
Reference Design Values for Visually Graded Dimension Lumber (2–4 in. Breadth) (All Species Except Southern Pine)

Species and Commercial Grade	Size Classification	Bending, F_b	Tension Parallel to Grain, F_t	Shear Parallel to Grain, F_v	Compression Perpendicular to Grain, $F_{c\perp}$	Compression Parallel to Grain, F_c	Modulus of Elasticity E	Modulus of Elasticity E_{min}	Grading Rules Agency
					Design Values in Pounds per Square Inch (psi)				
Hem-Fir									
Select structural	2 in. and wider	1400	925	150	405	1500	1,600,000	580,000	WCLIB
No.1 and Btr		1100	725	150	405	1350	1,500,000	550,000	WWPA
No.1		975	625	150	405	1350	1,500,000	550,000	
No.2		850	525	150	405	1300	1,300,000	470,000	
No.3		500	300	150	405	725	1,200,000	440,000	
Stud	2 in. and wider	675	400	150	405	800	1,200,000	440,000	
Construction	2–4 in. wide	975	600	150	405	1550	1,300,000	470,000	
Standard		550	325	150	405	1300	1,200,000	440,000	
Utility		250	150	150	405	850	1,100,000	400,000	
Hem-Fir (North)									
Select structural	2 in. and wider	1300	775	145	405	1700	1,700,000	620,000	NLGA
No.1 and Btr		1200	725	145	405	1550	1,700,000	620,000	
No.1/No.2		1000	575	145	405	1450	1,600,000	580,000	

No.3		575	325	145	405	850	1,400,000	510,000	NELMA
Stud	2 in. and wider	775	450	145	405	925	1,400,000	510,000
Construction	2–4 in. wide	1150	650	145	405	1750	1,500,000	550,000
Standard		650	350	145	405	1500	1,400,000	510,000
Utility		300	175	145	405	975	1,300,000	470,000
Mixed Maple								
Select structural	2 in. and wider	1000	600	195	620	875	1,300,000	470,000
No.1		725	425	195	620	700	1,200,000	440,000
No.2		700	425	195	620	550	1,100,000	400,000
No.3		400	250	195	620	325	1,000,000	370,000
Stud	2 in. and wider	550	325	195	620	350	1,000,000	370,000
Construction	2–4 in. wide	800	475	195	620	725	1,100,000	400,000
Standard		450	275	195	620	575	1,000,000	370,000
Utility		225	125	195	620	375	900,000	330,000

Source: Courtesy of the American Forest & Paper Association, Washington, DC.

Notes: Tabulated design values are for normal load duration and dry service conditions. See NDS 4.3 for a comprehensive description of design value adjustment factors.

TABLE B.3 SIZE FACTOR, C_F (SOUTHERN PINE)

Appropriate size adjustment factors have already been incorporated in the tabulated design values for most thicknesses of Southern Pine and Mixed Southern Pine dimension lumber. For dimension lumber 4 in. breadth, 8 in. and deeper (all grades except Dense Structural 86, Dense Structural 72, and Dense Structural 65), tabulated bending design values, F_b, shall be permitted to be multiplied by the size factor, $C_F = 1.1$. For dimension lumber deeper than 12 in. (all grades except Dense Structural 86, Dense Structural 72, and Dense Structural 65), tabulated bending, tension and compression parallel to grain design values of 12 in. depth lumber shall be multiplied by the size factor, $C_F = 0.9$. When the depth, d, of Dense Structural 86, Dense Structural 72, or Dense Structural 65 dimension lumber exceeds 12 in., the tabulated bending design value, F_b, shall be multiplied by the following size factor:

$$C_F = (12/d)^{1/9}$$

TABLE B.3
Reference Design Values for Visually Graded Southern Pine Dimension Lumber (2–4 in. Breadth)

Species and Commercial Grade	Size Classification	Design Values in Pounds per Square Inch (psi)					Modulus of Elasticity		Grading Rules Agency
		Bending, F_b	Tension Parallel to Grain, F_t	Shear Parallel to Grain, F_v	Compression Perpendicular to Grain, $F_{c\perp}$	Compression Parallel to Grain, F_c	E	E_{min}	
Southern Pine									
Dense select structural	2–4 in. wide	3050	1650	175	660	2250	1,900,000	690,000	SPIB
Select structural	(depth)	2850	1600	175	565	2100	1,800,000	660,000	
Non-dense select structural		2650	1350	175	480	1950	1,700,000	620,000	
No.1 dense		2000	1100	175	660	2000	1,800,000	660,000	
No.1		1850	1050	175	565	1850	1,700,000	620,000	
No.1 non-dense		1700	900	175	480	1700	1,600,000	580,000	
No.2 dense		1700	875	175	660	1850	1,700,000	620,000	
No.2		1500	825	175	565	1650	1,600,000	580,000	
No.2 non-dense		1350	775	175	480	1600	1,400,000	510,000	
No.3 and stud		850	475	175	565	975	1,400,000	510,000	
Construction	4 in. wide	1100	625	175	565	1800	1,500,000	550,000	
Standard	(depth)	625	350	175	565	1500	1,300,000	470,000	
Utility		300	175	175	565	975	1,300,000	470,000	
Dense select structural	5–6 in. wide	2700	1500	175	660	2150	1,900,000	690,000	
Select structural	(depth)	2550	1400	175	565	2000	1,800,000	660,000	
Non-dense select structural		2350	1200	175	480	1850	1,700,000	620,000	
No.1 dense		1750	950	175	660	1900	1,800,000	660,000	
No.1		1650	900	175	565	1750	1,700,000	620,000	
No.1 non-dense		1500	800	175	480	1600	1,600,000	580,000	
No.2 dense		1450	775	175	660	1750	1,700,000	620,000	
No.2		1250	725	175	565	1600	1,600,000	580,000	
No.2 non-dense		1150	675	175	480	1500	1,400,000	510,000	
No.3 and stud		750	425	175	565	925	1,400,000	510,000	

(continued)

TABLE B.3 (continued)
Reference Design Values for Visually Graded Southern Pine Dimension Lumber (2–4 in. Breadth)

Species and Commercial Grade	Size Classification	Design Values in Pounds per Square Inch (psi)							Grading Rules Agency
		Bending, F_b	Tension Parallel to Grain, F_t	Shear Parallel to Grain, F_v	Compression Perpendicular to Grain, $F_{c\perp}$	Compression Parallel to Grain, F_c	Modulus of Elasticity		
							E	E_{min}	
Dense select structural	8 in. wide[a]	2450	1350	175	660	2050	1,900,000	690,000	
Select structural	(depth)	2300	1300	175	565	1900	1,800,000	660,000	
Non-dense select structural		2100	1100	175	480	1750	1,700,000	620,000	
No.1 dense		1650	875	175	660	1800	1,800,000	660,000	
No.1		1500	825	175	565	1650	1,700,000	620,000	
No.1 non-dense		1350	725	175	480	1550	1,600,000	580,000	
No.2 dense		1400	675	175	660	1700	1,700,000	620,000	
No.2		1200	650	175	565	1550	1,600,000	580,000	
No.2 non-dense		1100	600	175	480	1450	1,400,000	510,000	
No.3 and stud		700	400	175	565	875	1,400,000	510,000	
Dense select structural	10 in. wide[a]	2150	1200	175	660	2000	1,900,000	690,000	
Select structural	(depth)	2050	1100	175	565	1850	1,800,000	660,000	
Non-dense select structural		1850	950	175	480	1750	1,700,000	620,000	
No.1 dense		1450	775	175	660	1750	1,800,000	660,000	
No.1		1300	725	175	565	1600	1,700,000	620,000	
No.1 non-dense		1200	650	175	480	1500	1,600,000	580,000	

Grade								
No.2 dense		1200	625	175	660	1650	1,700,000	620,000
No.2		1050	575	175	565	1500	1,600,000	580,000
No.2 non-dense		950	550	175	480	1400	1,400,000	510,000
No.3 and stud		600	325	175	565	850	1,400,000	510,000
Dense select structural	12 in. wide[a]	2050	1100	175	660	1950	1,900,000	690,000
Select Structural	(depth)	1900	1050	175	565	1800	1,800,000	660,000
Non-dense select structural		1750	900	175	480	1700	1,700,000	620,000
No.1 dense		1350	725	175	660	1700	1,800,000	660,000
No.1		1250	675	175	565	1600	1,700,000	620,000
No.1 non-dense		1150	600	175	480	1500	1,600,000	580,000
No.2 dense		1150	575	175	660	1600	1,700,000	620,000
No.2		975	550	175	565	1450	1,600,000	580,000
No.2 non-dense		900	525	175	480	1350	1,400,000	510,000
No.3 and stud		575	325	175	565	825	1,400,000	510,000
Dense structural	>12 in. wide[b]	$C_F = (12/d)^{1/9}$						
All except dense structural	(depth)	Multiply values of 12 in. wide by a size factor, $C_F = 0.9$						

Source: Courtesy of the American Wood Council, Washington, DC.

Notes: Tabulated design values are for normal load duration and dry service conditions, unless specified otherwise. See NDS 4.3 for a comprehensive description of design value adjustment factors.

[a] All, except dense structural, for 4 in. thick and 8 in. or more width (depth) up to 12 in. depth multiply F_b (bending) only by a size factor of 1.1.

[b] For more than 12 in. depth, for all species except dense structural, multiply the values of 12 in. depth by a size factor, $C_F = 0.9$. For dense structural of more than 12 in. depth, multiply the bending value of 12 in. depth by the size factor $C_F = (12/d)^{1/9}$.

TABLE B.4
Reference Design Values for Visually Graded Timbers (5 in. × 5 in. and Larger)

Species and Commercial Grade	Size Classification	Design Values in Pounds per Square Inch (psi)							Grading Rules Agency
		Bending, F_b	Tension Parallel to Grain, F_t	Shear Parallel to Grain, F_v	Compression Perpendicular to Grain, $F_{c\perp}$	Compression Parallel to Grain, F_c	Modulus of Elasticity E	E_{min}	
Alaska Cedar									
Select structural	Beams and stringers	1400	675	155	525	925	1,200,000	440,000	WCLIB
No.1		1150	475	155	525	775	1,200,000	440,000	
No.2		750	300	155	525	500	1,000,000	370,000	
Select structural	Posts and timbers	1300	700	155	525	975	1,200,000	440,000	
No.1		1050	575	155	525	850	1,200,000	440,000	
No.2		625	350	155	525	600	1,000,000	370,000	
Baldcypress									
Select structural	5 in. × 5 in. and larger	1150	750	200	615	1050	1,300,000	470,000	SPIB
No.1		1000	675	200	615	925	1,300,000	470,000	
No.2		625	425	175	615	600	1,000,000	370,000	
Balsam Fir									
Select structural	Beams and stringers	1350	900	125	305	950	1,400,000	510,000	NELMA
No.1		1100	750	125	305	800	1,400,000	510,000	NSLB
No.2		725	350	125	305	500	1,100,000	400,000	
Select structural	Posts and timbers	1250	825	125	305	1000	1,400,000	510,000	
No.1		1000	675	125	305	875	1,400,000	510,000	
No.2		575	375	125	305	400	1,100,000	400,000	
Beech-Birch-Hickory									
Select structural	Beams and stringers	1650	975	180	715	975	1,500,000	550,000	NELMA
No.1		1400	700	180	715	825	1,500,000	550,000	

Grade	Size classification							Agency
No.2		900	450	180	715	525	1,200,000	440,000	
Select structural	Posts and timbers	1550	1050	180	715	1050	1,500,000	550,000	NLGA
No.1	timbers	1250	850	180	715	900	1,500,000	550,000	
No.2		725	475	180	715	425	1,200,000	440,000	
Coast Sitka Spruce								
Select structural	Beams and stringers	1150	675	115	455	775	1,500,000	550,000	
No.1	stringers	950	475	115	455	650	1,500,000	550,000	
No.2		625	325	115	455	425	1,200,000	440,000	
Select structural	Posts and timbers	1100	725	115	455	825	1,500,000	550,000	
No.1	timbers	875	575	115	455	725	1,500,000	550,000	
No.2		525	350	115	455	500	1,200,000	440,000	
Douglas Fir-Larch								
Dense select structural	Beams and stringers	1900	1100	170	730	1300	1,700,000	620,000	WCLIB
Select structural	stringers	1600	950	170	625	1100	1,600,000	580,000	
Dense No.1		1550	775	170	730	1100	1,700,000	620,000	
No.1		1350	675	170	625	925	1,600,000	580,000	
No.2		875	425	170	625	600	1,300,000	470,000	
Dense select structural	Posts and timbers	1750	1150	170	730	1350	1,700,000	620,000	
Select structural	timbers	1500	1000	170	625	1150	1,600,000	580,000	
Dense No.1		1400	950	170	730	1200	1,700,000	620,000	
No.1		1200	825	170	625	1000	1,600,000	580,000	
No.2		750	475	170	625	700	1,300,000	470,000	
Dense select structural	Beams and stringers	1900	1100	170	730	1300	1,700,000	620,000	WWPA
Select structural	stringers	1600	950	170	625	1100	1,600,000	580,000	
Dense No.1		1550	775	170	730	1100	1,700,000	620,000	
No.1		1350	675	170	625	925	1,600,000	580,000	
No.2 dense		1000	500	170	730	700	1,400,000	510,000	
No.2		875	425	170	625	600	1,300,000	470,000	

(continued)

TABLE B.4 (continued)
Reference Design Values for Visually Graded Timbers (5 in. × 5 in. and Larger)

| Species and Commercial Grade | Size Classification | Design Values in Pounds per Square Inch (psi) | | | | | Modulus of Elasticity | | Grading Rules Agency |
		Bending, F_b	Tension Parallel to Grain, F_t	Shear Parallel to Grain, F_v	Compression Perpendicular to Grain, $F_{c\perp}$	Compression Parallel to Grain, F_c	E	E_{min}	
Dense select structural	Posts and	1750	1150	170	730	1350	1,700,000	620,000	
Select structural	timbers	1500	1000	170	625	1150	1,600,000	580,000	
Dense No.1		1400	950	170	730	1200	1,700,000	620,000	
No.1		1200	825	170	625	1000	1,600,000	580,000	
No.2 dense		850	550	170	730	825	1,400,000	510,000	
No.2		750	475	170	625	700	1,300,000	470,000	

Source: Courtesy of the American Forest & Paper Association, Washington, DC.

Notes: Tabulated design values are for normal load duration and dry service conditions, unless specified otherwise. See NDS 4.3 for a comprehensive description of design value adjustment factors.

TABLE B.5
Section Properties of *Western Species* Structural Glued Laminated Timber (GLULAM)

Depth d (in.)	Area A (in.²)	I_x (in.⁴)	S_x (in.³)	r_x (in.)	I_y (in.⁴)	S_y (in.³)
		X–X Axis			**Y–Y Axis**	
		3-1/8 in. Width			**(r_y = 0.902 in.)**	
6	18.75	56.25	18.75	1.732	15.26	9.766
7-1/2	23.44	109.9	29.30	2.165	19.07	12.21
9	28.13	189.8	42.19	2.598	22.89	14.65
10-1/2	32.81	301.5	57.42	3.031	26.70	17.09
12	37.50	450.0	75.00	3.464	30.52	19.53
13-1/2	42.19	640.7	94.92	3.897	34.33	21.97
15	46.88	878.9	117.2	4.330	38.15	24.41
16-1/2	51.56	1,170	141.8	4.763	41.96	26.86
18	56.25	1,519	168.8	5.196	45.78	29.30
19-1/2	60.94	1,931	198.0	5.629	49.59	31.74
21	65.63	2,412	229.7	6.062	53.41	34.18
22-1/2	70.31	2,966	263.7	6.495	57.22	36.62
24	75.00	3,600	300.0	6.928	61.04	39.06
		5-1/8 in. Width			**(r_y = 1.479 in.)**	
6	30.75	92.25	30.75	1.732	67.31	26.27
7-1/2	38.44	180.2	48.05	2.165	84.13	32.83
9	46.13	311.3	69.19	2.598	101.0	39.40
10-1/2	53.81	494.4	94.17	3.031	117.8	45.96
12	61.50	738.0	123.0	3.464	134.6	52.53
13-1/2	69.19	1,051	155.7	3.897	151.4	59.10
15	76.88	1,441	192.2	4.330	168.3	65.66
16-1/2	84.56	1,919	232.5	4.763	185.1	72.23
18	92.25	2,491	276.8	5.196	201.9	78.80
19-1/2	99.94	3,167	324.8	5.629	218.7	85.36
21	107.6	3,955	376.7	6.062	235.6	91.93
22-1/2	115.3	4,865	432.4	6.495	252.4	98.50
24	123.0	5,904	492.0	6.928	269.2	105.1
25-1/2	130.7	7,082	555.4	7.361	286.0	111.6
27	138.4	8,406	622.7	7.794	302.9	118.2
28-1/2	146.1	9,887	693.8	8.227	319.7	124.8
30	153.8	11,530	768.8	8.660	336.5	131.3
31-1/2	161.4	13,350	847.5	9.093	353.4	137.9
33	169.1	15,350	930.2	9.526	370.2	144.5
34-1/2	176.8	17,540	1017	9.959	387.0	151.0
36	184.5	19,930	1107	10.39	403.8	157.6
		6-3/4 in. Width			**(r_y = 1.949 in.)**	
7-1/2	50.63	237.3	63.28	2.165	192.2	56.95
9	60.75	410.1	91.13	2.598	230.7	68.34
10-1/2	70.88	651.2	124.0	3.031	269.1	79.73
12	81.00	972.0	162.0	3.464	307.5	91.13
13-1/2	91.13	1384	205.0	3.897	346.0	102.5
15	101.3	1898	253.1	4.330	384.4	113.9

(*continued*)

TABLE B.5 (continued)
Section Properties of *Western Species* Structural Glued Laminated Timber (GLULAM)

Depth d (in.)	Area A (in.²)	X–X Axis			Y–Y Axis	
		I_x (in.⁴)	S_x (in.³)	r_x (in.)	I_y (in.⁴)	S_y (in.³)
16-1/2	111.4	2527	306.3	4.763	422.9	125.3
18	121.5	3281	364.5	5.196	461.3	136.7
19-1/2	131.6	4171	427.8	5.629	499.8	148.1
21	141.8	5209	496.1	6.062	538.2	159.5
22-1/2	151.9	6,407	569.5	6.495	576.7	170.9
24	162.0	7,776	648.0	6.928	615.1	182.3
25-1/2	172.1	9,327	731.5	7.361	653.5	193.6
27	182.3	11,070	820.1	7.794	692.0	205.0
28-1/2	192.4	13,020	913.8	8.227	730.4	216.4
30	202.5	15,190	1013	8.660	768.9	227.8
31-1/2	212.6	17,580	1116	9.093	807.3	239.2
33	222.8	20,210	1225	9.526	845.8	250.6
34-1/2	232.9	23,100	1339	9.959	884.2	262.0
36	243.0	26,240	1458	10.39	922.6	273.4
37-1/2	253.1	29,660	1582	10.83	961.1	284.8
39	263.3	33,370	1711	11.26	999.5	296.2
40-1/2	273.4	37,370	1845	11.69	1038	307.5
42	283.5	41,670	1985	12.12	1076	318.9
43-1/2	293.6	46,300	2129	12.56	1115	330.3
45	303.8	51,260	2278	12.99	1153	341.7
46-1/2	313.9	56,560	2433	13.42	1192	353.1
48	324.0	62,210	2592	13.86	1230	364.5
49-1/2	334.1	68,220	2757	14.29	1269	375.9
51	344.3	74,620	2926	14.72	1307	387.3
52-1/2	354.4	81,400	3101	15.16	1346	398.7
54	364.5	88,570	3281	15.59	1384	410.1
55-1/2	374.6	96,160	3465	16.02	1422	421.5
57	384.8	104,200	3655	16.45	1461	432.8
58-1/2	394.9	112,600	3850	16.89	1499	444.2
60	405.0	121,500	4050	17.32	1538	455.6
		8-3/4 in. Width			**(r_y = 2.526 in.)**	
9	78.75	531.6	118.1	2.598	502.4	114.8
10-1/2	91.88	844.1	160.8	3.031	586.2	134.0
12	105.0	1260	210.0	3.464	669.9	153.1
13-1/2	118.1	1794	265.8	3.897	753.7	172.3
15	131.3	2461	328.1	4.330	837.4	191.4
16-1/2	144.4	3,276	397.0	4.763	921.1	210.5
18	157.5	4,253	472.5	5.196	1005	229.7
19-1/2	170.6	5,407	554.5	5.629	1089	248.8
21	183.8	6,753	643.1	6.062	1172	268.0
22-1/2	196.9	8,306	738.3	6.495	1256	287.1
24	210.0	10,080	840.0	6.928	1340	306.3
25-1/2	223.1	12,090	948.3	7.361	1424	325.4
27	236.3	14,350	1063	7.794	1507	344.5
28-1/2	249.4	16,880	1185	8.227	1591	363.7

TABLE B.5 (continued)
Section Properties of *Western Species* Structural Glued
Laminated Timber (GLULAM)

Depth d (in.)	Area A (in.²)	X–X Axis			Y–Y Axis	
		I_x (in.⁴)	S_x (in.³)	r_x (in.)	I_y (in.⁴)	S_y (in.³)
30	262.5	19,690	1313	8.660	1675	382.8
31-1/2	275.6	22,790	1447	9.093	1759	402.0
33	288.8	26,200	1588	9.526	1842	421.1
34-1/2	301.9	29,940	1736	9.959	1926	440.2
36	315.0	34,020	1890	10.39	2010	459.4
37-1/2	328.1	38,450	2051	10.83	2094	478.5
39	341.3	43,250	2218	11.26	2177	497.7
40-1/2	354.4	48,440	2392	11.69	2261	516.8
42	367.5	54,020	2573	12.12	2345	535.9
43-1/2	380.6	60,020	2760	12.56	2428	555.1
45	393.8	66,450	2953	12.99	2512	574.2
46-1/2	406.9	73,310	3153	13.42	2596	593.4
48	420.0	80,640	3360	13.86	2680	612.5
49-1/2	433.1	88,440	3573	14.29	2763	631.6
51	446.3	96,720	3793	14.72	2847	650.8
52-1/2	459.4	105,500	4020	15.16	2931	669.9
54	472.5	114,800	4253	15.59	3015	689.1
55-1/2	485.6	124,700	4492	16.02	3098	708.2
57	498.8	135,000	4738	16.45	3182	727.3
58-1/2	511.9	146,000	4991	16.89	3266	746.5
60	525.0	157,500	5250	17.32	3350	765.6
		10-3/4 in. Width			**(r_y = 3.103 in.)**	
12	129.0	1,548	258.0	3.464	1242	231.1
13-1/2	145.1	2,204	326.5	3.897	1398	260.0
15	161.3	3,023	403.1	4.330	1553	288.9
16-1/2	177.4	4,024	487.8	4.763	1708	317.8
18	193.5	5,225	580.5	5.196	1863	346.7
19-1/2	209.6	6,642	681.3	5.629	2019	375.6
21	225.8	8,296	790.1	6.062	2174	404.5
22-1/2	241.9	10,200	907.0	6.495	2329	433.4
24	258.0	12,380	1032	6.928	2485	462.3
25-1/2	274.1	14,850	1165	7.361	2640	491.1
27	290.3	17,630	1306	7.794	2795	520.0
28-1/2	306.4	20,740	1455	8.227	2950	548.9
30	322.5	24,190	1613	8.660	3106	577.8
31-1/2	338.6	28,000	1778	9.093	3261	606.7
33	354.8	32,190	1951	9.526	3416	635.6
34-1/2	370.9	36,790	2133	9.959	3572	664.5
36	387.0	41,800	2322	10.39	3727	693.4
37-1/2	403.1	47,240	2520	10.83	3882	722.3
39	419.3	53,140	2725	11.26	4037	751.2
40-1/2	435.4	59,510	2939	11.69	4193	780.0
42	451.5	66,370	3161	12.12	4348	808.9
43-1/2	467.6	73,740	3390	12.56	4503	837.8

(*continued*)

TABLE B.5 (continued)
Section Properties of *Western Species* Structural Glued
Laminated Timber (GLULAM)

Depth d (in.)	Area A (in.^{2})	X–X Axis			Y–Y Axis	
		I_x (in.^{4})	S_x (in.^{3})	r_x (in.)	I_y (in.^{4})	S_y (in.^{3})
45	483.8	81,630	3628	12.99	4659	866.7
46-1/2	499.9	90,070	3874	13.42	4814	895.6
48	516.0	99,070	4128	13.86	4969	924.5
49-1/2	532.1	108,700	4390	14.29	5124	953.4
51	548.3	118,800	4660	14.72	5280	982.3
52-1/2	564.4	129,600	4938	15.16	5435	1011
54	580.5	141,100	5225	15.59	5590	1040
55-1/2	596.6	153,100	5519	16.02	5746	1069
57	612.8	165,900	5821	16.45	5901	1098
58-1/2	628.9	179,300	6132	16.89	6056	1127
60	645.0	193,500	6450	17.32	6211	1156

Source: Courtesy of the American Forest & Paper Association, Washington, DC.

TABLE B.6
Section Properties of *Southern Pine* Structural Glued
Laminated Timber (GLULAM)

Depth d (in.)	Area A (in.^{2})	X–X Axis			Y–Y Axis	
		I_x (in.^{4})	S_x (in.^{3})	r_x (in.)	I_y (in.^{4})	S_y (in.^{3})
		3 in. Width			(r_y = 0.866 in.)	
5-1/2	16.50	41.59	15.13	1.588	12.38	8.250
6-7/8	20.63	81.24	23.63	1.985	15.47	10.31
8-1/4	24.75	140.4	34.03	2.382	18.56	12.38
9-5/8	28.88	222.9	46.32	2.778	21.66	14.44
11	33.00	332.8	60.50	3.175	24.75	16.50
12-3/8	37.13	473.8	76.57	3.572	27.84	18.56
13-3/4	41.25	649.9	94.53	3.969	30.94	20.63
15-1/8	45.38	865.0	114.4	4.366	34.03	22.69
16-1/2	49.50	1,123	136.1	4.763	37.13	24.75
17-7/8	53.63	1,428	159.8	5.160	40.22	26.81
19-1/4	57.75	1,783	185.3	5.557	43.31	28.88
20-5/8	61.88	2,193	212.7	5.954	46.41	30.94
22	66.00	2,662	242.0	6.351	49.50	33.00
23-3/8	70.13	3193	273.2	6.748	52.59	35.06
		5 in. Width			(r_y = 1.443 in.)	
6-7/8	34.38	135.4	39.39	1.985	71.61	28.65
8-1/4	41.25	234.0	56.72	2.382	85.94	34.38
9-5/8	48.13	371.5	77.20	2.778	100.3	40.10
11	55.00	554.6	100.8	3.175	114.6	45.83
12-3/8	61.88	789.6	127.6	3.572	128.9	51.56
13-3/4	68.75	1,083	157.6	3.969	143.2	57.29
15-1/8	75.63	1,442	190.6	4.366	157.6	63.02

TABLE B.6 (continued)
Section Properties of *Southern Pine* Structural Glued
Laminated Timber (GLULAM)

Depth d (in.)	Area A (in.²)	X–X Axis			Y–Y Axis	
		I_x (in.⁴)	S_x (in.³)	r_x (in.)	I_y (in.⁴)	S_y (in.³)
16-1/2	82.50	1,872	226.9	4.763	171.9	68.75
17-7/8	89.38	2,380	266.3	5.160	186.2	74.48
19-1/4	96.25	2,972	308.8	5.557	200.5	80.21
20-5/8	103.1	3,656	354.5	5.954	214.8	85.94
22	110.0	4,437	403.3	6.351	229.2	91.67
23-3/8	116.9	5,322	455.3	6.748	243.5	97.40
24-3/4	123.8	6,317	510.5	7.145	257.8	103.1
26-1/8	130.6	7,429	568.8	7.542	272.1	108.9
27-1/2	137.5	8,665	630.2	7.939	286.5	114.6
28-7/8	144.4	10,030	694.8	8.335	300.8	120.3
30-1/4	151.3	11,530	762.6	8.732	315.1	126.0
31-5/8	158.1	13,180	833.5	9.129	329.4	131.8
33	165.0	14,970	907.5	9.526	343.8	137.5
34-3/8	171.9	16,920	984.7	9.923	358.1	143.2
35-3/4	178.8	19,040	1065	10.32	372.4	149.0
		6-3/4 in. Width			**($r_y = 1.949$ in.)**	
6-7/8	46.41	182.8	53.17	1.985	176.2	52.21
8-1/4	55.69	315.9	76.57	2.382	211.4	62.65
9-5/8	64.97	501.6	104.2	2.778	246.7	73.09
11	74.25	748.7	136.1	3.175	281.9	83.53
12-3/8	83.53	1,066	172.3	3.572	317.2	93.97
13-3/4	92.81	1,462	212.7	3.969	352.4	104.4
15-1/8	102.1	1,946	257.4	4.366	387.6	114.9
16-1/2	111.4	2,527	306.3	4.763	422.9	125.3
17-7/8	120.7	3,213	359.5	5.160	458.1	135.7
19-1/4	129.9	4,012	416.9	5.557	493.4	146.2
20-5/8	139.2	4,935	478.6	5.954	528.6	156.6
22	148.5	5,990	544.5	6.351	563.8	167.1
23-3/8	157.8	7,184	614.7	6.748	599.1	177.5
24-3/4	167.1	8,528	689.1	7.145	634.3	187.9
26-1/8	176.3	10,030	767.8	7.542	669.6	198.4
27-1/2	185.6	11,700	850.8	7.939	704.8	208.8
28-7/8	194.9	13,540	938.0	8.335	740.0	219.3
30-1/4	204.2	15,570	1029	8.732	775.3	229.7
31-5/8	213.5	17,790	1125	9.129	810.5	240.2
33	222.8	20,210	1225	9.526	845.8	250.6
34-3/8	232.0	22,850	1329	9.923	881.0	261.0
35-3/4	241.3	25,700	1438	10.32	916.2	271.5
37-1/8	250.6	28,780	1551	10.72	951.5	281.9
38-1/2	259.9	32,100	1668	11.11	986.7	292.4
39-7/8	269.2	35,660	1789	11.51	1022	302.8
41-1/4	278.4	39,480	1914	11.91	1057	313.2
42-5/8	287.7	43,560	2044	12.30	1092	323.7
44	297.0	47,920	2178	12.70	1128	334.1

(*continued*)

TABLE B.6 (continued)
Section Properties of *Southern Pine* Structural Glued
Laminated Timber (GLULAM)

Depth *d* (in.)	Area *A* (in.²)	X–X Axis I_x (in.⁴)	S_x (in.³)	r_x (in.)	Y–Y Axis I_y (in.⁴)	S_y (in.³)
45-3/8	306.3	52,550	2316	13.10	1163	344.6
46-3/4	315.6	57,470	2459	13.50	1198	355.0
48-1/8	324.8	62,700	2606	13.89	1233	365.4
49-1/2	334.1	68,220	2757	14.29	1269	375.9
50-7/8	343.4	74,070	2912	14.69	1304	386.3
52-1/4	352.7	80,240	3071	15.08	1339	396.8
53-5/8	362.0	86,740	3235	15.48	1374	407.2
55	371.3	93,590	3403	15.88	1410	417.7
56-3/8	380.5	100,800	3575	16.27	1445	428.1
57-3/4	389.8	108,300	3752	16.67	1480	438.5
59-1/8	399.1	116,300	3933	17.07	1515	449.0
60-1/2	408.4	124,600	4118	17.46	1551	459.4
		8-1/2 in. Width			**($r_y = 2.454$ in.)**	
9-5/8	81.81	631.6	131.2	2.778	492.6	115.9
11	93.50	942.8	171.4	3.175	562.9	132.5
12-3/8	105.2	1,342	216.9	3.572	633.3	149.0
13-3/4	116.9	1,841	267.8	3.969	703.7	165.6
15-1/8	128.6	2,451	324.1	4.366	774.1	182.1
16-1/2	140.3	3,182	385.7	4.763	844.4	198.7
17-7/8	151.9	4,046	452.6	5.160	914.8	215.2
19-1/4	163.6	5,053	525.0	5.557	985.2	231.8
20-5/8	175.3	6,215	602.6	5.954	1056	248.4
22	187.0	7,542	685.7	6.351	1126	264.9
23-3/8	198.7	9,047	774.1	6.748	1196	281.5
24-3/4	210.4	10,740	867.8	7.145	1267	298.0
26-1/8	222.1	12,630	966.9	7.542	1337	314.6
27-1/2	233.8	14,730	1071	7.939	1407	331.1
28-7/8	245.4	17,050	1181	8.335	1478	347.7
30-1/4	257.1	19,610	1296	8.732	1548	364.3
31-5/8	268.8	22,400	1417	9.129	1618	380.8
33	280.5	25,460	1543	9.526	1689	397.4
34-3/8	292.2	28,770	1674	9.923	1759	413.9
35-3/4	303.9	32,360	1811	10.32	1830	430.5
37-1/8	315.6	36,240	1953	10.72	1900	447.0
38-1/2	327.3	40,420	2100	11.11	1970	463.6
39-7/8	338.9	44,910	2253	11.51	2041	480.2
41-1/4	350.6	49,720	2411	11.91	2111	496.7
42-5/8	362.3	54,860	2574	12.30	2181	513.3
44	374.0	60,340	2743	12.70	2252	529.8
45-3/8	385.7	66,170	2917	13.10	2322	546.4
46-3/4	397.4	72,370	3096	13.50	2393	562.9
48-1/8	409.1	78,950	3281	13.89	2463	579.5
49-1/2	420.8	85,910	3471	14.29	2533	596.1
50-7/8	432.4	93,270	3667	14.69	2604	612.6
52-1/4	444.1	101,000	3868	15.08	2674	629.2

TABLE B.6 (continued)
Section Properties of *Southern Pine* Structural Glued
Laminated Timber (GLULAM)

Depth d (in.)	Area A (in.²)	X–X Axis			Y–Y Axis	
		I_x (in.⁴)	S_x (in.³)	r_x (in.)	I_y (in.⁴)	S_y (in.³)
53-5/8	455.8	109,200	4074	15.48	2744	645.7
55	467.5	117,800	4285	15.88	2815	662.3
56-3/8	479.2	126,900	4502	16.27	2885	678.8
57-3/4	490.9	136,400	4725	16.67	2955	695.4
59-1/8	502.6	146,400	4952	17.07	3026	712.0
60-1/2	514.3	156,900	5185	17.46	3096	728.5
		10-½ in. Width			**(r_y = 3.031 in.)**	
11	115.5	1,165	211.8	3.175	1061	202.1
12-3/8	129.9	1,658	268.0	3.572	1194	227.4
13-3/4	144.4	2,275	330.9	3.969	1326	252.7
15-1/8	158.8	3,028	400.3	4.366	1459	277.9
16-1/2	173.3	3,931	476.4	4.763	1592	303.2
17-7/8	187.7	4,997	559.2	5.160	1724	328.5
19-1/4	202.1	6,242	648.5	5.557	1857	353.7
20-5/8	216.6	7,677	744.4	5.954	1990	379.0
22	231.0	9,317	847.0	6.351	2122	404.3
23-3/8	245.4	11,180	956.2	6.748	2255	429.5
24-3/4	259.9	13,270	1072	7.145	2388	454.8
26-1/8	274.3	15,600	1194	7.542	2520	480.0
27-1/2	288.8	18,200	1323	7.939	2653	505.3
28-7/8	303.2	21,070	1459	8.335	2786	530.6
30-1/4	317.6	24,220	1601	8.732	2918	555.8
31-5/8	332.1	27,680	1750	9.129	3051	581.1
33	346.5	31,440	1906	9.526	3183	606.4
34-3/8	360.9	35,540	2068	9.923	3316	631.6
35-3/4	375.4	39,980	2237	10.32	3449	656.9
37-1/8	389.8	44,770	2412	10.72	3581	682.2
38-1/2	404.3	49,930	2594	11.11	3714	707.4
39-7/8	418.7	55,480	2783	11.51	3847	732.7
41-1/4	433.1	61,420	2978	11.91	3979	758.0
42-5/8	447.6	67,760	3180	12.30	4112	783.2
44	462.0	74,540	3388	12.70	4245	808.5
45-3/8	476.4	81,740	3603	13.10	4377	833.8
46-3/4	490.9	89,400	3825	13.50	4510	859.0
48-1/8	505.3	97,530	4053	13.89	4643	884.3
49-1/2	519.8	106,100	4288	14.29	4775	909.6
50-7/8	534.2	115,200	4529	14.69	4908	934.8
52-1/4	548.6	124,800	4778	15.08	5040	960.1
53-5/8	563.1	134,900	5032	15.48	5173	985.4
55	577.5	145,600	5294	15.88	5306	1011
56-3/8	591.9	156,800	5562	16.27	5438	1036
57-3/4	606.4	168,500	5836	16.67	5571	1061
59-1/8	620.8	180,900	6118	17.07	5704	1086
60-1/2	635.3	193,800	6405	17.46	5836	1112

Source: Courtesy of the American Forest & Paper Association, Washington, DC.

TABLE B.7
Reference Design Values for Structural Glued Laminated Softwood Timber (Members Stressed Primarily in Bending)

Stress Class	Bending about X–X Axis Loaded Perpendicular to Wide Faces of Laminations						Bending about Y–Y Axis Loaded Parallel to Wide Faces of Laminations					Axially Loaded			Fasteners
	Extreme Fiber In Bending		Compression Perpendicular to Grain $F_{c\perp x}$ (psi)	Shear Parallel to Grain (Horizontal) F_{vx}^{d} (psi)	Modulus of Elasticity E_x (10^6 psi)	Modulus of Elasticity for Beam and Column Stability $E_{x\,min}$ (10^6 psi)	Extreme Fiber in Bending F_{by} (psi)	Compression Perpendicular to Grain $F_{c\perp y}$ (psi)	Shear Parallel to Grain (Horizontal) $F_{vy}^{d,e}$ (psi)	Modulus of Elasticity E_y (10^6 psi)	Modulus of Elasticity for Beam and Column Stability $E_{y\,min}$ (10^6 psi)	Tension Parallel to Grain F_t (psi)	Compression Parallel to Grain F_c (psi)	Modulus of Elasticity E_{axial} (10^6 psi)	Specific Gravity for Fastener Design G
	Tension Zone Stressed in Tension (Positive Bending) F_{bx}^{+} (psi)	Compression Zone Stressed in Tension (Negative Bending) F_{bx}^{-}[a] (psi)													
16F-1.3E	1600	925	315	195	1.3	0.67	800	315	170	1.1	0.57	675	925	1.2	0.42
20F-1.5E	2000	1100	425	210[f]	1.5	0.78	800	315	185	1.2	0.62	725	925	1.3	0.42
24F-1.7E	2400	1450	500	210[f]	1.7	0.88	1050	315	185	1.3	0.67	775	1000	1.4	0.42
24F-1.8E	2400	1450[b]	650	265[c]	1.8	0.93	1450	560	230[c]	1.6	0.83	1100	1600	1.7	0.50[i]
26F-1.9E[g]	2600	1950	650	265[c]	1.9	0.98	1600	560	230[c]	1.6	0.83	1150	1600	1.7	0.50[i]
28F-2.1E SP[g]	2800	2300	740	300	2.1[i]	1.09[i]	1600	650	260	1.7	0.88	1250	1750	1.7	0.55
30F-2.1E SP[g,h]	3000	2400	740	300	2.1[i]	1.09[i]	1750	650	260	1.7	0.88	1250	1750	1.7	0.55

Source: Courtesy of the American Forest & Paper Association, Washington, DC.

Design values in this table represent design values for groups of similar structural glued laminated timber combinations. Higher design values for some properties may be specified by manufacturer.

a   For balanced layups, F_{bx}^- shall be equal to F_{bx}^+ for the stress class. Designer shall specify when balanced layup is required.

b   Negative bending stress, F_{bx}^+, is permitted to be increased to 1850 psi for Douglas Fir and to 1950 psi for Southern Pine for specific combinations. Designer shall specify when these increased stresses are required.

c   For structural glued laminated timber of Southern Pine, the basic shear design values, F_{vx} and F_{vy}, are permitted to be increased to 300 and 260 psi, respectively.

d   The design value for shear, F_{vx} and F_{vy}, shall be decreased by multiplying by a factor of 0.72 for non-prismatic members, notched members, and for all members subject to impact or cyclic loading. The reduced design value shall be used for design of members at connections that transfer shear by mechanical fasteners (NDS 3.4.3.3). The reduced design value shall also be used for determination of design values for radial tension (NDS 5.2.2).

e   Design values are for timbers with laminations made from a single piece of lumber across the width or multiple pieces that have been edge bonded. For timbers manufactured from multiple piece laminations (across width) that are not edge bonded, value shall be multiplied by 0.4 for members with five, seven, or nine laminations or by 0.5 for all other members. This reduction shall be cumulative with the adjustment in footnote (d).

f   Certain Southern Pine combinations may contain lumber with wane. If lumber with wane is used, the design value for shear parallel to grain, F_{vx}, shall be multiplied by 0.67 if wane is allowed on both sides. If wane is limited to one side, F_{vx} shall be multiplied by 0.83. This reduction shall be cumulative with the adjustment in footnote (d).

g   26F, 28F, and 30F beams are not produced by all manufacturers, therefore, availability may be limited. Contact supplier or manufacturer for details.

h   30F combinations are restricted to a maximum 6 in. nominal width.

i   For 28F and 30F members with more than 15 laminations, $E_x = 2.0$ million psi and $E_{x,min} = 1.04$ million psi.

j   For structural glued laminated timber of Southern Pine, specific gravity for fastener design is permitted to be increased to 0.55.

TABLE B.8
Reference Design Values for Structural Glued Laminated Softwood Timber (Members Stressed Primarily in Axial Tension or Compression)

Identification Number	Species	Grade	Modulus of Elasticity E (10⁶ psi)	Modulus of Elasticity for Beam and Column Stability, E_{min} (10⁶ psi)	Compression Perpendicular to Grain, $F_{c\perp}$ (psi)	Tension Parallel to Grain, Two or More Laminations, F_t (psi)	Compression Parallel to Grain, Four or More Laminations, F_c (psi)	Compression Parallel to Grain, Two or Three Laminations, F_c (psi)	Bending, Four or More Laminations, F_{by} (psi)	Bending, Three Laminations, F_{by} (psi)	Bending, Two Laminations, F_{by} (psi)	Shear Parallel to Grain[a,b,c], F_{vy} (psi)	Bending[d], Two Laminations to 15 in. Deep,[e] F_{bx} (psi)	Shear Parallel to Grain[c], F_{vx} (psi)
Visually Graded Western Species														
1	DF	L3	1.5	0.78	560	900	1550	1200	1450	1250	1000	230	1250	265
2	DF	L2	1.6	0.83	560	1250	1950	1600	1800	1600	1300	230	1700	265
3	DF	L2D	1.9	0.98	650	1450	2300	1850	2100	1850	1550	230	2000	265
4	DF	L1CL	1.9	0.98	590	1400	2100	1900	2200	2000	1650	230	1900	265
5	DF	L1D	2.0	1.04	650	1600	2400	2100	2400	2100	1800	230	2200	265
14	HF	L3	1.3	0.67	375	800	1100	975	1200	1050	850	190	1100	215
15	HF	L2	1.4	0.73	375	1050	1350	1300	1500	1350	1100	190	1450	215
16	HF	L1	1.6	0.83	375	1200	1500	1450	1750	1550	1300	190	1600	215
17	HF	L1D	1.7	0.88	500	1400	1750	1700	2000	1850	1550	190	1900	215
22	SW	L3	1.0	0.52	315	525	850	675	800	700	550	170	725	195

#	Ratio															
69		AC	L3	1.2	0.62	470	725	1150	1100	1100	975	775	230	1000	265
70		AC	L2	1.3	0.67	470	975	1450	1450	1400	1250	1000	230	1350	265
71		AC	L1D	1.6	0.83	560	1250	1900	1900	1850	1650	1400	230	1700	265
72		AC	L1S	1.6	0.83	560	1250	1900	1900	1850	1650	1400	230	1900	265
Visually Graded Southern Pine															
47		SP	N2M14	1.4	0.73	650	1200	1900	1150	1750	1550	1300	260	1400	300
47	1:10	SP	N2M10	1.4	0.73	650	1150	1700	1150	1750	1550	1300	260	1400	300
47	1:8	SP	N2M	1.4	0.73	650	1000	1500	1150	1600	1550	1300	260	1350	300
48		SP	N2D14	1.7	0.88	740	1400	2200	1350	2000	1800	1500	260	1600	300
48	1:10	SP	N2D10	1.7	0.88	740	1350	2000	1350	2000	1800	1500	260	1600	300
48	1:8	SP	N2D	1.7	0.88	740	1150	1750	1350	1850	1800	1500	260	1600	300
49		SP	N1M16	1.7	0.88	650	1350	2100	1450	1950	1750	1500	260	1800	300
49	1:12	SP	N1M12	1.7	0.88	650	1300	1900	1450	1950	1750	1500	260	1750	300
49	1:10	SP	N1M	1.7	0.88	650	1150	1700	1450	1850	1750	1500	260	1550	300
50		SP	N1D14	1.9	0.98	740	1550	2300	1700	2300	2100	1750	260	2100	300
50	1:12	SP	N1D12	1.9	0.98	740	1550	2200	1700	2300	2100	1750	260	2100	300
50	1:10	SP	N1D	1.9	0.98	740	1350	2000	1700	2100	2100	1750	260	1800	300

Source: Courtesy of the American Forest & Paper Association, Washington, DC.

a For members with two or three laminations, the shear design value for transverse loads parallel to the wide faces of the laminations, F_{vy}, shall be reduced by multiplying by a factor of 0.84 or 0.95, respectively.

b The shear design value for transverse loads applied parallel to the wide faces of the laminations, F_{vy}, shall be multiplied by 0.4 for members with five, seven, or nine laminations manufactured from multiple piece laminations (across width) that are not edge bonded. The shear design value, F_{vy}, shall be multiplied by 0.5 for all other members manufactured from multiple piece laminations with unbonded edge joints. This reduction shall be cumulative with the adjustment in footnotes (a) and (c).

c The design value for shear, F_{vx} and F_{vy}, shall be decreased by multiplying by a factor of 0.72 for non-prismatic members, notched members, and for all members subject to impact or cyclic loading. The reduced design value shall be used for design of members at connections (NDS 3.4.3.3) that transfer shear by mechanical fasteners. The reduced design value shall also be used for determination of design values for radial tension (NDS 5.2.2).

d Tabulated design values are for members without special tension laminations. If special tension laminations are used, the design value for bending, F_{bx}, shall be permitted to be increased by multiplying by 1.18. This factor shall not be applied cumulatively with the adjustment in footnote (e).

e For members greater than 15 in. deep and without special tension laminations, the bending design value, F_{bx}, shall be reduced by multiplying by a factor of 0.88. This factor shall not be applied cumulatively with the adjustment in footnote (d).

TABLE B.9
Reference Design Values for Structural Composite Lumber

Grade	Orientation	G Shear of Elasticity psi	E Modulus of Elasticity psi	F_b^a Flexural Stress psi	F_t^b Tension Stress psi	$F_{c\perp}^c$ Compression Perpendicular to Grain psi	$F_c\|$ Compression Parallel to Grain psi	F_v Horizontal Shear Parallel to Grain psi
TimberStrand LSL								
1.3E	Beam/Column	81,250	1.3×10^6	3,140	1,985	1,240	2,235	745
	Plank	81,250	1.3×10^6	3,510		790	2,235	280
1.55E	Beam	96,875	1.55×10^6	4,295	1,975	1,455	3,270	575
Microllam LVL								
1.9E	Beam	118,750	1.9×10^6	4,805	2,870	1,365	4,005	530
Parallam PSL								
1.8E			1.8×10^6					
and 2.0E	Column	112,500	2.0×10^6	4,435	3,245	775	3,990	355
2.0E	Beam	125,000	2.0×10^6	5,360	3,750	1,365	4,630	540

[a] For 12 in., depth, For other depths, multiply, F_b by the factors as follows:

For TimberStrand LSL, multiply by $[12/d]^{0.092}$

For Microllam LVL, multiply by $[12/d]^{0.136}$

For Parallam, PSL, multiply by $[12/d]^{0.111}$

[b] F_t has been adjusted to reflect the volume effects for most standard applications.

[c] $F_{c\perp}$ shall not be increased for duration of load.

TABLE B.10
Common Wire, Box, or Sinker Nails: Reference Lateral Design Values (Z) for Single Shear (Two Member) Connections

Side Member Thickness, t_s (in.)	Nail Diameter, D (in.)	Common Wire Nail Pennyweight	Box Nail Pennyweight	Sinker Nail Pennyweight	G = 0.67 Red Oak (lb)	G = 0.55 Mixed Maple Southern Pine (lb)	G = 0.5 Douglas Fir-Larch (lb)	G = 0.49 Douglas Fir-Larch (N) (lb)	G = 0.46 Douglas Fir (S) Hem-Fir (N) (lb)	G = 0.43 Hem-Fir (lb)	G = 0.42 Spruce-Pine-Fir (lb)	G = 0.37 Redwood (Open Grain) (lb)	G = 0.36 Eastern Softwoods Spruce-Pine-Fir (S) Western Cedars Western Woods (lb)	G = 0.35 Northern Species (lb)
3/4	0.099	6d		7d	73	61	55	54	51	48	47	39	38	36
	0.113	6d	8d	8d	94	79	72	71	65	58	57	47	46	44
	0.120			10d	107	89	80	77	71	64	62	52	50	48
	0.128		10d		121	101	87	84	78	70	68	57	56	54
	0.131	8d			127	104	90	87	80	73	70	60	58	56
	0.135		16d	12d	135	108	94	91	84	76	74	63	61	58
	0.148	10d	20d	16d	154	121	105	102	94	85	83	70	69	66
	0.162	16d	40d		183	138	121	117	108	99	96	82	80	77
	0.177			20d	200	153	134	130	121	111	107	92	90	87
	0.192	20d		30d	206	157	138	134	125	114	111	96	93	90
	0.207	30d		40d	216	166	147	143	133	122	119	103	101	97
	0.225	40d			229	178	158	154	144	132	129	112	110	106
	0.244	50d		60d	234	182	162	158	147	136	132	115	113	109
1	0.099	6d		7d	73	61	55	54	51	48	47	42	41	40
	0.113	6d[b]	8d	8d	94	79	72	71	67	63	61	55	54	51
	0.120			10d	107	89	81	80	76	71	69	60	59	56
	0.128		10d		121	101	93	91	86	80	79	66	64	61
	0.131	8d			127	106	97	95	90	84	82	68	66	63
	0.135		16d	12d	135	113	103	101	96	89	86	71	69	66
	0.148	10d	20d	16d	154	128	118	115	109	99	96	80	77	74
	0.162	16d	40d		184	154	141	137	125	113	109	91	89	85
	0.177	20d		20d	213	178	155	150	138	125	121	102	99	95
	0.192	20d		30d	222	183	159	154	142	128	124	105	102	98

(continued)

TABLE B.10 (continued)
Common Wire, Box, or Sinker Nails: Reference Lateral Design Values (Z) for Single Shear (Two Member) Connections

Side Member Thickness, t_s (in.)	Nail Diameter, D (in.)	Pennyweight			G = 0.67 Red Oak (lb)	G = 0.55 Mixed Maple Southern Pine (lb)	G = 0.5 Douglas Fir-Larch (lb)	G = 0.49 Douglas Fir-Larch (N) (lb)	G = 0.46 Douglas Fir (S) Hem-Fir (N) (lb)	G = 0.43 Hem-Fir (lb)	G = 0.42 Spruce-Pine-Fir (lb)	G = 0.37 Redwood (Open Grain) (lb)	G = 0.36 Eastern Softwoods Spruce-Pine-Fir (S) Western Cedars Western Woods (lb)	G = 0.35 Northern Species (lb)
		Common Wire Nail	Box Nail	Sinker Nail										
	0.207	30d		40d	243	192	167	162	149	135	131	111	109	104
	0.225	40d		60d	268	202	177	171	159	144	140	120	117	112
	0.244	50d			274	207	181	175	162	148	143	123	120	115
1-1/4	0.099		6d	7d	73	61	55	54	51	48	47	42	41	40
	0.113	6d	8d	8d	94	79	72	71	67	63	61	55	54	52
	0.120			10d	107	89	81	80	76	71	69	62	60	59
	0.128		10d		121	101	93	91	86	80	79	70	69	67
	0.131	8d			127	106	97	95	90	84	82	73	72	70
	0.135		16d	12d	135	113	103	101	96	89	88	78	76	74
	0.148	10d	20d	16d	154	128	118	115	109	102	100	89	87	84
	0.162	16d	40d		184	154	141	138	131	122	120	103	100	95
	0.177			20d	213	178	163	159	151	141	136	113	110	105
	0.192	20d		30d	222	185	170	166	157	145	140	116	113	108
	0.207	30d		40d	243	203	186	182	169	152	147	123	119	114
	0.225	40d		60d	268	224	200	193	177	160	155	130	127	121
	0.244	50d			276	230	204	197	181	163	158	133	129	124
1-1/2	0.099			7d	73	61	55	54	51	48	47	42	41	40
	0.113		8d	8d	94	79	72	71	67	63	61	56	54	52
	0.120			10d	107	89	81	80	76	71	69	62	60	59
	0.128		10d		121	101	93	91	86	80	79	70	69	67
	0.131	8d			127	106	97	95	90	84	82	73	72	70

Length (in.)	Nail size	Nail size	Diameter (in.)									
	16d		0.135	135	113	103	101	96	89	88	78	76	74
	20d	10d	0.148	154	128	118	115	109	102	100	89	87	84
	40d	16d	0.162	184	154	141	138	131	122	120	106	104	101
		20d	0.177	213	178	163	159	151	141	138	123	121	117
		30d	0.192	222	185	170	166	157	147	144	128	126	120
		40d	0.207	243	203	186	182	172	161	158	135	131	125
			0.225	268	224	205	201	190	178	172	143	138	132
		50d	0.244	276	230	211	206	196	181	175	146	141	135
1-3/4		8*d*	0.113	94	79	72	71	67	63	61	55	54	52
		10*d*	0.120	107	89	81	80	76	71	69	62	60	59
	10d	12d	0.128	121	101	93	91	86	80	79	70	69	67
	16d	16d	0.135	135	113	103	101	96	89	88	78	76	74
	20d	20d	0.148	154	128	118	115	109	102	100	89	87	84
	40d	40d	0.162	184	154	141	138	131	122	120	106	104	101
		20d	0.177	213	178	163	159	151	141	138	123	121	117
		30d	0.192	222	185	170	166	157	147	144	128	126	122
		40d	0.207	243	203	186	182	172	161	158	140	137	133
			0.225	268	224	205	201	190	178	174	155	151	144
		60d	0.244	276	230	211	206	196	183	179	159	154	147

Source: Courtesy of the American Forest & Paper Association, Washington, DC.

a Single shear connection.

b Italic *d* indicates that the nail length is insufficient to provide 10 diameter penetration. Multiply the tabulated values by the ratio (penetration/10 diameter).

TABLE B.11
Nail and Spike Reference Withdrawal Design Values (*W*)

Pounds Per Inch of Penetration

Specific Gravity, *G*	Common Wire Nails, Box Nails, and Common Wire Spikes Diameter, *D*									
	0.099 in.	0.113 in.	0.128 in.	0.131 in.	0.135 in.	0.148 in.	0.162 in.	0.192 in.	0.207 in.	0.225 in.
0.73	62	71	80	82	85	93	102	121	130	141
0.71	58	66	75	77	79	87	95	113	121	132
0.68	52	59	67	69	71	78	85	101	109	118
0.67	50	57	65	66	68	75	82	97	105	114
0.58	35	40	45	46	48	52	57	68	73	80
0.55	31	35	40	41	42	46	50	59	64	70
0.51	25	29	33	34	35	38	42	49	53	58
0.50	24	28	31	32	33	36	40	47	50	55
0.49	23	26	30	30	31	34	38	45	48	52
0.47	21	24	27	27	28	31	34	40	43	47
0.46	20	22	25	26	27	29	32	38	41	45
0.44	18	20	23	23	24	26	29	34	37	40
0.43	17	19	21	22	23	25	27	32	35	38
0.42	16	18	20	21	21	23	26	30	33	35
0.41	15	17	19	19	20	22	24	29	31	33
0.40	14	16	18	18	19	21	23	27	29	31
0.39	13	15	17	17	18	19	21	25	27	29
0.38	12	14	16	16	17	18	20	24	25	28
0.37	11	13	15	15	16	17	19	22	24	26
0.36	11	12	14	14	14	16	17	21	22	24
0.35	10	11	13	13	14	15	16	19	21	23
0.31	7	8	9	10	10	11	12	14	15	17

Source: Courtesy of the American Forest & Paper Association, Washington, DC.

Common Nails/Spikes, Box Nails					Threaded Nails Wire Diameter, D				
0.244 in.	0.263 in.	0.283 in.	0.312 in.	0.375 in.	0.120 in.	0.135 in.	0.148 in.	0.177 in.	0.207 in.
153	165	178	196	236	82	93	102	121	141
143	154	166	183	220	77	87	95	113	132
128	138	149	164	197	69	78	85	101	118
124	133	144	158	190	66	75	82	97	114
86	93	100	110	133	46	52	57	68	80
76	81	88	97	116	41	46	50	59	70
63	67	73	80	96	34	38	42	49	58
60	64	69	76	91	32	36	40	47	55
57	61	66	72	87	30	34	38	45	52
51	55	59	65	78	27	31	34	40	47
48	52	56	62	74	26	29	32	38	45
43	47	50	55	66	23	26	29	34	40
41	44	47	52	63	22	25	27	32	38
38	41	45	49	59	21	23	26	30	35
36	39	42	46	56	19	22	24	29	33
34	37	40	44	52	18	21	23	27	31
32	34	37	41	49	17	19	21	25	29
30	32	35	38	46	16	18	20	24	28
28	30	33	36	43	15	17	19	22	26
26	28	30	33	40	14	16	17	21	24
24	26	28	31	38	13	15	16	19	23
18	19	21	23	28	10	11	12	14	17

TABLE B.12
Wood Screws: Reference Lateral Design Values (Z) for Single Shear (Two Member) Connections

Side Member Thickness, t_s (in.)	Wood Screw Diameter, D (in.)	Wood Screw Number	G = 0.67 Red Oak (lb)	G = 0.55 Mixed Maple Southern Pine (lb)	G = 0.5 Douglas Fir-Larch (lb)	G = 0.49 Douglas Fir-Larch (N) (lb)	G = 0.46 Douglas Fir (S) Hem-Fir (N) (lb)	G = 0.43 Hem-Fir (lb)	G = 0.42 Spruce-Pine-Fir (lb)	G = 0.37 Redwood (Open Grain) (lb)	G = 0.36 Eastern Softwoods Spruce-Pine-Fir (S) Western Cedars Western Woods (lb)	G = 0.35 Northern Species (lb)
1/2	0.138	6	88	67	59	57	53	49	47	41	40	38
	0.151	7	96	74	65	63	59	54	52	45	44	42
	0.164	8	107	82	73	71	66	61	59	51	50	48
	0.177	9	121	94	83	81	76	70	68	59	58	56
	0.190	10	130	101	90	87	82	75	73	64	63	60
	0.216	12	156	123	110	107	100	93	91	79	78	75
	0.242	14	168	133	120	117	110	102	99	87	86	83
5/8	0.138	6	94	76	66	64	59	53	52	44	43	41
	0.151	7	104	83	72	70	64	58	56	48	47	45
	0.164	8	120	92	80	77	72	65	63	54	53	51
	0.177	9	136	103	91	88	81	74	72	62	61	58
	0.190	10	146	111	97	94	88	80	78	67	65	63
	0.216	12	173	133	117	114	106	97	95	82	80	77
	0.242	14	184	142	126	123	115	106	103	89	87	84
3/4	0.138	6	94	79	72	71	65	58	57	47	46	44
	0.151	7	104	87	80	77	71	64	62	52	50	48
	0.164	8	120	101	88	85	78	71	69	58	56	54
	0.177	9	142	114	99	96	88	80	78	66	64	61
	0.190	10	153	122	107	103	95	86	83	71	69	66
	0.216	12	192	144	126	122	113	103	100	86	84	80
	0.242	14	203	154	135	131	122	111	108	93	91	87

1	0.138	6	94	79	72	71	67	63	61	55	54	51
	0.151	7	104	87	80	78	74	69	68	60	59	56
	0.164	8	120	101	92	90	85	80	78	67	65	62
	0.177	9	142	118	108	106	100	94	90	75	73	70
	0.190	10	153	128	117	114	108	101	97	81	78	75
	0.216	12	193	161	147	143	131	118	114	96	93	89
	0.242	14	213	178	157	152	139	126	122	102	100	95
1-1/4	0.138	6	94	79	72	71	67	63	61	55	54	52
	0.151	7	104	87	80	78	74	69	68	60	59	57
	0.164	8	120	101	92	90-	85	80	78	70	68	66
	0.177	9	142	118	108	106	100	94	92	82	80	78
	0.190	10	153	128	117	114	108	101	99	88	87	84
	0.216	12	193	161	147	144	137	128	125	108	105	100
	0.242	14	213	178	163	159	151	141	138	115	111	106
1-1/2	0.138	6	94	79	72	71	67	63	61	55	54	52
	0.151	7	104	87	80	78	74	69	68	60	59	57
	0.164	8	120	101	92	90	85	80	78	70	68	66
	0.177	9	142	118	108	106	100	94	92	82	80	78
	0.190	10	153	128	117	114	108	101	99	88	87	84
	0.216	12	193	161	147	144	137	128	125	111	109	106
	0.242	14	213	178	163	159	151	141	138	123	120	117
1-3/4	0.138	6	94	79	72	71	67	63	61	55	54	52
	0.151	7	104	87	80	78	74	69	68	60	59	57
	0.164	8	120	101	92	90	85	80	78	70	68	66
	0.177	9	142	118	108	106	100	94	92	82	80	78
	0.190	10	153	128	117	114	108	101	99	88	87	84
	0.216	12	193	161	147	144	137	128	125	111	109	106
	0.242	14	213	178	163	159	151	141	138	123	120	117

Source: Courtesy of the American Forest & Paper Association, Washington, DC.

[a] Single shear connection.

TABLE B.13
Cut Thread or Rolled Thread Wood Screw Reference Withdrawal Design Values (W)

Pounds Per Inch of Thread Penetration

Specific	Wood Screw Number										
Gravity, G	6	7	8	9	10	12	14	16	18	20	24
0.73	209	229	249	268	288	327	367	406	446	485	564
0.71	198	216	235	254	272	310	347	384	421	459	533
0.68	181	199	216	233	250	284	318	352	387	421	489
0.67	176	193	209	226	243	276	309	342	375	409	475
0.58	132	144	157	169	182	207	232	256	281	306	356
0.55	119	130	141	152	163	186	208	231	253	275	320
0.51	102	112	121	131	141	160	179	198	217	237	275
0.50	98	107	117	126	135	154	172	191	209	228	264
0.49	94	103	112	121	130	147	165	183	201	219	254
0.47	87	95	103	111	119	136	152	168	185	201	234
0.46	83	91	99	107	114	130	146	161	177	193	224
0.44	76	83	90	97	105	119	133	148	162	176	205
0.43	73	79	86	93	100	114	127	141	155	168	196
0.42	69	76	82	89	95	108	121	134	147	161	187
0.41	66	72	78	85	91	103	116	128	141	153	178
0.40	63	69	75	81	86	98	110	122	134	146	169
0.39	60	65	71	77	82	93	105	116	127	138	161
0.38	57	62	67	73	78	89	99	110	121	131	153
0.37	54	59	64	69	74	84	94	104	114	125	145
0.36	51	56	60	65	70	80	89	99	108	118	137
0.35	48	53	57	62	66	75	84	93	102	111	130
0.31	38	41	45	48	52	59	66	73	80	87	102

Source: Courtesy of the American Forest & Paper Association, Washington, DC.

Note: Tabulated withdrawal design values (*W*) are in pounds per inch of thread penetration into side grain of main member. Thread length is approximately 2/3 the total wood screw length.

TABLE B.14
Bolts: Reference Lateral Design Values (Z) for Single Shear (Two Member) Connections

Main Member, t_m (in.)	Side Member, t_s (in.)	Bolt Diameter, D (in.)	G = 0.67 Red Oak				G = 0.55 Mixed Maple Southern Pine				G = 0.50 Douglas Fir-Larch				G = 0.49 Douglas Fir-Larch (N)				G = 0.46 Douglas Fir (S) Hem-Fir (N)			
			$Z_\parallel$ (lb)	$Z_{s\perp}$ (lb)	$Z_{m\perp}$ (lb)	$Z_\perp$ (lb)	$Z_\parallel$ (lb)	$Z_{s\perp}$ (lb)	$Z_{m\perp}$ (lb)	$Z_\perp$ (lb)	$Z_\parallel$ (lb)	$Z_{s\perp}$ (lb)	$Z_{m\perp}$ (lb)	$Z_\perp$ (lb)	$Z_\parallel$ (lb)	$Z_{s\perp}$ (lb)	$Z_{m\perp}$ (lb)	$Z_\perp$ (lb)	$Z_\parallel$ (lb)	$Z_{s\perp}$ (lb)	$Z_{m\perp}$ (lb)	$Z_\perp$ (lb)
1-1/2	1-1/2	1/2	650	420	420	330	530	330	330	250	480	300	300	220	470	290	290	210	440	270	270	190
		5/8	810	500	500	370	660	400	400	280	600	360	360	240	590	350	350	240	560	320	320	220
		3/4	970	580	580	410	800	460	460	310	720	420	420	270	710	400	400	260	670	380	380	240
		7/8	1130	660	660	440	930	520	520	330	850	470	470	290	830	460	460	280	780	420	420	250
		1	1290	740	740	470	1060	580	580	350	970	530	530	310	950	510	510	300	890	480	480	280
1-3/4	1-3/4	1/2	760	490	490	390	620	390	390	290	560	350	350	250	550	340	340	250	520	320	320	230
		5/8	940	590	590	430	770	470	470	330	700	420	420	280	690	410	410	280	650	380	380	250
		3/4	1130	680	680	480	930	540	540	360	850	480	480	310	830	470	470	300	780	440	440	280
		7/8	1320	770	770	510	1080	610	610	390	990	550	550	340	970	530	530	320	910	500	500	300
		1	1510	860	860	550	1240	680	680	410	1130	610	610	360	1110	600	600	350	1040	560	560	320
2-1/2	1-1/2	1/2	770	480	540	440	660	400	420	350	610	370	370	310	610	360	360	300	580	340	330	270
		5/8	1070	660	630	520	930	560	490	390	850	520	430	340	830	520	420	330	780	470	390	300
		3/4	1360	890	720	570	1120	660	560	430	1020	590	500	380	1000	560	480	360	940	520	450	330
		7/8	1590	960	800	620	1300	720	620	470	1190	630	550	410	1170	600	540	390	1090	550	500	360
		1	1820	1020	870	660	1490	770	680	490	1360	680	610	440	1330	650	590	420	1250	600	550	390
3-1/2	1-1/2	1/2	770	480	560	440	660	400	470	360	610	370	430	330	610	360	420	320	580	340	400	310
		5/8	1070	660	760	590	940	560	620	500	880	520	540	460	870	520	530	450	830	470	490	410
		3/4	1450	890	900	770	1270	660	690	580	1200	590	610	510	1190	560	590	490	1140	520	550	450
		7/8	1890	960	990	830	1680	720	770	630	1590	630	680	550	1570	600	650	530	1470	550	600	480
		1	2410	1020	1080	890	2010	770	830	670	1830	680	740	590	1790	650	710	560	1680	600	660	520
3-1/2	1-3/4	1/2	830	510	590	480	720	420	510	390	670	380	470	350	660	380	460	340	620	360	440	320
		5/8	1160	680	820	620	1000	580	640	520	930	530	560	460	920	530	550	450	880	500	510	410

(*continued*)

TABLE B.14 (continued)
Bolts: Reference Lateral Design Values (Z) for Single Shear (Two Member) Connections

Main Member, t_m (in.)	Side Member, t_s (in.)	Bolt Diameter, D (in.)	$G = 0.67$ Red Oak				$G = 0.55$ Mixed Maple Southern Pine				$G = 0.50$ Douglas Fir-Larch				$G = 0.49$ Douglas Fir-Larch (N)				$G = 0.46$ Douglas Fir (S) Hem-Fir (N)			
			$Z_\parallel$ (lb)	$Z_{s\perp}$ (lb)	$Z_{m\perp}$ (lb)	$Z_\perp$ (lb)	$Z_\parallel$ (lb)	$Z_{s\perp}$ (lb)	$Z_{m\perp}$ (lb)	$Z_\perp$ (lb)	$Z_\parallel$ (lb)	$Z_{s\perp}$ (lb)	$Z_{m\perp}$ (lb)	$Z_\perp$ (lb)	$Z_\parallel$ (lb)	$Z_{s\perp}$ (lb)	$Z_{m\perp}$ (lb)	$Z_\perp$ (lb)	$Z_\parallel$ (lb)	$Z_{s\perp}$ (lb)	$Z_{m\perp}$ (lb)	$Z_\perp$ (lb)
		3/4	1530	900	940	780	1330	770	720	580	1250	680	640	520	1240	660	620	500	1190	600	580	460
		7/8	1970	1120	1040	840	1730	840	810	640	1620	740	710	550	1590	700	690	530	1490	640	640	490
		1	2480	1190	1130	900	2030	890	880	670	1850	790	780	590	1820	750	760	570	1700	700	700	530
3-1/2	3-1/2	1/2	830	590	590	530	750	520	520	460	720	490	490	430	710	480	480	420	690	460	460	410
		5/8	1290	880	880	780	1170	780	780	650	1120	700	700	560	1110	690	690	550	1070	650	650	500
		3/4	1860	1190	1190	950	1690	960	960	710	1610	870	870	630	1600	850	850	600	1540	800	800	560
		7/8	2540	1410	1410	1030	2170	1160	1160	780	1970	1060	1060	680	1940	1040	1040	650	1810	980	980	590
		1	3020	1670	1670	1100	2480	1360	1360	820	2260	1230	1230	720	2210	1190	1190	690	2070	1110	1110	640
5-1/4	1-1/2	5/8	1070	660	760	590	940	560	640	500	880	520	590	460	870	520	590	450	830	470	560	430
		3/4	1450	890	990	780	1270	660	850	660	1200	590	790	590	1190	560	780	560	1140	520	740	520
		7/8	1890	960	1260	960	1680	720	1060	720	1590	630	940	630	1570	600	900	600	1520	550	830	550
		1	2410	1020	1500	1020	2150	770	1140	770	2050	680	1010	680	2030	650	970	650	1930	600	910	600
5-1/4	1-3/4	5/8	1160	680	820	620	1000	580	690	520	930	530	630	470	920	530	630	470	880	500	590	440
		3/4	1530	900	1050	800	1330	770	890	680	1250	680	830	630	1240	660	810	620	1190	600	780	590
		7/8	1970	1120	1320	1020	1730	840	1090	840	1640	740	960	740	1620	700	920	700	1550	640	850	640
		1	2480	1190	1530	1190	2200	890	1170	890	2080	790	1040	790	2060	750	1000	750	1990	700	930	700
5-1/4	3-1/2	5/8	1290	880	880	780	1170	780	780	680	1120	700	730	630	1110	690	720	620	1070	650	690	580
		3/4	1860	1190	1240	1080	1690	960	1090	850	1610	870	1030	780	1600	850	1010	750	1540	800	970	710

Member	Side	Size[a]																				
5-1/2	1-1/2	7/8	2540	1410	1640	1260	2300	1160	1380	1000	2190	1060	1230	870	2170	1040	1190	840	2060	980	1100	770
		1	3310	1670	1940	1420	2870	1390	1520	1060	2660	1290	1360	940	2630	1260	1320	900	2500	1210	1230	830
5-1/2	3-1/2	5/8	1070	660	760	590	940	560	640	500	880	520	590	460	870	520	590	450	830	470	560	430
		3/4	1450	890	990	780	1270	660	850	660	1200	590	790	590	1190	560	780	560	1140	520	740	520
		7/8	1890	960	1260	960	1680	720	1090	720	1590	630	980	630	1570	600	940	600	1520	550	860	550
		1	2410	1020	1560	1020	2150	770	1190	770	2050	680	1060	680	2030	650	1010	650	1930	600	940	600
7-1/2	1-1/2	5/8	1290	880	880	780	1170	780	780	680	1120	700	730	630	1110	690	720	620	1070	650	690	580
		3/4	1860	1190	1240	1080	1690	960	1090	850	1610	870	1030	780	1600	850	1010	750	1540	800	970	710
		7/8	2540	1410	1640	1260	2300	1160	1410	1020	2190	1060	1260	910	2170	1040	1220	870	2060	980	1130	790
		1	3310	1670	1980	1470	2870	1390	1550	1100	2660	1290	1390	970	2630	1260	1340	930	2500	1210	1250	860
7-1/2	3-1/2	5/8	1070	660	760	590	940	560	640	500	880	520	590	460	870	520	590	450	830	470	560	430
		3/4	1450	890	990	780	1270	660	850	660	1200	590	790	590	1190	560	780	560	1140	520	740	520
		7/8	1890	960	1260	960	1680	720	1090	720	1590	630	1010	630	1570	600	990	600	1520	550	950	550
		1	2410	1020	1560	1020	2150	770	1350	770	2050	680	1270	680	2030	650	1240	650	1930	600	1190	600
7-1/2	3-1/2	5/8	1290	880	880	780	1170	780	780	680	1120	700	730	630	1110	690	720	620	1070	650	690	580
		3/4	1860	1190	1240	1080	1690	960	1090	850	1610	870	1030	780	1600	850	1010	750	1540	800	970	710
		7/8	2540	1410	1640	1260	2300	1160	1450	1020	2190	1060	1360	930	2170	1040	1340	900	2060	980	1280	850
		1	3310	1670	2090	1470	2870	1390	1830	1210	2660	1290	1630	1110	2630	1260	1570	1080	2500	1210	1470	1030

Source: Courtesy of the American Forest & Paper Association, Washington, DC.

[a] Single shear connection.

TABLE B.15
Lag Screws: Reference Lateral Design Values (Z) for Single Shear (Two Member) Connections

Side Member Thickness, t_s (in.)	Lag Screw Diameter, D (in.)	G = 0.67 Red Oak				G = 0.55 Mixed Maple Southern Pine				G = 0.50 Douglas Fir-Larch				G = 0.49 Douglas Fir-Larch (N)				G = 0.46 Douglas Fir (S) Hem-Fir (N)			
		$Z_{\parallel}$ (lb)	$Z_{s\perp}$ (lb)	$Z_{m\perp}$ (lb)	$Z_{\perp}$ (lb)	$Z_{\parallel}$ (lb)	$Z_{s\perp}$ (lb)	$Z_{m\perp}$ (lb)	$Z_{\perp}$ (lb)	$Z_{\parallel}$ (lb)	$Z_{s\perp}$ (lb)	$Z_{m\perp}$ (lb)	$Z_{\perp}$ (lb)	$Z_{\parallel}$ (lb)	$Z_{s\perp}$ (lb)	$Z_{m\perp}$ (lb)	$Z_{\perp}$ (lb)	$Z_{\parallel}$ (lb)	$Z_{s\perp}$ (lb)	$Z_{m\perp}$ (lb)	$Z_{\perp}$ (lb)
1/2	1/4	150	110	110	110	130	90	100	90	120	90	90	80	120	90	90	80	110	80	90	80
	5/16	170	130	130	120	150	110	120	100	150	100	110	100	140	100	110	90	140	100	100	90
	3/8	180	130	130	120	160	110	110	100	150	100	110	90	150	90	110	90	140	90	100	90
5/8	1/4	160	120	130	120	140	100	110	100	130	90	100	90	130	90	100	90	120	90	90	80
	5/16	190	140	140	130	160	110	120	110	150	110	110	100	150	100	110	100	150	100	110	90
	3/8	190	130	140	120	170	110	120	100	160	100	110	100	160	100	110	90	150	100	100	90
3/4	1/4	180	140	140	130	150	110	120	100	140	100	110	100	140	100	110	90	130	90	100	90
	5/16	210	150	160	140	180	120	130	120	170	110	120	100	160	110	120	100	160	100	110	100
	3/8	210	140	160	130	180	120	130	110	170	110	120	100	170	110	120	100	160	100	110	90
1	1/4	180	140	140	140	160	120	120	120	150	120	120	110	150	110	110	110	150	110	110	100
	5/16	230	170	170	160	210	140	150	130	190	130	140	120	190	120	140	120	180	120	130	110
	3/8	230	160	170	160	210	130	150	120	200	120	140	110	190	120	140	110	180	110	130	100
1-1/4	1/4	180	140	140	140	160	120	120	120	150	120	120	110	150	110	110	110	150	110	110	100
	5/16	230	170	170	160	210	150	150	140	200	140	140	130	200	140	140	130	190	130	140	120
	3/8	230	170	170	160	210	150	150	140	200	140	140	130	200	130	140	120	190	120	140	120
1-1/2	1/4	180	140	140	140	160	120	120	120	150	120	120	110	150	110	110	110	150	110	110	100
	5/16	230	170	170	160	210	150	150	140	200	140	140	130	200	140	140	130	190	140	140	130
	3/8	230	170	170	160	210	150	150	140	200	140	140	130	200	140	140	130	190	140	140	120
	7/16	360	260	260	240	320	230	230	200	310	200	210	180	310	190	210	180	300	180	200	160
	1/2	460	310	320	280	410	250	290	230	390	220	270	200	390	220	260	200	370	210	250	190
	5/8	700	410	500	370	600	340	420	310	560	310	380	280	550	310	380	270	530	290	360	260
	3/4	950	550	660	490	830	470	560	410	770	440	510	380	760	430	510	370	730	400	480	360
	7/8	1240	720	830	630	1080	560	710	540	1020	490	660	490	1010	470	650	470	970	430	610	430
	1	1550	800	1010	780	1360	600	870	600	1290	530	810	530	1280	500	790	500	1230	470	760	470

1-3/4	1/4	180	140	140	140	160	120	120	120	150	120	120	110	150	110	110	110	150	110	110	100
	5/16	230	170	170	160	210	150	150	140	200	140	140	130	200	140	140	130	190	140	140	130
	3/8	230	170	170	160	210	150	150	140	200	140	140	130	200	140	140	130	190	140	140	120
	7/16	360	260	260	240	320	230	230	210	310	210	210	190	300	210	210	190	300	200	200	180
	1/2	460	320	320	290	410	270	270	250	390	240	240	220	380	240	260	220	380	220	250	200
	5/8	740	500	500	400	660	360	360	320	600	330	320	290	570	320	410	290	570	300	390	270
	3/4	1030	720	720	520	890	480	440	430	820	450	450	390	780	440	540	380	780	420	510	360
	7/8	1320	890	890	650	1150	630	600	550	1060	570	550	510	1010	550	680	490	1010	500	650	470
	1	1630	1070	1070	790	1420	700	750	670	1320	610	590	610	1270	590	830	590	1270	550	790	550
2-1/2	1/4	180	140	140	140	160	120	120	120	150	120	120	110	150	110	110	110	150	110	110	100
	5/16	230	170	170	160	210	150	150	140	200	140	140	130	200	140	140	130	190	140	140	130
	3/8	230	170	170	160	210	150	150	140	200	140	140	130	200	140	140	130	190	140	140	120
	7/16	360	260	260	240	320	230	230	210	310	210	210	190	300	210	210	190	300	200	200	180
	1/2	460	320	320	290	410	290	290	250	390	270	270	240	390	260	260	230	380	250	250	220
	5/8	740	440	500	450	670	430	440	390	640	390	420	350	630	380	410	340	610	360	390	320
	3/4	1110	680	740	610	1010	550	650	490	960	500	610	450	950	490	600	430	920	460	580	410
	7/8	1550	830	1000	740	1370	690	880	600	1280	630	830	550	1260	620	810	530	1190	580	770	500
	1	1940	980	1270	860	1660	830	1080	720	1550	770	990	660	1520	750	970	640	1450	720	920	620
3-1/2	1/4	180	140	140	140	160	120	120	120	150	120	120	110	150	110	110	110	150	110	110	100
	5/16	230	170	170	160	210	150	150	140	200	140	140	130	200	140	140	130	190	140	140	130
	3/8	230	170	170	160	210	150	150	140	200	140	140	130	200	140	140	130	190	140	140	120
	7/16	360	260	260	240	320	230	230	210	310	210	210	190	300	210	210	190	300	200	200	180
	1/2	460	320	320	290	410	290	290	250	390	270	270	240	390	260	260	230	380	250	250	220
	5/8	740	500	500	450	670	440	440	390	640	420	420	360	630	410	410	360	610	390	390	340
	3/4	1110	740	740	650	1010	650	650	560	960	600	610	520	950	580	600	510	920	550	580	490
	7/8	1550	990	1000	860	1400	800	880	710	1340	720	830	640	1320	700	810	620	1280	660	780	570
	1	2020	1140	1270	1010	1830	930	1120	810	1740	850	1060	740	1730	830	1040	720	1670	790	1000	680

Source: Courtesy of the American Forest & Paper Association, Washington, DC.

[a] Single shear connection.

TABLE B.16
Lag Screw Reference Withdrawal Design Values (W)

Pounds Per Inch of Thread Penetration

Specific Gravity, G	Lag Screw Unthreaded Shank Diameter, D											
	1/4 in.	5/16 in.	3/8 in.	7/16 in.	1/2 in.	5/8 in.	3/4 in.	7/8 in.	1 in.	1-1/8 in.	1-1/4 in.
0.73	397	469	538	604	668	789	905	1016	1123	1226	1327
0.71	381	450	516	579	640	757	868	974	1077	1176	1273
0.68	357	422	484	543	600	709	813	913	1009	1103	1193
0.67	349	413	473	531	587	694	796	893	987	1078	1167
0.58	281	332	381	428	473	559	641	719	795	869	940
0.55	260	307	352	395	437	516	592	664	734	802	868
0.51	232	274	314	353	390	461	528	593	656	716	775
0.50	225	266	305	342	378	447	513	576	636	695	752
0.49	218	258	296	332	367	434	498	559	617	674	730
0.47	205	242	278	312	345	408	467	525	580	634	686
0.46	199	235	269	302	334	395	453	508	562	613	664
0.44	186	220	252	283	312	369	423	475	525	574	621
0.43	179	212	243	273	302	357	409	459	508	554	600
0.42	173	205	235	264	291	344	395	443	490	535	579
0.41	167	198	226	254	281	332	381	428	473	516	559
0.40	161	190	218	245	271	320	367	412	455	497	538
0.39	155	183	210	236	261	308	353	397	438	479	518
0.38	149	176	202	227	251	296	340	381	422	461	498
0.37	143	169	194	218	241	285	326	367	405	443	479
0.36	137	163	186	209	231	273	313	352	389	425	460
0.35	132	156	179	200	222	262	300	337	373	407	441
0.31	110	130	149	167	185	218	250	281	311	339	367

Source: Courtesy of the American Forrest & Paper Association, Washington, DC.
Notes: Tabulated withdrawal design values (W) are in pounds per inch of thread penetration into side grain of main member. Length of thread penetration in main member shall not include the length of the tapered tip.

Appendix C: Steel

TABLE C.1
Properties Data of Each Shape Follow the Dimensions Data Sheets

W Shapes

(a) Dimensions

Shape	Area, A (in.²)	Depth, d (in.)		Web Thickness, t_w (in.)		$\frac{t_w}{2}$ (in.)	Flange Width, b_t (in.)		Thickness, t_f (in.)		Distance k k_{des} (in.)	k_{det} (in.)	k_1 (in.)	T (in.)	Workable Gage (in.)
W21×93	27.3	21.6	21-5/8	0.580	9/16	5/16	8.42	8-3/8	0.930	15/16	1.43	1-5/8	15/16	18-3/8	5-1/2
×83[a]	24.3	21.4	21-3/8	0.515	1/2	1/4	8.36	8-3/8	0.835	13/16	1.34	1-1/2	7/8	⟶	⟶
×73[a]	21.5	21.2	21-1/4	0.455	7/16	1/4	8.30	8-1/4	0.740	3/4	1.24	1-7/16	7/8		
×68[a]	20.0	21.1	21-1/8	0.430	7/16	1/4	8.27	8-1/4	0.685	11/16	1.19	1-3/8	7/8		
×62[a]	18.3	21.0	21	0.400	3/8	3/16	8.24	8-1/4	0.615	5/8	1.12	1-5/16	13/16		
×55[a]	16.2	20.8	20-3/4	0.375	3/8	3/16	8.22	8-1/4	0.522	1/2	1.02	1-3/16	13/16		
×48[a,b]	14.1	20.6	20-5/8	0.350	3/8	3/16	8.14	8-1/8	0.430	7/16	0.930	1-1/8	13/16	⟶	⟶
W21×57[a]	16.7	21.1	21	0.405	3/8	3/16	6.56	6-1/2	0.650	5/8	1.15	1-5/16	13/16	18-3/8	3-1/2
×50[a]	14.7	20.8	20-7/8	0.380	3/8	3/16	6.53	6-1/2	0.535	9/16	1.04	1-1/4	13/16	⟶	⟶
×44[a]	13.0	20.7	20-5/8	0.350	3/8	3/16	6.50	6-1/2	0.450	7/16	0.950	1-1/8	13/16		

Shape														T	Gage
W18×311[d]	91.6	22.3	22-3/8	1.52	1-1/2	3/4	12.0	12	2.74	2-3/4	3.24	3-7/16	1-3/8	15-1/2 →	5-1/2 →
×283[d]	83.3	21.9	21-7/8	1.40	1-3/8	11/16	11.9	11-7/8	2.50	2-1/2	3.00	3-3/16	1-5/16		
×258[d]	75.9	21.5	21-1/2	1.28	1-1/4	5/8	11.8	11-3/4	2.30	2-5/16	2.70	3	1-1/4	15-1/8	
×234[d]	68.8	21.1	21	1.16	1-3/16	5/8	11.7	11-5/8	2.11	2-1/8	2.51	2-3/4	1-3/16		
×211	62.1	20.7	20-5/8	1.06	1-1/16	9/16	11.6	11-1/2	1.91	1-15/16	2.31	2-9/16	1-3/16		
×192	56.4	20.4	20-3/8	0.960	15/16	1/2	11.5	11-1/2	1.75	1-3/4	2.15	2-7/16	1-1/8		
×175	51.3	20.0	20	0.890	7/8	7/16	11.4	11-3/8	1.59	1-9/16	1.99	2-7/16	1-1/4		
×158	46.3	19.7	19-3/4	0.810	13/16	7/16	11.3	11-1/4	1.44	1-7/16	1.84	2-3/8	1-1/4		
×143	42.1	19.5	19-1/2	0.730	3/4	3/8	11.2	11-1/4	1.32	1-5/16	1.72	2-3/16	1-3/16		
×130	38.2	19.3	19-1/4	0.670	11/16	3/8	11.2	11-1/8	1.20	1-3/16	1.60	2-1/16	1-3/16		
×119	35.1	19.0	19	0.655	5/8	5/16	11.3	11-1/4	1.06	1-1/16	1.46	1-15/16	1-3/16		
×106	31.1	18.7	18-3/4	0.590	9/16	5/16	11.2	11-1/4	0.940	15/16	1.34	1-13/16	1-1/8		
×97	28.5	18.6	18-5/8	0.535	9/16	5/16	11.1	11-1/8	0.870	7/8	1.27	1-3/4	1-1/8		
×86	25.3	18.4	18-3/8	0.480	1/2	1/4	11.1	11-1/8	0.770	3/4	1.17	1-5/8	1-1/16		
×76[a]	22.3	18.2	18-1/4	0.425	7/16	1/4	11.0	11	0.680	11/16	1.08	1-9/16	1-1/16		
W18×71	20.8	18.5	18-1/2	0.495	1/2	1/4	7.64	7-5/8	0.810	13/16	1.21	1-1/2	7/8	15-1/2 →	3-1/2[c] →
×65	19.1	18.4	18-3/8	0.450	7/16	1/4	7.59	7-5/8	0.750	3/4	1.15	1-7/16	7/8		
×60[a]	17.6	18.2	18-1/4	0.415	7/16	1/4	7.56	7-1/2	0.695	11/16	1.10	1-3/8	13/16		
×55[a]	16.2	18.1	18-1/8	0.390	3/8	3/16	7.53	7-1/2	0.630	5/8	1.03	1-5/16	13/16		
×50[a]	14.7	18.0	18	0.355	3/8	3/16	7.50	7-1/2	0.570	9/16	0.972	1-1/4	13/16		
W18×46[a]	13.5	18.1	18	0.360	3/8	3/16	6.06	6	0.605	5/8	1.01	1-1/4	13/16	15-1/2 →	3-1/2[c] →
×40[a]	11.8	17.9	17-7/8	0.315	5/16	3/16	6.02	6	0.525	1/2	0.927	1-3/16	13/16		
×35[a]	10.3	17.7	17-3/4	0.300	5/16	3/16	6.00	6	0.425	7/16	0.827	1-1/8	3/4		
W16×100	29.5	17.0	17	0.585	9/16	5/16	10.4	10-3/8	0.985	1	1.39	1-7/8	1-1/8	13-1/4 →	5-1/2 →
×89	26.2	16.8	16-3/4	0.525	1/2	1/4	10.4	10-3/8	0.875	7/8	1.28	1-3/4	1-1/16		
×77	22.6	16.5	16-1/2	0.455	7/16	1/4	10.3	10-1/4	0.760	3/4	1.16	1-5/8	1-1/16		
×67[a]	19.7	16.3	16-3/8	0.395	3/8	3/16	10.2	10-1/4	0.665	11/16	1.07	1-9/16	1		
W16×57	16.8	16.4	16-3/8	0.430	7/16	1/4	7.12	7-1/8	0.715	11/16	1.12	1-3/8	7/8	13-5/8 →	3-1/2g →
×50[a]	14.7	16.3	16-1/4	0.380	3/8	3/16	7.07	7-1/8	0.630	5/8	1.03	1-5/16	13/16		
×45[a]	13.3	16.1	16-1/8	0.345	3/8	3/16	7.04	7	0.565	9/16	0.967	1-1/4	13/16		
×40[a]	11.8	16.0	16	0.305	5/15	3/16	7.00	7	0.505	1/2	0.907	1-3/16	13/16		
×36[a]	10.6	15.9	15-7/8	0.295	5/16	3/16	6.99	7	0.430	7/16	0.832	1-1/8	3/4		
W16×31[a]	9.13	15.9	15-7/8	0.275	1/4	1/8	5.53	5-1/2	0.440	7/16	0.842	1-1/8	3/4	13-5/8 →	3-1/2 →
×26[a,e]	7.68	15.7	15-3/4	0.250	1/4	1/8	5.50	5-1/2	0.345	3/8	0.747	1-1/16	3/4	13-5/8	3-1/2

(continued)

TABLE C.1 (continued)
Properties Data of Each Shape Follow the Dimensions Data Sheets

W Shapes

Shape	Area, A (in.²)	Depth, d (in.)	Depth, d (in.)	Web Thickness, t_w (in.)	Web Thickness, t_w (in.)	$\frac{t_w}{2}$ (in.)	Flange Width, b_f (in.)	Flange Width, b_f (in.)	Flange Thickness, t_f (in.)	Flange Thickness, t_f (in.)	k_{des} (in.)	k_{det} (in.)	k_1 (in.)	T (in.)	Workable Gage (in.)
W14×730[d]	215	22.4	22-3/8	3.07	3-1/16	1-9/16	17.9	17-7/8	4.91	4-15/16	5.51	6-3/16	2-3/4	10	3-7-1/2-3[e]
×665[d]	196	21.6	21-5/8	2.83	2-13/16	1-7/16	17.7	17-5/8	4.52	4-1/2	5.12	5-13/16	2-5/8		3-7-1/2-3[e]
×605[d]	178	20.9	20-7/8	2.60	2-5/8	1-5/16	17.4	17-3/8	4.16	4-3/16	4.76	5-7/16	2-1/2		3-7-1/2-3
×550[d]	162	20.2	20-1/4	2.38	2-3/8	1-3/16	17.2	17-1/4	3.82	3-13/16	4.42	5-1/8	2-3/8		
×500[d]	147	19.6	19-5/8	2.19	2-3/16	1-1/8	17.0	17	3.50	3-1/2	4.10	4-13/16	2-5/16		
×455[d]	134	19.0	19	2.02	2	1	16.8	16-7/8	3.21	3-3/16	3.81	4-1/2	2-1/4		
×426[d]	125	18.7	18-5/8	1.88	1-7/8	15/16	16.7	16-3/4	3.04	3-1/16	3.63	4-5/16	2-1/8		
×398[d]	117	18.3	18-1/4	1.77	1-3/4	7/8	16.6	16-5/8	2.85	2-7/8	3.44	4-1/8	2-1/8		
×370[d]	109	17.9	17-7/8	1.66	1-5/8	13/16	16.5	16-1/2	2.66	2-11/16	3.26	3-15/16	2-1/8		
×342[d]	101	17.5	17-1/2	1.54	1-9/16	13/16	16.4	16-3/8	2.47	2-1/2	3.07	3-3/4	2		
×311[d]	91.4	17.1	17-1/8	1.41	1-7/16	3/4	16.2	16-1/4	2.26	2-1/4	2.86	3-9/16	1-15/16		
×283[d]	83.3	16.7	16-3/4	1.29	1-5/16	11/16	16.1	16-1/8	2.07	2-1/16	2.67	3-3/8	1-7/8		
×257	75.6	16.4	16-3/8	1.18	1-3/16	5/8	16.0	16	1.89	1-7/8	2.49	3-3/16	1-13/16		
×233	68.5	16.0	16	1.07	1-1/16	9/16	15.9	15-7/8	1.72	1-3/4	2.32	3	1-3/4		
×211	62.0	15.7	15-3/4	0.980	1	1/2	15.8	15-3/4	1.56	1-9/16	2.16	2-7/8	1-11/16		
×193	56.8	15.5	15-1/2	0.890	7/8	7/16	15.7	15-3/4	1.44	1-7/16	2.04	2-3/4	1-11/16		
×176	51.8	15.2	15-1/4	0.830	13/16	7/16	15.7	15-5/8	1.31	1-5/16	1.91	2-5/8	1-5/8		
×159	46.7	15.0	15	0.745	3/4	3/8	15.6	15-5/8	1.19	1-3/16	1.79	2-1/2	1-9/16		
×145	42.7	14.8	14-3/4	0.680	11/16	3/8	15.5	15-1/2	1.09	1-1/16	1.69	2-3/8	1-9/16		
W14×132	38.8	14.7	14-5/8	0.645	5/8	5/16	14.7	14-3/4	1.03	1	1.63	2-5/16	1-9/16	10	5-1/2
×120	35.3	14.5	14-1/2	0.590	9/16	5/16	14.7	14-5/8	0.940	15/16	1.54	2-1/4	1-1/2		
×109	32.0	14.3	14-3/8	0.525	1/2	1/4	14.6	14-5/8	0.860	7/8	1.46	2-3/16	1-1/2		
×99[b]	29.1	14.2	14-1/8	0.485	1/2	1/4	14.6	14-5/8	0.780	3/4	1.38	2-1/16	1-7/16		
×90[b]	26.5	14.0	14	0.440	7/16	1/4	14.5	14-1/2	0.710	11/16	1.31	2	1-7/16		

Shape														T	g
W14×82	24.0	14.3	14-1/4	0.510	1/2	1/4	10.1	10-1/8	0.855	7/8	1.45	1-11/16	1-1/16	10-7/8	5-1/2
×74	21.8	14.2	14-1/8	0.450	7/16	1/4	10.1	10-1/8	0.785	13/16	1.38	1-5/8	1-1/16	→	→
×68	20.0	14.0	14	0.415	7/16	1/4	10.0	10	0.720	3/4	1.31	1-9/16	1-1/16		
×61	17.9	13.9	13-7/8	0.375	3/8	3/16	10.0	10	0.645	5/8	1.24	1-1/2	1		
W14×53	15.6	13.9	13-7/8	0.370	3/8	3/16	8.06	8	0.660	11/16	1.25	1-1/2	1	10-7/8	5-1/2
×48	14.1	13.8	13-3/4	0.340	5/16	3/16	8.03	8	0.595	5/8	1.19	1-7/16	1	→	→
×43[a]	12.6	13.7	13-5/8	0.305	5/16	3/16	8.00	8	0.530	1/2	1.12	1-3/8	1		
W14×38[a]	11.2	14.1	14-1/8	0.310	5/16	3/16	6.77	6-3/4	0.515	1/2	0.915	1-1/4	13/16	11-5/8	3-1/2[c]
×34[a]	10.0	14.0	14	0.285	5/16	3/16	6.75	6-3/4	0.455	7/16	0.855	1-3/16	3/4	→	3-1/2
×30[a]	8.85	13.8	13-7/8	0.270	1/4	1/8	6.73	6-3/4	0.385	3/8	0.785	1-1/8	3/4		3-1/2
W14×26[a]	7.69	13.9	13-7/8	0.255	1/4	1/8	5.03	5	0.420	7/16	0.820	1-1/8	3/4	11-5/8	2-3/4[c]
×22[a]	6.49	13.7	13-3/4	0.230	1/4	1/8	5.00	5	0.335	5/16	0.735	1-1/16	3/4	11-5/8	2-3/4[c]
W12×336[d]	98.8	16.8	16-7/8	1.78	1-3/4	7/8	13.4	13-3/8	2.96	2-15/16	3.55	3-7/8	1-11/16	9-1/8	5-1/2
×305[d]	89.6	16.3	16-3/8	1.63	1-5/8	13/16	13.2	13-1/4	2.71	2-11/16	3.30	3-5/8	1-5/8		→
×279[d]	81.9	15.9	15-7/8	1.53	1-1/2	3/4	13.1	13-1/8	2.47	2-1/2	3.07	3-3/8	1-5/8		
×252[d]	74.0	15.4	15-3/8	1.40	1-3/8	11/16	13.0	13	2.25	2-1/4	2.85	3-1/8	1-1/2		
×230[d]	67.7	15.1	15	1.29	1-5/16	11/16	12.9	12-7/8	2.07	2-1/16	2.67	2-15/16	1-1/2		
×210	61.8	14.7	14-3/4	1.18	1-3/16	5/8	12.8	12-3/4	1.90	1-7/8	2.50	2-13/16	1-7/16		
×190	55.8	14.4	14-3/8	1.06	1-1/16	9/16	12.7	12-5/8	1.74	1-3/4	2.33	2-5/8	1-3/8		
×170	50.0	14.0	14	0.960	15/16	1/2	12.6	12-5/8	1.56	1-9/16	2.16	2-7/16	1-5/16		
×152	44.7	13.7	13-3/4	0.870	7/8	7/16	12.5	12-1/2	1.40	1-3/8	2.00	2-5/16	1-1/4		
×136	39.9	13.4	13-3/8	0.790	13/16	7/16	12.4	12-3/8	1.25	1-1/4	1.85	2-1/8	1-1/4		
×120	35.3	13.1	13-1/8	0.710	11/16	3/8	12.3	12-3/8	1.11	1-1/8	1.70	2	1-3/16		
×106	31.2	12.9	12-7/8	0.610	5/8	5/16	12.2	12-1/4	0.990	1	1.59	1-7/8	1-1/8		
×96	28.2	12.7	12-3/4	0.550	9/16	5/16	12.2	12-1/8	0.900	7/8	1.50	1-13/16	1-1/8		
×87	25.6	12.5	12-1/2	0.515	1/2	1/4	12.1	12-1/8	0.810	13/15	1.41	1-11/16	1-1/16		
×79	23.2	12.4	12-3/8	0.470	1/2	1/4	12.1	12-1/8	0.735	3/4	1.33	1-5/8	1-1/16		
×72	21.1	12.3	12-1/4	0.430	7/16	1/4	12.0	12	0.670	11/16	1.27	1-9/16	1-1/16		
×65[b]	19.1	12.1	12-1/8	0.390	3/8	3/16	12.0	12	0.605	5/8	1.20	1-1/2	1		
W12×58	17.0	12.2	12-1/4	0.360	3/8	3/16	10.0	10	0.640	5/8	1.24	1-1/2	15/16	9-1/4	5-1/2
×53	15.6	12.1	12	0.345	3/8	3/16	10.0	10	0.575	9/16	1.18	1-3/8	15/16	9-1/4	5-1/2
W12×50	14.6	12.2	12-1/4	0.370	3/8	3/16	8.08	8-1/8	0.640	5/8	1.14	1-1/2	15/16	9-1/4	5-1/2
×45	13.1	12.1	12	0.335	5/16	3/16	8.05	8	0.575	9/16	1.08	1-3/8	15/16	→	→
×40	11.7	11.9	12	0.295	5/16	3/16	8.01	8	0.515	1/2	1.02	1-3/8	7/8		

(continued)

TABLE C.1 (continued)
Properties Data of Each Shape Follow the Dimensions Data Sheets

W Shapes

Shape	Area, A (in.²)	Depth, d (in.)		Web Thickness, t_w (in.)		$\frac{t_w}{2}$ (in.)	Flange Width, b_t (in.)		Thickness, t_f (in.)		Distance k k_{des} (in.)	k_{det} (in.)	k_1 (in.)	T (in.)	Workable Gage (in.)
W12×35[a]	10.3	12.5	12-1/2	0.300	5/16	3/16	6.56	6-1/2	0.520	1/2	0.820	1-3/16	3/4	10-1/8	3-1/2
×30[a]	8.79	12.3	12-3/8	0.260	1/4	1/8	6.52	6-1/2	0.440	7/16	0.740	1-1/8	3/4	→	→
×26[a]	7.65	12.2	12-1/4	0.230	1/8	1/8	6.49	6-1/2	0.380	3/8	0.680	1-1/16	3/4	10-3/8	2-1/4[e]
W12×22[a]	6.48	12.3	12-1/4	0.260	1/4	1/8	4.03	4	0.425	7/16	0.725	15/16	5/8	→	→
×19[a]	5.57	12.2	12-1/8	0.235	1/4	1/8	4.01	4	0.350	3/8	0.650	7/8	9/16	10-3/8	
×16[a]	4.71	12.0	12	0.220	1/4	1/8	3.99	4	0.265	1/4	0.565	13/16	9/16	→	
×14[a,c]	4.16	11.9	11-7/8	0.200	3/16	1/8	3.97	4	0.225	1/4	0.525	3/4	9/16	→	
W10×112	32.9	11.4	11-3/8	0.755	3/4	3/8	10.4	10-3/8	1.25	1-1/4	1.75	1-15/16	1	7-1/2	5-1/2
×100	29.4	11.1	11-1/8	0.680	11/16	3/8	10.3	10-3/8	1.12	1-1/8	1.62	1-13/16	1	→	→
×88	25.9	10.8	10-7/8	0.605	5/8	5/16	10.3	10-1/4	0.990	1	1.49	1-11/16	15/16		
×77	22.6	10.6	10-5/8	0.530	1/2	1/4	10.2	10-1/4	0.870	7/8	1.37	1-9/16	7/8		
×68	20.0	10.4	10-3/8	0.470	1/2	1/4	10.1	10-1/8	0.770	3/4	1.27	1-7/16	7/8		
×60	17.6	10.2	10-1/4	0.420	7/16	1/4	10.1	10-1/8	0.680	11/16	1.18	1-3/8	13/16		
×54	15.8	10.1	10-1/8	0.370	3/8	3/16	10.0	10	0.615	5/8	1.12	1-5/16	13/16		
×49	14.4	10.0	10	0.340	5/16	3/16	10.0	10	0.560	9/16	1.06	1-1/4	13/16		

W10×45	13.3	10.1	10-1/8	0.350	3/8	3/16	8.02	8	0.620	5/8	1.12	1-5/16	13/16	7-1/2	5-1/2
×39	11.5	9.92	9-7/8	0.315	5/16	3/16	7.99	8	0.530	1/2	1.03	1-3/16	13/16	→	→
×33	9.71	9.73	9-3/4	0.290	5/16	3/16	7.96	8	0.435	7/16	0.935	1-1/8	3/4	→	→
W10×30	8.84	10.5	10-1/2	0.300	5/16	3/16	5.81	5-3/4	0.510	1/2	0.810	1-1/8	11/16	8-1/4	2-3/4[e]
×26	7.61	10.3	10-3/8	0.260	1/4	1/8	5.77	5-3/4	0.440	7/16	0.740	1-1/16	11/16	→	→
×22[a]	6.49	10.2	10-1/8	0.240	1/4	1/8	5.75	5-3/4	0.360	3/8	0.660	15/16	5/8	→	→
W10×19	5.62	10.2	10-1/4	0.250	1/4	1/8	4.02	4	0.395	3/8	0.695	15/16	5/8	8-3/8	2-1/4[e]
×17[a]	4.99	10.1	10-1/8	0.240	1/4	1/8	4.01	4	0.330	5/16	0.630	7/8	9/16	→	→
×15[a]	4.41	10.0	10	0.230	1/4	1/8	4.00	4	0.270	1/4	0.570	13/16	9/16	→	→
×12[a,b]	3.54	9.87	9-7/8	0.190	3/16	1/8	3.96	4	0.210	3/16	0.510	3/4	9/16	→	

Source: Courtesy of the American Institute of Steel Construction, Chicago, IL.

[a] Shape is slender for compression with $F_y = 50$ ksi.
[b] Shape exceeds compact limit for flexure with $F_y = 50$ ksi.
[c] The actual size, combination, and orientation of fastener components should be compared with the geometry of the cross section to ensure compatibility.
[d] Flange thickness greater than 2 in. Special requirements may apply per AISC Specification Section A3.1c.
[e] Shape does not meet the h/t_w limit for shear in Specification Section G2.1a with $F_y = 50$ ksi.

TABLE C.1
W Shapes

(b) Properties

Shape	Compact Section Criteria		Axis X-X				Axis Y-Y				r_{ts} (in.)	h_o (in.)	Torsional Properties	
	$\frac{b_f}{2t_f}$	$\frac{h}{t_w}$	I (in.^{4})	S (in.^{3})	r (in.)	Z (in.^{3})	I (in.^{4})	S (in.^{3})	r (in.)	Z (in.^{3})			J (in.^{4})	C_w (in.^{6})
W21×93	4.53	32.3	2070	192	8.70	221	92.9	22.1	1.84	34.7	2.24	20.7	6.03	9,940
83	5.00	36.4	1830	171	8.67	196	81.4	19.5	1.83	30.5	2.21	20.6	4.34	8,630
73	5.60	41.2	1600	151	8.64	172	70.6	17.0	1.81	26.6	2.19	20.5	3.02	7,410
68	6.04	43.6	1480	140	8.60	160	64.7	15.7	1.80	24.4	2.17	20.4	2.45	6,760
62	6.70	46.9	1330	127	8.54	144	57.5	14.0	1.77	21.7	2.15	20.4	1.83	5,960
55	7.87	50.0	1140	110	8.40	126	48.4	11.8	1.73	18.4	2.11	20.3	1.24	4,980
48	9.47	53.6	959	93.0	8.24	107	38.7	9.52	1.66	14.9	2.05	20.2	0.803	3,950
W21×57	5.04	46.3	1170	111	8.36	129	30.6	9.35	1.35	14.8	1.68	20.4	1.77	3,190
50	6.10	49.4	984	94.5	8.18	110	24.9	7.64	1.30	12.2	1.64	20.3	1.14	2,570
44	7.22	53.6	843	81.6	8.06	95.4	20.7	6.37	1.26	10.2	1.60	20.2	0.770	2,110
W18×311	2.19	10.4	6970	624	8.72	754	795	132	2.95	207	3.53	19.6	176	76,200
283	2.38	11.3	6170	565	8.61	676	704	118	2.91	185	3.47	19.4	134	65,900
258	2.56	12.5	5510	514	8.53	611	628	107	2.88	166	3.42	19.2	103	57,600
234	2.76	13.8	4900	466	8.44	549	558	95.8	2.85	149	3.37	19.0	78.7	50,100
211	3.02	15.1	4330	419	8.35	490	493	85.3	2.82	132	3.32	18.8	58.6	43,400
192	3.27	16.7	3870	380	8.28	442	440	76.8	2.79	119	3.28	18.6	44.7	38,000
175	3.58	18.0	3450	344	8.20	398	391	68.8	2.76	106	3.24	18.5	33.8	33,300

Section														
158	3.92	19.8	3060	310	8.12	356	347	61.4	2.74	94.8	3.20	18.3	25.2	29,000
143	4.25	22.0	2750	282	8.09	322	311	55.5	2.72	85.4	3.17	18.2	19.2	25,700
130	4.65	23.9	2460	256	8.03	290	278	49.9	2.70	76.7	3.13	18.1	14.5	22,700
119	5.31	24.5	2190	231	7.90	262	253	44.9	2.69	69.1	3.13	17.9	10.6	20,300
106	5.96	27.2	1910	204	7.84	230	220	39.4	2.66	60.5	3.10	17.8	7.48	17,400
97	6.41	30.0	1750	188	7.82	211	201	36.1	2.65	55.3	3.08	17.7	5.86	15,800
86	7.20	33.4	1530	166	7.77	186	175	31.6	2.63	48.4	3.05	17.6	4.10	13,600
76	8.11	37.8	1330	146	7.73	163	152	27.6	2.61	42.2	3.02	17.5	2.83	11,700
W18×71	4.71	32.4	1170	127	7.50	146	60.3	15.8	1.70	24.7	2.05	17.7	3.49	4,700
65	5.06	35.7	1070	117	7.49	133	54.8	14.4	1.69	22.5	2.03	17.6	2.73	4,240
60	5.44	38.7	984	108	7.47	123	50.1	13.3	1.68	20.6	2.02	17.5	2.17	3,850
55	5.98	41.1	890	98.3	7.41	112	44.9	11.9	1.67	18.5	2.00	17.5	1.66	3,430
50	6.57	45.2	800	88.9	7.38	101	40.1	10.7	1.65	16.6	1.98	17.4	1.24	3,040
W18×46	5.01	44.6	712	78.8	7.25	90.7	22.5	7.43	1.29	11.7	1.58	17.5	1.22	1,720
40	5.73	50.9	612	68.4	7.21	78.4	19.1	6.35	1.27	10.0	1.56	17.4	0.810	1,440
35	7.06	53.5	510	57.6	7.04	66.5	15.3	5.12	1.22	8.06	1.52	17.3	0.506	1,140
100	5.29	24.3	1,490	175	7.10	198	186	35.7	2.51	54.9	2.92	16.0	7.73	11,900
89	5.92	27.0	1,300	155	7.05	175	163	31.4	2.49	48.1	2.88	15.9	5.45	10,200
77	6.77	31.2	1,110	134	7.00	150	138	26.9	2.47	41.1	2.85	15.8	3.57	8,590
67	7.70	35.9	954	117	6.96	130	119	23.2	2.46	35.5	2.82	15.7	2.39	7,300
W16×57	4.98	33.0	758	92.2	6.72	105	43.1	12.1	1.60	18.9	1.92	15.7	2.22	2,660
50	5.61	37.4	659	81.0	6.68	92.0	37.2	10.5	1.59	16.3	1.89	15.6	1.52	2,270
45	6.23	41.1	586	72.7	6.65	82.3	32.8	9.34	1.57	14.5	1.88	15.6	1.11	1,990
40	6.93	46.5	518	64.7	6.63	73.0	28.9	8.25	1.57	12.7	1.86	15.5	0.794	1,730
36	8.12	48.1	448	56.5	6.51	64.0	24.5	7.00	1.52	10.8	1.83	15.4	0.545	1,460
W16×31	6.28	51.6	375	47.2	6.41	54.0	12.4	4.49	1.17	7.03	1.42	15.4	0.461	739
26	7.97	56.8	301	38.4	6.26	44.2	9.59	3.49	1.12	5.48	1.38	15.3	0.262	565
W14×730	1.82	3.71	14,300	1280	8.17	1660	4720	527	4.69	816	5.68	17.5	1450	362,000
665	1.95	4.03	12,400	1150	7.98	1480	4170	472	4.62	730	5.57	17.1	1120	305,000
605	2.09	4.39	10,800	1040	7.80	1320	3680	423	4.55	652	5.46	16.8	869	258,000
550	2.25	4.79	9,430	931	7.63	1180	3250	378	4.49	583	5.36	16.4	669	219,000
500	2.43	5.21	8,210	838	7.48	1050	2880	339	4.43	522	5.26	16.1	514	187,000

(continued)

TABLE C.1 (continued)
W Shapes

Shape	Compact Section Criteria $\frac{b_f}{2t_f}$	Compact Section Criteria $\frac{h}{t_w}$	Axis X–X I (in.⁴)	Axis X–X S (in.³)	Axis X–X r (in.)	Axis X–X Z (in.³)	Axis Y–Y I (in.⁴)	Axis Y–Y S (in.³)	Axis Y–Y r (in.)	Axis Y–Y Z (in.³)	r_{ts} (in.)	h_o (in.)	Torsional Properties J (in.⁴)	Torsional Properties C_w (in.⁶)
455	2.62	5.66	7,190	756	7.33	936	2560	304	4.38	468	5.17	15.8	395	160,000
426	2.75	6.08	6,600	706	7.26	869	2360	283	4.34	434	5.11	15.6	331	144,000
398	2.92	6.44	6,000	656	7.16	801	2170	262	4.31	402	5.06	15.4	273	129,000
370	3.10	6.89	5,440	607	7.07	736	1990	241	4.27	370	5.00	15.3	222	116,000
342	3.31	7.41	4,900	558	6.98	672	1810	221	4.24	338	4.94	15.1	178	103,000
311	3.59	8.09	4,330	506	6.88	603	1610	199	4.20	304	4.87	14.9	136	89,100
283	3.89	8.84	3,840	459	6.79	542	1440	179	4.17	274	4.81	14.7	104	77,700
257	4.23	9.71	3,400	415	6.71	487	1290	161	4.13	246	4.75	14.5	79.1	67,800
233	4.62	10.7	3,010	375	6.63	436	1150	145	4.10	221	4.69	14.3	59.5	59,000
211	5.06	11.6	2,660	338	6.55	390	1030	130	4.07	198	4.64	14.2	44.6	51,500
193	5.45	12.8	2,400	310	6.50	355	931	119	4.05	180	4.59	14.0	34.8	45,900
176	5.97	13.7	2,140	281	6.43	320	838	107	4.02	163	4.55	13.9	26.5	40,500
159	6.54	15.3	1,900	254	6.38	287	748	96.2	4.00	146	4.51	13.8	19.7	35,600
145	7.11	16.8	1,710	232	6.33	260	677	87.3	3.98	133	4.47	13.7	15.2	31,700
W14×132	7.15	17.7	1530	209	6.28	234	548	74.5	3.76	113	4.23	13.6	12.3	255
120	7.80	19.3	1380	190	6.24	212	495	67.5	3.74	102	4.20	13.5	9.37	227
109	8.49	21.7	1240	173	6.22	192	447	61.2	3.73	92.7	4.17	13.5	7.12	202
99	9.34	23.5	1110	157	6.17	173	402	55.2	3.71	83.6	4.14	13.4	5.37	1800
90	10.2	25.9	999	143	6.14	157	362	49.9	3.70	75.6	4.11	13.3	4.06	1600
W14×82	5.92	22.4	881	123	6.05	139	148	29.3	2.48	44.8	2.85	13.5	5.07	671
74	6.41	25.4	795	112	6.04	126	134	26.6	2.48	40.5	2.82	13.4	3.87	599
68	6.97	27.5	722	103	6.01	115	121	24.2	2.46	36.9	2.80	13.3	3.01	538
61	7.75	30.4	640	92.1	5.98	102	107	21.5	2.45	32.8	2.78	13.2	2.19	471
W14×53	6.11	30.9	541	77.8	5.89	87.1	57.7	14.3	1.92	22.0	2.22	13.3	1.94	2540
48	6.75	33.6	484	70.2	5.85	78.4	51.4	12.8	1.91	19.6	2.20	13.2	1.45	2240

43	7.54	37.4	428	62.6	5.82	69.6	45.2	11.3	1.89	17.3	2.18	13.1	1.05	1950
W14×38	6.57	39.6	385	54.6	5.87	61.5	26.7	7.88	1.55	12.1	1.82	13.6	0.798	1,230
34	7.41	43.1	340	48.6	5.83	54.6	23.3	6.91	1.53	10.6	1.80	13.5	0.569	1,070
30	8.74	45.4	291	42.0	5.73	47.3	19.6	5.82	1.49	8.99	1.77	13.5	0.380	887
W14×26	5.98	48.1	245	35.3	5.65	40.2	8.91	3.55	1.08	5.54	1.31	13.5	0.358	405
22	7.46	53.3	199	29.0	5.54	33.2	7.00	2.80	1.04	4.39	1.27	13.4	0.208	314
W12×336	2.26	5.47	4060	483	6.41	603	1190	177	3.47	274	4.13	13.9	243	57,000
305	2.45	5.98	3550	435	6.29	537	1050	159	3.42	244	4.05	13.6	185	48,600
279	2.66	6.35	3110	393	6.16	481	937	143	3.38	220	4.00	13.4	143	42,000
252	2.89	6.96	2720	353	6.06	428	828	127	3.34	196	3.93	13.2	108	35,800
230	3.11	7.56	2420	321	5.97	386	742	115	3.31	177	3.87	13.0	83.8	31,200
210	3.37	8.23	2140	292	5.89	348	664	104	3.28	159	3.82	12.8	64.7	27,200
190	3.65	9.16	1890	263	5.82	311	589	93.0	3.25	143	3.76	12.6	48.8	23,600
170	4.03	10.1	1650	235	5.74	275	517	82.3	3.22	126	3.71	12.5	35.6	20,100
152	4.46	11.2	1430	209	5.66	243	454	72.8	3.19	111	3.66	12.3	25.8	17,200
136	4.96	12.3	1240	186	5.58	214	398	64.2	3.16	98.0	3.61	12.2	18.5	14,700
120	5.57	13.7	1070	163	5.51	186	345	56.0	3.13	85.4	3.56	12.0	12.9	12,400
106	6.17	15.9	933	145	5.47	164	301	49.3	3.11	75.1	3.52	11.9	9.13	10,700
96	6.76	17.7	833	131	5.44	147	270	44.4	3.09	67.5	3.49	11.8	6.85	9,410
87	7.48	18.9	740	118	5.38	132	241	39.7	3.07	60.4	3.46	11.7	5.10	8,270
79	8.22	20.7	662	107	5.34	119	216	35.8	3.05	54.3	3.43	11.6	3.84	7,330
72	8.99	22.6	597	97.4	5.31	108	195	32.4	3.04	49.2	3.40	11.6	2.93	6,540
65	9.92	24.9	533	87.9	5.28	96.8	174	29.1	3.02	44.1	3.38	11.5	2.18	5,780
W12×58	7.82	27.0	475	78.0	5.28	86.4	107	21.4	2.51	32.5	2.82	11.6	2.10	3,570
53	8.69	28.1	425	70.6	5.23	77.9	95.8	19.2	2.48	29.1	2.79	11.5	1.58	3,160
W12×50	6.31	26.8	391	64.2	5.18	71.9	56.3	13.9	1.96	21.3	2.25	11.6	1.71	1,880
45	7.00	29.6	348	57.7	5.15	64.2	50.0	12.4	1.95	19.0	2.23	11.5	1.26	1,650
40	7.77	33.6	307	51.5	5.13	57.0	44.1	11.0	1.94	16.8	2.21	11.4	0.906	1,440
W12×35	6.31	36.2	285	45.6	5.25	51.2	24.5	7.47	1.54	11.5	1.79	12.0	0.741	879
30	7.41	41.8	238	38.6	5.21	43.1	20.3	6.24	1.52	9.56	1.77	11.9	0.457	720
26	8.54	47.2	204	33.4	5.17	37.2	17.3	5.34	1.51	8.17	1.75	11.8	0.300	607
W12×22	4.74	41.8	156	25.4	4.91	29.3	4.66	2.31	0.848	3.66	1.04	11.9	0.293	164

(continued)

TABLE C.1 (continued)
W Shapes

Shape	bf/2tf	h/tw	I (in.⁴) X	S (in.³) X	r (in.) X	Z (in.³) X	I (in.⁴) Y	S (in.³) Y	r (in.) Y	Z (in.³) Y	rts (in.)	hₒ (in.)	J (in.⁴)	Cw (in.⁶)
19	5.72	46.2	130	21.3	4.82	24.7	3.76	1.88	0.822	2.98	1.02	11.8	0.180	131
16	7.53	49.4	103	17.1	4.67	20.1	2.82	1.41	0.773	2.26	0.982	11.7	0.103	96.9
14	8.82	54.3	88.6	14.9	4.62	17.4	2.36	1.19	0.753	1.90	0.962	11.7	0.0704	80.4
W10×112	4.17	10.4	716	126	4.66	147	236	45.3	2.68	69.2	3.07	10.1	15.1	6,020
100	4.62	11.6	623	112	4.60	130	207	40.0	2.65	61.0	3.03	10.0	10.9	5,150
88	5.18	13.0	534	98.5	4.54	113	179	34.8	2.63	53.1	2.99	9.85	7.53	4,330
77	5.86	14.8	455	85.9	4.49	97.6	154	30.1	2.60	45.9	2.95	9.73	5.11	3,630
68	6.58	16.7	394	75.7	4.44	85.3	134	26.4	2.59	40.1	2.91	9.63	3.56	3,100
60	7.41	18.7	341	66.7	4.39	74.6	116	23.0	2.57	35.0	2.88	9.54	2.48	2,640
54	8.15	21.2	303	60.0	4.37	66.6	103	20.6	2.56	31.3	2.86	9.48	1.82	2,320
49	8.93	23.1	272	54.6	4.35	60.4	93.4	18.7	2.54	28.3	2.84	9.42	1.39	2,070
W10×45	6.47	22.5	248	49.1	4.32	54.9	53.4	13.3	2.01	20.3	2.27	9.48	1.51	1,200
39	7.53	25.0	209	42.1	4.27	46.8	45.0	11.3	1.98	17.2	2.24	9.39	0.976	992
33	9.15	27.1	171	35.0	4.19	38.8	36.6	9.20	1.94	14.0	2.20	9.30	0.583	791
W10×30	5.70	29.5	170	32.4	4.38	36.6	16.7	5.75	1.37	8.84	1.60	10.0	0.622	414
26	6.56	34.0	144	27.9	4.35	31.3	14.1	4.89	1.36	7.50	1.58	9.89	0.402	345
22	7.99	36.9	118	23.2	4.27	26.0	11.4	3.97	1.33	6.10	1.55	9.81	0.239	275
W10×19	5.09	35.4	96.3	18.8	4.14	21.6	4.29	2.14	0.874	3.35	1.06	9.85	0.233	104
17	6.08	36.9	81.9	16.2	4.05	18.7	3.56	1.78	0.845	2.80	1.04	9.78	0.156	85.1
15	7.41	38.5	68.9	13.8	3.95	16.0	2.89	1.45	0.810	2.30	1.01	9.72	0.104	68.3
12	9.43	46.6	53.8	10.9	3.90	12.6	2.18	1.10	0.785	1.74	0.983	9.66	0.0547	50.9

Source: Courtesy of the American Institute of Steel Construction, Chicago, IL.

TABLE C.2
Properties Data of Each Shape Follow the Dimensions Data Sheets

C Shapes

(a) Dimensions

Shape	Area, A (in.²)	Depth, d (in.)	Web Thickness, t_w (in.)	$\frac{t_w}{2}$ (in.)	Flange Width, b_f (in.)	Thickness, t_f (in.)		Distance k (in.)	T (in.)	Workable Gage (in.)	r_{ts} (in.)	h_0 (in.)
C15×50	14.7	15.0	0.716	11/16	3/8	3.72	3-3/4	0.650	5/8	1-7/16	12-1/8	2-1/4	1.17	14.4
×40	11.8	15.0	0.520	1/2	1/4	3.52	3-1/2	0.650	5/8	1-7/16	12-1/8	2	1.15	14.4
×33.9	10.0	15.0	0.400	3/8	3/16	3.40	3-3/8	0.650	5/8	1-7/16	12-1/8	2	1.13	14.4
C12×30	8.81	12.0	0.510	1/2	1/4	3.17	3-1/8	0.501	1/2	1-1/8	9-3/4	1-3/4ᵃ	1.01	11.5
×25	7.34	12.0	0.387	3/8	3/16	3.05	3	0.501	1/2	1-1/8	9-3/4	1-3/4ᵃ	1.00	11.5
×20.7	6.08	12.0	0.282	5/16	3/16	2.94	3	0.501	1/2	1-1/8	9-3/4	1-3/4ᵃ	0.983	11.5
C10×30	8.81	10.0	0.673	11/16	3/8	3.03	3	0.436	7/16	1	8	1-3/4ᵃ	0.925	9.56
×25	7.34	10.0	0.526	1/2	1/4	2.89	2-7/8	0.436	7/16	1	8	1-3/4ᵃ	0.911	9.56
×20	5.87	10.0	0.379	3/8	3/16	2.74	2-3/4	0.436	7/16	1	8	1-1/2ᵃ	0.894	9.56
×15.3	4.48	10.0	0.240	1/4	1/8	2.60	2-5/8	0.436	7/16	1	8	1-1/2ᵃ	0.869	9.56
C9×20	5.87	9.00	0.448	7/16	1/4	2.65	2-5/8	0.413	7/16	1	7	1-1/2ᵃ	0.848	8.59
×15	4.41	9.00	0.285	5/16	3/16	2.49	2-1/2	0.413	7/16	1	7	1-3/8ᵃ	0.824	8.59
×13.4	3.94	9.00	0.233	1/4	1/8	2.43	2-3/8	0.413	7/16	1	7	1-3/8ᵃ	0.813	8.59
C8×18.7	5.51	8.00	0.487	1/2	1/4	2.53	2-1/2	0.390	3/8	15/16	6-1/8	1-1/2ᵃ	0.800	7.61
×13.7	4.04	8.00	0.303	5/16	3/16	2.34	2-3/8	0.390	3/8	15/16	6-1/8	1-3/8ᵃ	0.774	7.61
×11.5	3.37	8.00	0.220	1/4	1/8	2.26	2-1/4	0.390	3/8	15/16	6-1/8	1-3/8ᵃ	0.756	7.61

(continued)

TABLE C.2 (continued)
Properties Data of Each Shape Follow the Dimensions Data Sheets

C Shapes

Shape	Area, A (in.²)	Depth, d (in.)	Web Thickness, t_w (in.)	Web $\frac{t_w}{2}$ (in.)	Flange Width, b_f (in.)	Flange Thickness, t_f (in.)	Distance k (in.)	Distance T (in.)	Distance Workable Gage (in.)	r_{ts} (in.)	h_0 (in.)
C7×14.7	4.33	7.00	0.419 7/16	1/4	2.30 2-1/4	0.366 3/8	7/8	5-1/4	1-1/4[a]	0.738	6.63
×12.2	3.60	7.00	0.314 5/16	3/16	2.19 2-1/4	0.366 3/8	7/8	5-1/4	1-1/4[a]	0.721	6.63
×9.8	2.87	7.00	0.210 3/16	1/8	2.09 2-1/8	0.366 3/8	7/8	5-1/4	1-1/4[a]	0.698	6.63
C6×13	3.81	6.00	0.437 7/16	1/4	2.16 2-1/8	0.343 5/16	13/16	4/8	1-3/8[a]	0.689	5.66
×10.5	3.08	6.00	0.314 5/16	3/16	2.03 2	0.343 5/16	13/16	4-3/8	1-1/8[a]	0.669	5.66
×8.2	2.39	6.00	0.200 3/16	1/8	1.92 1-7/8	0.343 5/16	13/16	4-3/8	1-1/8[a]	0.643	5.66
C5×9	2.64	5.00	0.325 5/16	3/16	1.89 1-7/8	0.320 5/16	3/4	3-1/2	1-1/8[a]	0.617	4.68
×6.7	1.97	5.00	0.190 3/16	1/8	1.75 1-3/4	0.320 5/16	3/4	3-1/2	—	0.584	4.68
C4×7.2	2.13	4.00	0.321 5/16	3/16	1.72 1-3/4	0.296 5/16	3/4	2-1/2	1[a]	0.563	3.70
×5.4	1.58	4.00	0.184 3/16	1/8	1.58 1-5/8	0.296 5/16	3/4	2-1/2	—	0.528	3.70
×4.5	1.38	4.00	0.125 1/8	1/16	1.58 1-5/8	0.296 5/16	3/4	2-1/2	—	0.524	3.70
C3×6	1.76	3.00	0.356 3/8	3/16	1.60 1-5/8	0.273 1/4	11/16	1-5/8	—	0.519	2.73
×5	1.47	3.00	0.258 1/4	1/8	1.50 1-1/2	0.273 1/4	11/16	1-5/8	—	0.495	2.73
×4.1	1.20	3.00	0.170 3/16	1/8	1.41 1-3/8	0.273 1/4	11/16	1-5/8	—	0.469	2.73
×3.5	1.09	3.00	0.132 1/8	1/16	1.37 1-3/8	0.273 1/4	11/16	1-5/8	—	0.455	2.73

Source: Courtesy of the American Institute of Steel Construction, Chicago, IL.

[a] The actual size, combination, and orientation of fastener components should be compared with the geometry of the cross section to ensure compatibility.

—, Flange is too narrow to establish a workable gage.

TABLE C.2
C Shapes

(b) Properties

Shape	Shear Ctr, e_0 (in.)	Axis X–X				Axis Y–Y						Torsional Properties			
		I (in.⁴)	S (in.³)	r (in.)	Z (in.³)	I (in.⁴)	S (in.³)	r (in.)	$\bar{x}$ (in.)	Z (in.³)	x_p (in.)	J (in.⁴)	C_w (in.⁶)	$\bar{r}_0$ (in.)	H
C15×50	0.583	404	53.8	5.24	68.5	11.0	3.77	0.865	0.799	8.14	0.490	2.65	492	5.49	0.937
40	0.767	348	46.5	5.45	57.5	9.17	3.34	0.883	0.778	6.84	0.392	1.45	410	5.73	0.927
33.9	0.896	315	42.0	5.62	50.8	8.07	3.09	0.901	0.788	6.19	0.332	1.01	358	5.94	0.920
C12×30	0.618	162	27.0	4.29	33.8	5.12	2.05	0.762	0.674	4.32	0.367	0.861	151	4.54	0.919
25	0.746	144	24.0	4.43	29.4	4.45	1.87	0.779	0.674	3.82	0.306	0.538	130	4.72	0.909
20.7	0.870	129	21.5	4.61	25.6	3.86	1.72	0.797	0.698	3.47	0.253	0.369	112	4.93	0.899
C10×30	0.368	103	20.7	3.42	26.7	3.93	1.65	0.668	0.649	3.78	0.441	1.22	79.5	3.63	0.922
25	0.494	91.1	18.2	3.52	23.1	3.34	1.47	0.675	0.617	3.18	0.367	0.687	68.3	3.75	0.912
20	0.636	78.9	15.8	3.66	19.4	2.80	1.31	0.690	0.606	2.70	0.294	0.368	56.9	3.93	0.900
15.3	0.796	67.3	13.5	3.87	15.9	2.27	1.15	0.711	0.634	2.34	0.224	0.209	45.5	4.19	0.884
C9×20	0.515	60.9	13.5	3.22	16.9	2.41	1.17	0.640	0.583	2.46	0.326	0.427	39.4	3.46	0.899
15	0.681	51.0	11.3	3.40	13.6	1.91	1.01	0.659	0.586	2.04	0.245	0.208	31.0	3.69	0.882
13.4	0.742	47.8	10.6	3.49	12.6	1.75	0.954	0.666	0.601	1.94	0.219	0.168	28.2	3.79	0.875
C8×18.7	0.431	43.9	11.0	2.82	13.9	1.97	1.01	0.598	0.565	2.17	0.344	0.434	25.1	3.05	0.894
13.7	0.604	36.1	9.02	2.99	11.0	1.52	0.848	0.613	0.554	1.73	0.252	0.186	19.2	3.26	0.874
11.5	0.697	32.5	8.14	3.11	9.63	1.31	0.775	0.623	0.572	1.57	0.211	0.130	16.5	3.41	0.862

(continued)

TABLE C.2 (continued)
C Shapes

Shape	Shear Ctr, e_0 (in.)	Axis X-X				Axis Y-Y						Torsional Properties			
		I (in.^{4})	S (in.^{3})	r (in.)	Z (in.^{3})	I (in.^{4})	S (in.^{3})	r (in.)	$\bar{x}$ (in.)	Z (in.^{3})	x_p (in.)	J (in.^{4})	C_w (in.^{6})	$\bar{r}_0$ (in.)	H
C7×14.7	0.441	27.2	7.78	2.51	9.75	1.37	0.772	0.561	0.532	1.63	0.309	0.267	13.1	2.75	0.875
12.2	0.538	24.2	6.92	2.60	8.46	1.16	0.696	0.568	0.525	1.42	0.257	0.161	11.2	2.86	0.862
9.8	0.647	21.2	6.07	2.72	7.19	0.957	0.617	0.578	0.541	1.26	0.205	0.0996	9.15	3.03	0.846
C6×13	0.380	17.3	5.78	2.13	7.29	1.05	0.638	0.524	0.514	1.35	0.318	0.237	7.19	2.37	0.858
10.5	0.486	15.1	5.04	2.22	6.18	0.860	0.561	0.529	0.500	1.14	0.256	0.128	5.91	2.48	0.842
8.2	0.599	13.1	4.35	2.34	5.16	0.687	0.488	0.536	0.512	0.987	0.199	0.0736	4.70	2.64	0.823
C5×9	0.427	8.89	3.56	1.83	4.39	0.624	0.444	0.486	0.478	0.913	0.264	0.109	2.93	2.10	0.815
6.7	0.552	7.48	2.99	1.95	3.55	0.470	0.372	0.489	0.484	0.757	0.215	0.0549	2.22	2.26	0.791
C4×7.2	0.386	4.58	2.29	1.47	2.84	0.425	0.337	0.447	0.459	0.695	0.266	0.0817	1.24	1.75	0.767
5.4	0.501	3.85	1.92	1.56	2.29	0.312	0.277	0.444	0.457	0.565	0.231	0.0399	0.921	1.88	0.741
4.5	0.587	3.65	1.83	1.63	2.12	0.289	0.265	0.457	0.493	0.531	0.321	0.0322	0.871	2.00	0.710
C3×6	0.322	2.07	1.38	1.08	1.74	0.300	0.263	0.413	0.455	0.543	0.294	0.0725	0.462	1.40	0.690
5	0.392	1.85	1.23	1.12	1.52	0.241	0.228	0.405	0.439	0.464	0.245	0.0425	0.379	1.45	0.674
4.1	0.461	1.65	1.10	1.17	1.32	0.191	0.196	0.398	0.437	0.399	0.262	0.0269	0.307	1.53	0.655
3.5	0.493	1.57	1.04	1.20	1.24	0.169	0.182	0.394	0.443	0.364	0.296	0.0226	0.276	1.57	0.645

Source: Courtesy of the American Institute of Steel Construction, Chicago, IL.

TABLE C.3
Other Properties of Each of these Shape Continue in "(b) Properties" Section following up this "(a) Properties" Section

Angles

(a) Properties

Shape	k (in.)	Wt. (lb/ft)	Area, A (in.²)	Axis X–X						Flexural-Torsional Properties		
				I (in.⁴)	S (in.³)	r (in.)	$\bar{y}$ (in.)	Z (in.³)	y_p (in.)	J (in.⁴)	C_w (in.⁶)	$\bar{r}_0$ (in.)
L4×3-1/2×1/2	7/8	11.9	3.50	5.30	1.92	1.23	1.24	3.46	0.497	0.301	0.302	2.03
×3/8	3/4	9.10	2.67	4.15	1.48	1.25	1.20	2.66	0.433	0.132	0.134	2.06
×5/16	11/16	7.70	2.25	3.53	1.25	1.25	1.17	2.24	0.401	0.0782	0.0798	2.08
×1/4	5/8	6.20	1.81	2.89	1.01	1.26	1.14	1.81	0.368	0.0412	0.0419	2.09
L4×3×5/8	1	13.6	3.89	6.01	2.28	1.23	1.37	4.08	0.810	0.529	0.472	1.91
×1/2	7/8	11.1	3.25	5.02	1.87	1.24	1.32	3.36	0.747	0.281	0.255	1.94
×3/8	3/4	8.50	2.48	3.94	1.44	1.26	1.27	2.60	0.683	0.123	0.114	1.97
×5/16	11/16	7.20	2.09	3.36	1.22	1.27	1.25	2.19	0.651	0.0731	0.0676	1.98
×1/4	5/8	5.80	1.69	2.75	0.988	1.27	1.22	1.77	0.618	0.0386	0.0356	1.99
L3-1/2×3-1/2×3 1/2	7/8	11.1	3.25	3.63	1.48	1.05	1.05	2.66	0.466	0.281	0.238	1.87
×7/16	13/16	9.80	2.87	3.25	1.32	1.06	1.03	2.36	0.412	0.192	0.164	1.89
×3/8	3/4	8.50	2.48	2.86	1.15	1.07	1.00	2.06	0.357	0.123	0.106	1.90
×5/16	11/16	7.20	2.09	2.44	0.969	1.08	0.979	1.74	0.301	0.0731	0.0634	1.92
×1/4	5/8	5.80	1.69	2.00	0.787	1.09	0.954	1.41	0.243	0.0386	0.0334	1.93

(continued)

TABLE C.3 (continued)
Other Properties of Each of these Shape Continue in "(b) Properties" Section following up this "(a) Properties" Section

Angles

Shape	k (in.)	Wt. (lb/ft)	Area, A (in.²)	Axis X–X						Flexural-Torsional Properties		
				I (in.⁴)	S (in.³)	r (in.)	$\bar{y}$ (in.)	Z (in.³)	y_p (in.)	J (in.⁴)	C_w (in.⁶)	$\bar{r}_0$ (in.)
L3-1/2×3×1/2	7/8	10.2	3.00	3.45	1.45	1.07	1.12	2.61	0.480	0.260	0.191	1.75
×7/16	13/16	9.10	2.65	3.10	1.29	1.08	1.09	2.32	0.446	0.178	0.132	1.76
×3/8	3/4	7.90	2.30	2.73	1.12	1.09	1.07	2.03	0.411	0.114	0.0858	1.78
×5/16	11/16	6.60	1.93	2.33	0.951	1.09	1.05	1.72	0.375	0.0680	0.0512	1.79
×1/4	5/8	5.40	1.56	1.92	0.773	1.10	1.02	1.39	0.336	0.0360	0.0270	1.80
L3-1/2×2-1/2×1/2	7/8	9.40	2.75	3.24	1.41	1.08	1.20	2.52	0.736	0.234	0.159	1.66
×3/8	3/4	7.20	2.11	2.56	1.09	1.10	1.15	1.96	0.668	0.103	0.0714	1.69
×5/16	11/16	6.10	1.78	2.20	0.925	1.11	1.13	1.67	0.633	0.0611	0.0426	1.71
×1/4	5/8	4.90	1.44	1.81	0.753	1.12	1.10	1.36	0.596	0.0322	0.0225	1.72
L3×3×1/2	7/8	9.40	2.75	2.20	1.06	0.895	0.929	1.91	0.458	0.230	0.144	1.59
×7/16	13/16	8.30	2.43	1.98	0.946	0.903	0.907	1.70	0.405	0.157	0.100	1.60
×3/8	3/4	7.20	2.11	1.75	0.825	0.910	0.884	1.48	0.351	0.101	0.0652	1.62
×5/16	11/16	6.10	1.78	1.50	0.699	0.918	0.860	1.26	0.296	0.0597	0.0390	1.64
×1/4	5/8	4.90	1.44	1.23	0.569	0.926	0.836	1.02	0.239	0.0313	0.0206	1.65
×3/16	9/16	3.71	1.09	0.948	0.433	0.933	0.812	0.774	0.181	0.0136	0.00899	1.67
L3×2-1/2×1/2	7/8	8.50	2.50	2.07	1.03	0.910	0.995	1.86	0.494	0.213	0.112	1.46
×7/16	13/16	7.60	2.21	1.87	0.921	0.917	0.972	1.66	0.462	0.146	0.0777	1.48
×3/8	3/4	6.60	1.92	1.65	0.803	0.924	0.949	1.45	0.430	0.0943	0.0507	1.49
×5/16	11/16	5.60	1.67	1.41	0.681	0.932	0.925	1.23	0.397	0.0560	0.0304	1.51
×1/4	5/8	4.50	1.31	1.16	0.555	0.940	0.900	1.000	0.363	0.0296	0.0161	1.52
×3/16	9/16	3.39	0.996	0.899	0.423	0.947	0.874	0.761	0.328	0.0130	0.00705	1.54

L3×2×1/2	13/16	7.70	2.25	1.92	1.00	0.922	1.08	1.78	0.736	0.192	0.0908	1.39
×3/8	11/16	5.90	1.73	1.54	0.779	0.937	1.03	1.39	0.668	0.0855	0.0413	1.42
×5/16	5/8	5.00	1.46	1.32	0.662	0.945	1.01	1.19	0.633	0.0510	0.0248	1.43
×1/4	9/16	4.10	1.19	1.09	0.541	0.953	0.980	0.969	0.596	0.0270	0.0132	1.45
×3/16	1/2	3.07	0.902	0.847	0.414	0.961	0.952	0.743	0.556	0.0119	0.00576	1.46
L2-1/2×2-1/2×1/2	3/4	7.70	2.25	1.22	0.716	0.735	0.803	1.29	0.450	0.188	0.0791	1.30
×3/8	5/8	5.90	1.73	0.972	0.558	0.749	0.758	1.01	0.347	0.0833	0.0362	1.33
×5/16	9/16	5.00	1.46	0.837	0.474	0.756	0.735	0.853	0.293	0.0495	0.0218	1.35
×1/4	1/2	4.10	1.19	0.692	0.387	0.764	0.711	0.695	0.237	0.0261	0.0116	1.36
×3/16	7/16	3.07	0.900	0.535	0.295	0.771	0.687	0.529	0.180	0.0114	0.00510	1.38
L2-1/2×2×3/8	5/8	5.30	1.55	0.914	0.546	0.766	0.826	0.982	0.425	0.0746	0.0268	1.22
×5/16	9/16	4.50	1.31	0.790	0.465	0.774	0.803	0.839	0.391	0.0444	0.0162	1.23
×1/4	1/2	3.62	1.06	0.656	0.381	0.782	0.779	0.688	0.356	0.0235	0.00868	1.25
×3/16	7/16	2.75	0.809	0.511	0.293	0.790	0.754	0.529	0.318	0.0103	0.00382	1.26
L2-1/2×1-1/2×1/4	1/2	3.22	0.938	0.594	0.364	0.792	0.866	0.644	0.606	0.0209	0.00694	1.19
×3/16	7/16	2.47	0.715	0.464	0.280	0.801	0.839	0.497	0.568	0.00921	0.00306	1.20
L2×2×3/8	5/8	4.70	1.36	0.476	0.348	0.591	0.632	0.629	0.342	0.0658	0.0174	1.05
×5/16	9/16	3.92	1.15	0.414	0.298	0.598	0.609	0.537	0.290	0.0393	0.0106	1.06
×1/4	1/2	3.19	0.938	0.346	0.244	0.605	0.586	0.440	0.236	0.0209	0.00572	1.08
×3/16	7/16	2.44	0.715	0.271	0.188	0.612	0.561	0.338	0.180	0.00921	0.00254	1.09
×1/8	3/8	1.65	0.484	0.189	0.129	0.620	0.534	0.230	0.123	0.00293	0.000789	1.10

Source: Courtesy of the American Institute of Steel Construction, Chicago, IL.

Note: For compactness criteria, refer to end of this table.

Workable Gages in Angle Legs (in.)

Leg	8	7	6	5	4	3-1/2	3	2-1/2	2	1-3/4	1-1/2	1-3/8	1-1/4	1
g	4-1/2	4	3-1/2	3	2-1/2	2	1-3/4	1-3/8	1-1/8	1	7/8	7/8	3/4	5/8
g_1	3	2-1/2	2-1/4	2										
g_2	3	3	2-1/2	1-3/4										

Note: Other gages are permitted to suit specific requirements subject to clearances and edge distance limitations.

TABLE C.3
Angles Angles: Other Properties of Each of these Shape Continue in "(b) Properties" Section following up this "(a) Properties" Section

(b) Properties

Shape	Axis Y–Y						Axis Z–Z				Q_s
	I (in.⁴)	S (in.³)	r (in.)	$\bar{x}$ (in.)	Z (in.³)	x_p (in.)	I (in.⁴)	S (in.³)	r (in.)	Tan α	$F_y = 36$ ksi
L4×3-1/2×1/2	3.76	1.50	1.04	0.994	2.69	0.438	1.80	0.719	0.716	0.750	1.00
×3/8	2.96	1.16	1.05	0.947	2.06	0.334	1.38	0.555	0.719	0.755	1.00
×5/16	2.52	0.980	1.06	0.923	1.74	0.281	1.17	0.470	0.721	0.757	0.997
×1/4	2.07	0.794	1.07	0.897	1.40	0.227	0.950	0.382	0.723	0.759	0.912
L4×3×5/8	2.85	1.34	0.845	0.867	2.45	0.498	1.59	0.720	0.631	0.534	1.00
×1/2	2.40	1.10	0.858	0.822	1.99	0.407	1.30	0.592	0.633	0.542	1.00
×3/8	1.89	0.851	0.873	0.775	1.52	0.311	1.01	0.460	0.636	0.551	1.00
×5/16	1.62	0.721	0.880	0.750	1.28	0.262	0.851	0.390	0.638	0.554	0.997
×1/4	1.33	0.585	0.887	0.725	1.03	0.211	0.691	0.318	0.639	0.558	0.912
L3-1/2×3-1/2×1/2	3.63	1.48	1.05	1.05	2.66	0.466	1.51	0.609	0.679	1.00	1.00
×7/16	3.25	1.32	1.06	1.03	2.36	0.412	1.34	0.540	0.681	1.00	1.00
×3/8	2.86	1.15	1.07	1.00	2.05	0.357	1.17	0.471	0.683	1.00	1.00
×5/16	2.44	0.969	1.08	0.979	1.74	0.301	0.989	0.400	0.685	1.00	1.00
×1/4	2.00	0.787	1.09	0.954	1.41	0.243	0.807	0.326	0.688	1.00	0.965
L3-1/2×3×1/2	2.32	1.09	0.877	0.869	1.97	0.431	1.15	0.537	0.618	0.713	1.00
×7/16	2.09	0.971	0.885	0.846	1.75	0.382	1.03	0.478	0.620	0.717	1.00
×3/8	1.84	0.847	0.892	0.823	1.52	0.331	0.895	0.418	0.622	0.720	1.00
×5/16	1.58	0.718	0.900	0.798	1.28	0.279	0.761	0.356	0.624	0.722	1.00
×1/4	1.30	0.585	0.908	0.773	1.04	0.226	0.623	0.292	0.628	0.725	0.965
L3-1/2×2-1/2×1/2	1.36	0.756	0.701	0.701	1.39	0.395	0.782	0.420	0.532	0.485	1.00
×3/8	1.09	0.589	0.716	0.655	1.07	0.303	0.608	0.329	0.535	0.495	1.00
×5/16	0.937	0.501	0.723	0.632	0.900	0.256	0.518	0.281	0.538	0.500	1.00
×1/4	0.775	0.410	0.731	0.607	0.728	0.207	0.425	0.232	0.541	0.504	0.965

L3×3×1/2	2.20	1.06	0.895	0.929	1.91	0.458	0.924	0.436	0.580	1.00	1.00
×7/16	1.98	0.946	0.903	0.907	1.70	0.405	0.819	0.386	0.580	1.00	1.00
×3/8	1.75	0.825	0.910	0.884	1.48	0.351	0.712	0.336	0.581	1.00	1.00
×5/16	1.50	0.699	0.918	0.860	1.25	0.296	0.603	0.284	0.583	1.00	1.00
×1/4	1.23	0.569	0.926	0.836	1.02	0.239	0.491	0.231	0.585	1.00	1.00
×3/16	0.948	0.433	0.933	0.812	0.774	0.181	0.374	0.176	0.586	1.00	0.912
L3×2-1/2×1/2	1.29	0.736	0.718	0.746	1.34	0.418	0.666	0.370	0.516	0.666	1.00
×7/16	1.17	0.656	0.724	0.724	1.19	0.370	0.591	0.329	0.516	0.671	1.00
×3/8	1.03	0.573	0.731	0.701	1.03	0.321	0.514	0.287	0.517	0.675	1.00
×5/16	0.888	0.487	0.739	0.677	0.873	0.271	0.437	0.244	0.518	0.679	1.00
×1/4	0.734	0.397	0.746	0.653	0.707	0.220	0.356	0.199	0.520	0.683	1.00
×3/16	0.568	0.303	0.753	0.627	0.536	0.167	0.272	0.153	0.521	0.687	0.912
L3×2×1/2	0.667	0.470	0.543	0.580	0.887	0.377	0.409	0.266	0.425	0.413	1.00
×3/8	0.539	0.368	0.555	0.535	0.679	0.291	0.318	0.209	0.426	0.426	1.00
×5/16	0.467	0.314	0.562	0.511	0.572	0.247	0.271	0.179	0.428	0.432	1.00
×1/4	0.390	0.258	0.569	0.487	0.463	0.200	0.223	0.179	0.431	0.437	1.00
×3/16	0.305	0.198	0.577	0.462	0.351	0.153	0.173	0.116	0.435	0.442	0.912
L2-1/2×2-1/2×1/2	1.22	0.716	0.735	0.803	1.29	0.450	0.521	0.295	0.481	1.00	1.00
×3/8	0.972	0.558	0.749	0.758	1.00	0.347	0.400	0.226	0.481	1.00	1.00
×5/16	0.837	0.474	0.756	0.735	0.853	0.293	0.339	0.192	0.481	1.00	1.00
×1/4	0.692	0.387	0.764	0.711	0.694	0.237	0.275	0.156	0.482	1.00	1.00
×3/16	0.535	0.295	0.771	0.687	0.528	0.180	0.210	0.119	0.482	1.00	0.983
L2-1/2×2×3/8	0.513	0.361	0.574	0.578	0.657	0.311	0.273	0.189	0.419	0.612	1.00
×5/16	0.446	0.309	0.581	0.555	0.557	0.264	0.233	0.161	0.420	0.618	1.00
×1/4	0.372	0.253	0.589	0.532	0.454	0.214	0.191	0.133	0.423	0.624	1.00
×3/16	0.292	0.195	0.597	0.508	0.347	0.164	0.149	0.104	0.426	0.628	0.983
L2-1/2×1-1/2×1/4	0.160	0.142	0.411	0.372	0.261	0.189	0.0975	0.0818	0.321	0.354	1.00
×3/16	0.126	0.110	0.418	0.347	0.198	0.145	0.0760	0.0644	0.324	0.360	0.983
L2×2×3/8	0.476	0.348	0.591	0.632	0.628	0.342	0.203	0.144	0.386	1.00	1.00
×5/16	0.414	0.298	0.598	0.609	0.536	0.290	0.173	0.122	0.386	1.00	1.00
×1/4	0.346	0.244	0.605	0.586	0.440	0.236	0.141	0.1000	0.387	1.00	1.00
×3/16	0.271	0.188	0.612	0.561	0.338	0.180	0.109	0.0771	0.389	1.00	1.00
×1/8	0.189	0.129	0.620	0.534	0.230	0.123	0.0751	0.0531	0.391	1.00	0.912

Source: Courtesy of the American Institute of Steel Construction, Chicago, IL.

Note: For compactness criteria, refer to end of this table.

TABLE C.3 (continued)

Compactness Criteria for Angles

t	Compression	Flexure	
	Non-Slender up to	Compact up to	Non-Compact up to
	Width of Angle Leg (in.)		
1-1/8	8	8	—
1	↓	↓	—
7/8			—
3/4			—
5/8	↓		—
9/16	7	↓	—
1/2	6	7	8
7/16	5	6	↓
3/8	4	5	
5/16	4	4	↓
1/4	3	3-1/2	6
3/16	2	2-1/2	4
1/8	1-1/2	1-1/2	3

Note: Compactness criteria given for $F_y = 36$ ksi. $C_v = 1.0$ for all angles.

TABLE C.4
Other Properties for Each of these Shape Continue in "(b) Dimensions and Properties," following this "(a) Dimensions and Properties" Section

Rectangular HSS

(a) Dimensions and Properties

Shape	Design Wall Thickness, t (in.)	Nominal Wt. (lb/ft)	Area, A (in.²)	b/t	h/t	Axis X–X			
						I (in.⁴)	S (in.³)	r (in.)	Z (in.³)
HSS6×4×1/2	0.465	28.30	7.88	5.60	9.90	34.0	11.3	2.08	14.6
×3/8	0.349	22.30	6.18	8.46	14.2	28.3	9.43	2.14	11.9
×5/16	0.291	19.06	5.26	10.7	17.6	24.8	8.27	2.17	10.3
×1/4	0.233	15.58	4.30	14.2	22.8	20.9	6.96	2.20	8.53
×3/16	0.174	11.98	3.28	20.0	31.5	16.4	5.46	2.23	6.60
×1/8	0.116	8.15	2.23	31.5	48.7	11.4	3.81	2.26	4.56
HSS6×3×1/2	0.465	24.90	6.95	3.45	9.90	26.8	8.95	1.97	12.1
×3/8	0.349	19.75	5.48	5.60	14.2	22.7	7.57	2.04	9.90
×5/16	0.291	16.93	4.68	7.31	17.6	20.1	6.69	2.07	8.61
×1/4	0.233	13.88	3.84	9.88	22.8	17.0	5.66	2.10	7.19
×3/16	0.174	10.70	2.93	14.2	31.5	13.4	4.47	2.14	5.59
×1/8	0.116	7.30	2.00	22.9	48.7	9.43	3.14	2.17	3.87
HSS6×2×3/8	0.349	17.20	4.78	2.73	14.2	17.1	5.71	1.89	7.93
×5/16	0.291	14.80	4.10	3.87	17.6	15.3	5.11	1.93	6.95
×1/4	0.233	12.18	3.37	5.58	22.8	13.1	4.37	1.97	5.84
×3/16	0.174	9.43	2.58	8.49	31.5	10.5	3.49	2.01	4.58
×1/8	0.116	6.45	1.77	14.2	48.7	7.42	2.47	2.05	3.19

(continued)

TABLE C.4 (continued)

Other Properties for Each of these Shape Continue in "(b) Dimensions and Properties following this "(a) Dimensions and Properties" Section

Rectangular HSS

Shape	Design Wall Thickness, t (in.)	Nominal Wt. (lb/ft)	Area, A (in.²)	b/t	h/t	Axis X–X			
						I (in.⁴)	S (in.³)	r (in.)	Z (in.³)
HSS5×4×1/2	0.465	24.90	6.95	5.60	7.75	21.2	8.49	1.75	10.9
×3/8	0.349	19.75	5.48	8.46	11.3	17.9	7.17	1.81	8.96
×5/16	0.291	16.93	4.68	10.7	14.2	15.8	6.32	1.84	7.79
×1/4	0.233	13.88	3.84	14.2	18.5	13.4	5.35	1.87	6.49
×3/16	0.174	10.70	2.93	20.0	25.7	10.6	4.22	1.90	5.05
×1/8	0.116	7.30	2.00	31.5	40.1	7.42	2.97	1.93	3.50
HSS5×3×1/2	0.465	21.50	6.02	3.45	7.75	16.4	6.57	1.65	8.83
×3/8	0.349	17.20	4.78	5.60	11.3	14.1	5.65	1.72	7.34
×5/16	0.291	14.80	4.10	7.31	14.2	12.6	5.03	1.75	6.42
×1/4	0.233	12.18	3.37	9.88	18.5	10.7	4.29	1.78	5.38
×3/16	0.174	9.43	2.58	14.2	25.7	8.53	3.41	1.82	4.21
×1/8	0.116	6.45	1.77	22.9	40.1	6.03	2.41	1.85	2.93
HSS5×2-1/2×1/4	0.233	11.33	3.14	7.73	18.5	9.40	3.76	1.73	4.83
×3/16	0.174	8.79	2.41	11.4	25.7	7.51	3.01	1.77	3.79
×1/8	0.116	6.02	1.65	18.6	40.1	5.34	2.14	1.80	2.65
HSS5×2×3/8	0.349	14.65	4.09	2.73	11.3	10.4	4.14	1.59	5.71
×5/16	0.291	12.67	3.52	3.87	14.2	9.35	3.74	1.63	5.05
×1/4	0.233	10.48	2.91	5.58	18.5	8.08	3.23	1.67	4.27
×3/16	0.174	8.15	2.24	8.49	25.7	6.50	2.60	1.70	3.37
×1/8	0.116	5.60	1.54	14.2	40.1	4.65	1.86	1.74	2.37

HSS4×3×3/8	0.349	14.65	4.09	5.60	8.46	7.93	3.97	1.39	5.12
× 5/16	0.291	12.67	3.52	7.31	10.7	7.14	3.57	1.42	4.51
× 1/4	0.233	10.48	2.91	9.88	14.2	6.15	3.07	1.45	3.81
× 3/16	0.174	8.15	2.24	14.2	20.0	4.93	2.47	1.49	3.00
× 1/8	0.116	5.60	1.54	22.9	31.5	3.52	1.76	1.52	2.11
HSS4×2-1/2×3/8	0.349	13.37	3.74	4.16	8.46	6.77	3.38	1.35	4.48
× 5/16	0.291	11.60	3.23	5.59	10.7	6.13	3.07	1.38	3.97
× 1/4	0.233	9.63	2.67	7.73	14.2	5.32	2.66	1.41	3.38
× 3/16	0.174	7.51	2.06	11.4	20.0	4.30	2.15	1.44	2.67
× 1/8	0.116	5.17	1.42	18.6	31.5	3.09	1.54	1.47	1.88
HSS4×2×3/8	0.349	12.09	3.39	2.73	8.46	5.60	2.80	1.29	3.84
× 5/16	0.291	10.54	2.94	3.87	10.7	5.13	2.56	1.32	3.43
× 1/4	0.233	8.78	2.44	5.58	14.2	4.49	2.25	1.36	2.94
× 3/16	0.174	6.87	1.89	8.49	20.0	3.66	1.83	1.39	2.34
× 1/8	0.116	4.75	1.30	14.2	31.5	2.65	1.32	1.43	1.66
HSS3-1/2×2-1/2×3/8	0.349	12.09	3.39	4.16	7.03	4.75	2.72	1.18	3.59
× 5/16	0.291	10.54	2.94	5.59	9.03	4.34	2.48	1.22	3.20
× 1/4	0.233	8.78	2.44	7.73	12.0	3.79	2.17	1.25	2.74
× 3/16	0.174	6.87	1.89	11.4	17.1	3.09	1.76	1.28	2.18
× 1/8	0.116	4.75	1.30	18.6	27.2	2.23	1.28	1.31	1.54
HSS3-1/2×2×1/4	0.233	7.93	2.21	5.58	12.0	3.17	1.81	1.20	2.36
× 3/16	0.174	6.23	1.71	8.49	17.1	2.61	1.49	1.23	1.89
× 1/8	0.116	4.32	1.19	14.2	27.2	1.90	1.09	1.27	1.34

Source: Courtesy of the American Institute of Steel Construction, Chicago, IL.

TABLE C.4
Rectangular HSS

(b) Dimensions and Properties

Shape	Axis Y-Y				Workable Flat		Torsion		Surface Area
	I (in.^{4})	S (in.^{3})	r (in.)	Z (in.^{3})	Depth (in.)	Width (in.)	J (in.^{4})	C (in.^{3})	(ft^2/ft)
HSS6×4×1/2	17.8	8.89	1.50	11.0	3-3/4	—	40.3	17.8	1.53
×3/8	14.9	7.47	1.55	8.94	4-5/16	2-5/16	32.8	14.2	1.57
×5/16	13.2	6.58	1.58	7.75	4-5/8	2-5/8	28.4	12.2	1.58
×1/4	11.1	5.56	1.61	6.45	4-7/8	2-7/8	23.6	10.1	1.60
×3/16	8.76	4.38	1.63	5.00	5-3/16	3-3/16	18.2	7.74	1.62
×1/8	6.15	3.08	1.66	3.46	5-7/16	3-7/16	12.6	5.30	1.63
HSS6×3×1/2	8.69	5.79	1.12	7.28	3-3/4	—	23.1	12.7	1.37
×3/8	7.48	4.99	1.17	6.03	4-5/16	—	19.3	10.3	1.40
×5/16	6.67	4.45	1.19	5.27	4-5/8	—	16.9	8.91	1.42
×1/4	5.70	3.80	1.22	4.41	4-7/8	—	14.2	7.39	1.43
×3/16	4.55	3.03	1.25	3.45	5-3/16	2-3/16	11.1	5.71	1.45
×1/8	3.23	2.15	1.27	2.40	5-7/16	2-7/16	7.73	3.93	1.47
HSS6×2×3/8	2.77	2.77	0.760	3.46	4-5/16	—	8.42	6.35	1.23
×5/16	2.52	2.52	0.785	3.07	4-5/8	—	7.60	5.58	1.25
×1/4	2.21	2.21	0.810	2.61	4-7/8	—	6.55	4.70	1.27
×3/16	1.80	1.80	0.836	2.07	5-3/16	—	5.24	3.68	1.28
×1/8	1.31	1.31	0.861	1.46	5-7/16	—	3.72	2.57	1.30

Shape									
HSS5×4×1/2	14.9	7.43	1.46	9.35	2-3/4	—	30.3	14.5	1.37
×3/8	12.6	6.30	1.52	7.67	3-5/16	2-5/16	24.9	11.7	1.40
×5/16	11.1	5.57	1.54	6.67	3-5/8	2-5/8	21.7	10.1	1.42
×1/4	9.46	4.73	1.57	5.57	3-7/8	2-7/8	18.0	8.32	1.43
×3/16	7.48	3.74	1.60	4.34	4-3/16	3-3/16	14.0	6.41	1.45
×1/8	5.27	2.64	1.62	3.01	4-7/16	3-7/16	9.66	4.39	1.47
HSS5×3×1/2	7.18	4.78	1.09	6.10	2-3/4	—	17.6	10.3	1.20
×3/8	6.25	4.16	1.14	5.10	3-5/16	—	14.9	8.44	1.23
×5/16	5.60	3.73	1.17	4.48	3-5/8	—	13.1	7.33	1.25
×1/4	4.81	3.21	1.19	3.77	3-7/8	—	11.0	6.10	1.27
×3/16	3.85	2.57	1.22	2.96	4-3/16	2-3/16	8.64	4.73	1.28
×1/8	2.75	1.83	1.25	2.07	4-7/16	2-7/16	6.02	3.26	1.30
HSS5×2-1/2×1/4	3.13	2.50	0.999	2.95	3-7/8	—	7.93	4.99	1.18
×3/16	2.53	2.03	1.02	2.33	4-3/16	—	6.26	3.89	1.20
×1/8	1.82	1.46	1.05	1.64	4-7/16	—	4.40	2.70	1.22
HSS5×2×3/8	2.28	2.28	0.748	2.88	3-5/16	—	6.61	5.20	1.07
×5/16	2.10	2.10	0.772	2.57	3-5/8	—	5.99	4.59	1.08
×1/4	1.84	1.84	0.797	2.20	3-7/8	—	5.17	3.88	1.10
×3/16	1.51	1.51	0.823	1.75	4-3/16	—	4.15	3.05	1.12
×1/8	1.10	1.10	0.848	1.24	4-7/16	—	2.95	2.13	1.13
HSS4×3×3/8	5.01	3.34	1.11	4.18	2-5/16	—	10.6	6.59	1.07
×5/16	4.52	3.02	1.13	3.69	2-5/8	—	9.41	5.75	1.08
×1/4	3.91	2.61	1.16	3.12	2-7/8	—	7.96	4.81	1.10
×3/16	3.16	2.10	1.19	2.46	3-3/16	—	6.26	3.74	1.12
×1/8	2.27	1.51	1.21	1.73	3-7/16	—	4.38	2.59	1.13
HSS4×2-1/2×3/8	3.17	2.54	0.922	3.20	2-5/16	—	7.57	5.32	0.983
×5/16	2.89	2.32	0.947	2.85	2-5/8	—	6.77	4.67	1.00
×1/4	2.53	2.02	0.973	2.43	2-7/8	—	5.78	3.93	1.02
×3/16	2.06	1.65	0.999	1.93	3-1/8	—	4.59	3.08	1.03
×1/8	1.49	1.19	1.03	1.36	3-7/16	—	3.23	2.14	1.05

(continued)

TABLE C.4 (continued)
Rectangular HSS

Shape	Axis Y-Y				Workable Flat		Torsion		Surface Area (ft²/ft)
	I (in.⁴)	S (in.³)	r (in.)	Z (in.³)	Depth (in.)	Width (in.)	J (in.⁴)	C (in.³)	
HSS4×2×3/8	1.80	1.80	0.729	2.31	2-5/16	—	4.83	4.04	0.900
×5/16	1.67	1.67	0.754	2.08	2-5/8	—	4.40	3.59	0.917
×1/4	1.48	1.48	0.779	1.79	2-7/8	—	3.82	3.05	0.933
×3/16	1.22	1.22	0.804	1.43	3-3/16	—	3.08	2.41	0.950
×1/8	0.898	0.898	0.830	1.02	3-7/16	—	2.20	1.69	0.967
HSS3-1/2×2-1/2×3/8	2.77	2.21	0.904	2.82	—	—	6.16	4.57	0.900
×5/16	2.54	2.03	0.930	2.52	2-1/8	—	5.53	4.03	0.917
×1/4	2.23	1.78	0.956	2.16	2-3/8	—	4.75	3.40	0.933
×3/16	1.82	1.46	0.983	1.72	2-11/16	—	3.78	2.67	0.950
×1/8	1.33	1.06	1.01	1.22	2-15/16	—	2.67	1.87	0.967
HSS3-1/2×2×1/4	1.30	1.30	0.766	1.58	2-3/8	—	3.16	2.64	0.850
×3/16	1.08	1.08	0.792	1.27	2-11/16	—	2.55	2.09	0.867
×1/8	0.795	0.795	0.818	0.912	2-15/16	—	1.83	1.47	0.883

Source: Courtesy of the American Institute of Steel Construction, Chicago, IL.
—, Flat depth or width is too small to establish a workable flat.

TABLE C.5
Square HSS: Dimensions and Properties

HSS7–HSS4½

Shape	Design Wall Thickness, t (in.)	Nominal Wt. (lb/ft)	Area, A (in.²)	b/t	h/t	I (in.⁴)	S (in.³)	r (in.)	Z (in.³)	Workable Flat (in.)	Torsion J (in.⁴)	Torsion C (in.³)	Surface Area (ft²/ft)
HSS7×7×5/8	0.581	50.60	14.0	9.05	9.05	93.4	26.7	2.58	33.1	4-3/16	158	47.1	2.17
×1/2	0.465	41.91	11.6	12.1	12.1	80.5	23.0	2.63	27.9	4-3/4	133	39.3	2.20
×3/8	0.349	32.51	8.97	17.1	17.1	65.0	18.6	2.69	22.1	5-5/16	105	30.7	2.23
×5/16	0.291	27.54	7.59	21.1	21.1	56.1	16.0	2.72	18.9	5-5/8	89.7	26.1	2.25
×1/4	0.233	22.39	6.17	27.0	27.0	46.5	13.3	2.75	15.5	5-7/8	73.5	21.3	2.27
×3/16	0.174	17.06	4.67	37.2	37.2	36.0	10.3	2.77	11.9	6-3/16	56.1	16.2	2.28
×1/8	0.116	11.55	3.16	57.3	57.3	24.8	7.09	2.80	8.13	6-7/16	38.2	11.0	2.30
HSS6×6×5/8	0.581	42.10	11.7	7.33	7.33	55.2	18.4	2.17	23.2	3-3/16	94.9	33.4	1.83
×1/2	0.465	35.11	9.74	9.90	9.90	48.3	16.1	2.23	19.8	3-3/4	81.1	28.1	1.87
×3/8	0.349	27.41	7.58	14.2	14.2	39.5	13.2	2.28	15.8	4-5/16	64.6	22.1	1.90
×5/16	0.291	23.29	6.43	17.6	17.6	34.3	11.4	2.31	13.6	4-5/8	55.4	18.9	1.92
×1/4	0.233	18.99	5.24	22.8	22.8	28.6	9.54	2.34	11.2	4-7/8	45.6	15.4	1.93
×3/16	0.174	14.51	3.98	31.5	31.5	22.3	7.42	2.37	8.63	5-3/16	35.0	11.8	1.95
×1/8	0.116	9.85	2.70	48.7	48.7	15.5	5.15	2.39	5.92	5-7/16	23.9	8.03	1.97
HSS5-1/2×5-1/2×3/8	0.349	24.85	6.88	12.8	12.8	29.7	10.8	2.08	13.1	3-13/16	49.0	18.4	1.73
×5/16	0.291	21.16	5.85	15.9	15.9	25.9	9.43	2.11	11.3	4-1/8	42.2	15.7	1.75
×1/4	0.233	17.28	4.77	20.6	20.6	21.7	7.90	2.13	9.32	4-3/8	34.8	12.9	1.77
×3/16	0.174	13.23	3.63	28.6	28.6	17.0	6.17	2.16	7.19	4-11/16	26.7	9.85	1.78
×1/8	0.116	9.00	2.46	44.4	44.4	11.8	4.30	2.19	4.95	4-15/16	18.3	6.72	1.80
HSS5×5×1/2	0.465	28.30	7.88	7.75	7.75	26.0	10.4	1.82	13.1	2-3/4	44.6	18.7	1.53
×3/8	0.349	22.30	6.18	11.3	11.3	21.7	8.68	1.87	10.6	3-5/16	36.1	14.9	1.57
×5/16	0.291	19.03	5.26	14.2	14.2	19.0	7.62	1.90	9.16	3-5/8	31.2	12.8	1.58

(continued)

TABLE C.5 (continued)
Square HSS: Dimensions and Properties

Shape	Design Wall Thickness, t (in.)	Nominal Wt. (lb/ft)	Area, A (in.²)	b/t	h/t	I (in.⁴)	S (in.³)	r (in.)	Z (in.³)	Workable Flat (in.)	Torsion J (in.⁴)	Torsion C (in.³)	Surface Area (ft²/ft)
×1/4	0.233	15.58	4.30	18.5	18.5	16.0	6.41	1.93	7.61	3-7/8	25.8	10.5	1.60
×3/16	0.174	11.96	3.28	25.7	25.7	12.6	5.03	1.96	5.89	4-3/16	19.9	8.08	1.62
×1/8	0.116	8.15	2.23	40.1	40.1	8.80	3.52	1.99	4.07	4-7/16	13.7	5.53	1.63
HSS4-1/2×4-1/2×1/2	0.465	24.90	6.95	6.68	6.68	18.1	8.03	1.61	10.2	2-1/4	31.3	14.8	1.37
×3/8	0.349	19.75	5.48	9.89	9.89	15.3	6.79	1.67	8.36	2-13/16	25.7	11.9	1.40
×5/16	0.291	16.91	4.68	12.5	12.5	13.5	6.00	1.70	7.27	3-1/8	22.3	10.2	1.42
×1/4	0.233	13.88	3.84	16.3	16.3	11.4	5.08	1.73	6.06	3-3/8	18.5	8.44	1.43
×3/16	0.174	10.68	2.93	22.9	22.9	9.02	4.01	1.75	4.71	3-11/16	14.4	6.49	1.45
×1/8	0.116	7.30	2.00	35.8	35.8	6.35	2.82	1.78	3.27	3-15/16	9.92	4.45	1.47
HSS4×4×1/2	0.465	21.50	6.02	5.60	5.60	11.9	5.97	1.41	7.70	—	21.0	11.2	1.20
×3/8	0.349	17.20	4.78	8.46	8.46	10.3	5.13	1.47	6.39	2-5/16	17.5	9.14	1.23
×5/16	0.291	14.78	4.10	10.7	10.7	9.14	4.57	1.49	5.59	2-5/8	15.3	7.91	1.25
×1/4	0.233	12.18	3.37	14.2	14.2	7.80	3.90	1.52	4.69	2-7/8	12.8	6.56	1.27
×3/16	0.174	9.40	2.58	20.0	20.0	6.21	3.10	1.55	3.67	3-3/16	10.0	5.07	1.28
×1/8	0.116	6.45	1.77	31.5	31.5	4.40	2.20	1.58	2.56	3-7/16	6.91	3.49	1.30
HSS3-1/2×3-1/2×3/8	0.349	14.65	4.09	7.03	7.03	6.49	3.71	1.26	4.69	—	11.2	6.77	1.07
×5/16	0.291	12.65	3.52	9.03	9.03	5.84	3.34	1.29	4.14	2-1/8	9.89	5.90	1.08
×1/4	0.233	10.48	2.91	12.0	12.0	5.04	2.88	1.32	3.50	2-3/8	8.35	4.92	1.10
×3/16	0.174	8.13	2.24	17.1	17.1	4.05	2.31	1.35	2.76	2-11/16	6.56	3.83	1.12
×1/8	0.116	5.60	1.54	27.2	27.2	2.90	1.66	1.37	1.93	2-15/16	4.58	2.65	1.13

HSS3×3×3/8	0.349	12.09	3.39	5.60	5.60	3.78	2.52	1.06	3.25	—	6.64	4.74	0.900
×5/16	0.291	10.53	2.94	7.31	7.31	3.45	2.30	1.08	2.90	—	5.94	4.18	0.917
×1/4	0.233	8.78	2.44	9.88	9.88	3.02	2.01	1.11	2.48	—	5.08	3.52	0.933
×3/16	0.174	6.85	1.89	14.2	14.2	2.46	1.64	1.14	1.97	2-3/16	4.03	2.76	0.950
×1/8	0.116	4.75	1.30	22.9	22.9	1.78	1.19	1.17	1.40	2-7/16	2.84	1.92	0.967
HSS2-1/2×2-1/2×5/16	0.291	8.40	2.35	5.59	5.59	1.82	1.46	0.880	1.88	—	3.20	2.74	0.750
×1/4	0.233	7.08	1.97	7.73	7.73	1.63	1.30	0.908	1.63	—	2.79	2.35	0.767
×3/16	0.174	5.57	1.54	11.4	11.4	1.35	1.08	0.937	1.32	—	2.25	1.86	0.784
×1/8	0.116	3.90	1.07	18.6	18.6	0.998	0.799	0.965	0.947	—	1.61	1.31	0.800
HSS2-1/4×2-1/4×1/4	0.233	6.23	1.74	6.66	6.66	1.13	1.01	0.806	1.28	—	1.96	1.85	0.683
×3/16	0.174	4.94	1.37	9.93	9.93	0.953	0.847	0.835	1.04	—	1.60	1.48	0.700
×1/8	0.116	3.47	0.956	16.4	16.4	0.712	0.633	0.863	0.755	—	1.15	1.05	0.717
HSS2×2×1/4	0.233	5.38	1.51	5.58	5.58	0.747	0.747	0.704	0.964	—	1.31	1.41	0.600
×3/16	0.174	4.30	1.19	8.49	8.49	0.641	0.641	0.733	0.797	—	1.09	1.14	0.617
×1/8	0.116	3.04	0.840	14.2	14.2	0.486	0.486	0.761	0.584	—	0.796	0.817	0.633

Source: Courtesy of the American Institute of Steel Construction, Chicago, IL.

—, Flat depth or width is too small to establish a workable flat.

TABLE C.6
Round HSS: Dimensions and Properties

HSS6.625–HSS5.000

HSS4.500–HSS2.500

Shape	Design Wall Thickness, t (in.)	Nominal Wt. (lb/ft)	Area, A (in.²)	D/t	I (in.⁴)	S (in.³)	r (in.)	Z (in.³)	Torsion J (in.⁴)	Torsion C (in.³)
HSS6.625×0.500	0.465	32.74	9.00	14.2	42.9	13.0	2.18	17.7	85.9	25.9
×0.432	0.402	28.60	7.86	16.5	38.2	11.5	2.20	15.6	76.4	23.1
×0.375	0.349	25.06	6.88	19.0	34.0	10.3	2.22	13.8	68.0	20.5
×0.312	0.291	21.06	5.79	22.8	29.1	8.79	2.24	11.7	58.2	17.6
×0.280	0.260	18.99	5.20	25.5	26.4	7.96	2.25	10.5	52.7	15.9
×0.250	0.233	17.04	4.68	28.4	23.9	7.22	2.26	9.52	47.9	14.4
×0.188	0.174	12.94	3.53	38.1	18.4	5.54	2.28	7.24	36.7	11.1
×0.125ᵃ	0.116	8.69	2.37	57.1	12.6	3.79	2.30	4.92	25.1	7.59
HSS6.000×0.500	0.465	29.40	8.09	12.9	31.2	10.4	1.96	14.3	62.4	20.8
×0.375	0.349	22.55	6.20	17.2	24.8	8.28	2.00	11.2	49.7	16.6
×0.312	0.291	18.97	5.22	20.6	21.3	7.11	2.02	9.49	42.6	14.2
×0.280	0.260	17.12	4.69	23.1	19.3	6.45	2.03	8.57	38.7	12.9
×0.250	0.233	15.37	4.22	25.8	17.6	5.86	2.04	7.75	35.2	11.7
×0.188	0.174	11.68	3.18	34.5	13.5	4.51	2.06	5.91	27.0	9.02
×0.125ᵃ	0.116	7.85	2.14	51.7	9.28	3.09	2.08	4.02	18.6	6.19
HSS5.563×0.500	0.465	27.06	7.45	12.0	24.4	8.77	1.81	12.1	48.8	17.5
×0.375	0.349	20.80	5.72	15.9	19.5	7.02	1.85	9.50	39.0	14.0
×0.258	0.240	14.63	4.01	23.2	14.2	5.12	1.88	6.80	28.5	10.2
×0.188	0.174	10.80	2.95	32.0	10.7	3.85	1.91	5.05	21.4	7.70
×0.134	0.124	7.78	2.12	44.9	7.84	2.82	1.92	3.67	15.7	5.64

HSS5.500×0.500	0.465	26.73	7.36	11.8	23.5	8.55	1.79	11.8	47.0	17.1
×0.375	0.349	20.55	5.65	15.8	18.8	6.84	1.83	9.27	37.6	13.7
×0.258	0.240	14.46	3.97	22.9	13.7	5.00	1.86	6.64	27.5	10.0
HSS5.000×0.500	0.465	24.05	6.62	10.8	17.2	6.88	1.61	9.60	34.4	13.8
×0.375	0.349	18.54	5.10	14.3	13.9	5.55	1.65	7.56	27.7	11.1
×0.312	0.291	15.64	4.30	17.2	12.0	4.79	1.67	6.46	24.0	9.58
×0.258	0.240	13.08	3.59	20.8	10.2	4.08	1.69	5.44	20.4	8.15
×0.250	0.233	12.69	3.49	21.5	9.94	3.97	1.69	5.30	19.9	7.95
×0.188	0.174	9.67	2.64	28.7	7.69	3.08	1.71	4.05	15.4	6.15
×0.125	0.116	6.51	1.78	43.1	5.31	2.12	1.73	2.77	10.6	4.25
HSS4.500×0.375	0.349	16.54	4.55	12.9	9.87	4.39	1.47	6.03	19.7	8.78
×0.337	0.313	15.00	4.12	14.4	9.07	4.03	1.48	5.50	18.1	8.06
×0.237	0.220	10.80	2.96	20.5	6.79	3.02	1.52	4.03	13.6	6.04
×0.188	0.174	8.67	2.36	25.9	5.54	2.46	1.53	3.26	11.1	4.93
×0.125	0.116	5.85	1.60	38.8	3.84	1.71	1.55	2.23	7.68	3.41
HSS4.000×0.313	0.291	12.34	3.39	13.7	5.87	2.93	1.32	4.01	11.7	5.87
×0.250	0.233	10.00	2.76	17.2	4.91	2.45	1.33	3.31	9.82	4.91
×0.237	0.220	9.53	2.61	18.2	4.68	2.34	1.34	3.15	9.36	4.68
×0.226	0.210	9.12	2.50	19.0	4.50	2.25	1.34	3.02	9.01	4.50
×0.220	0.205	8.89	2.44	19.5	4.41	2.21	1.34	2.96	8.83	4.41
×0.188	0.174	7.66	2.09	23.0	3.83	1.92	1.35	2.55	7.67	3.83
×0.125	0.116	5.18	1.42	34.5	2.67	1.34	1.37	1.75	5.34	2.67
HSS3.500×0.313	0.291	10.66	2.93	12.0	3.81	2.18	1.14	3.00	7.61	4.35
×0.300	0.279	10.26	2.82	12.5	3.69	2.11	1.14	2.90	7.38	4.22
×0.250	0.233	8.69	2.39	15.0	3.21	1.83	1.16	2.49	6.41	3.66
×0.216	0.201	7.58	2.08	17.4	2.84	1.63	1.17	2.19	5.69	3.25
×0.203	0.189	7.15	1.97	18.5	2.70	1.54	1.17	2.07	5.41	3.09
×0.188	0.174	6.66	1.82	20.1	2.52	1.44	1.18	1.93	5.04	2.88
×0.125	0.116	4.51	1.23	30.2	1.77	1.01	1.20	1.33	3.53	2.02

(continued)

TABLE C.6 (continued)
Round HSS: Dimensions and Properties

Shape	Design Wall Thickness, t (in.)	Nominal Wt. (lb/ft)	Area, A (in.²)	D/t	I (in.⁴)	S (in.³)	r (in.)	Z (in.³)	Torsion J (in.⁴)	Torsion C (in.³)
HSS3.000×0.250	0.233	7.35	2.03	12.9	1.95	1.30	0.982	1.79	3.90	2.60
×0.216	0.201	6.43	1.77	14.9	1.74	1.16	0.992	1.58	3.48	2.32
×0.203	0.189	6.07	1.67	15.9	1.66	1.10	0.996	1.50	3.31	2.21
×0.188	0.174	5.65	1.54	17.2	1.55	1.03	1.00	1.39	3.10	2.06
×0.152	0.141	4.63	1.27	21.3	1.30	0.865	1.01	1.15	2.59	1.73
×0.134	0.124	4.11	1.12	24.2	1.16	0.774	1.02	1.03	2.32	1.55
×0.125	0.116	3.84	1.05	25.9	1.09	0.730	1.02	0.965	2.19	1.46
HSS2.875×0.250	0.233	7.02	1.93	12.3	1.70	1.18	0.938	1.63	3.40	2.37
×0.203	0.189	5.80	1.59	15.2	1.45	1.01	0.952	1.37	2.89	2.01
×0.188	0.174	5.40	1.48	16.5	1.35	0.941	0.957	1.27	2.70	1.88
×0.125	0.116	3.67	1.01	24.8	0.958	0.667	0.976	0.884	1.92	1.33
HSS2.500×0.250	0.233	6.01	1.66	10.7	1.08	0.862	0.806	1.20	2.15	1.72
×0.188	0.174	4.65	1.27	14.4	0.865	0.692	0.825	0.943	1.73	1.38
×0.125	0.116	3.17	0.869	21.6	0.619	0.495	0.844	0.660	1.24	0.990

Source: Courtesy of the American Institute of Steel Construction, Chicago, IL.
[a] Shape exceeds compact limit for flexure with $F_y = 42$ ksi.

TABLE C.7
Pipe: Dimensions and Properties

Pipe

Shape	Nominal Wt. (lb/ft)	Dimensions Outside Diameter (in.)	Inside Diameter (in.)	Nominal Wall Thickness (in.)	Design Wall Thickness (in.)	Area (in.²)	D/t	I (in.⁴)	S (in.³)	r (in.)	J (in.⁴)	Z (in.³)
Standard weight (Std.)												
Pipe 12 Std.	49.6	12.8	12.0	0.375	0.349	13.6	36.5	262	41.0	4.39	523	53.7
Pipe 10 Std.	40.5	10.8	10.0	0.365	0.340	11.1	31.6	151	28.1	3.68	302	36.9
Pipe 8 Std.	28.6	8.63	7.98	0.322	0.300	7.85	28.8	68.1	15.8	2.95	136	20.8
Pipe 6 Std.	19.0	6.63	6.07	0.280	0.261	5.22	25.4	26.5	7.99	2.25	52.9	10.6
Pipe 5 Std.	14.6	5.56	5.05	0.258	0.241	4.03	23.1	14.3	5.14	1.88	28.6	6.83
Pipe 4 Std.	10.8	4.50	4.03	0.237	0.221	2.97	20.4	6.82	3.03	1.51	13.6	4.05
Pipe 3-1/2 Std.	9.12	4.00	3.55	0.226	0.211	2.51	19.0	4.52	2.26	1.34	9.04	3.03
Pipe 3 Std.	7.58	3.50	3.07	0.216	0.201	2.08	17.4	2.85	1.63	1.17	5.69	2.19
Pipe 2-1/2 Std.	5.80	2.88	2.47	0.203	0.189	1.59	15.2	1.45	1.01	0.952	2.89	1.37
Pipe 2 Std.	3.66	2.38	2.07	0.154	0.143	1.00	16.6	0.627	0.528	0.791	1.25	0.713
Pipe 1-1/2 Std.	2.72	1.90	1.61	0.145	0.135	0.750	14.1	0.293	0.309	0.626	0.586	0.421
Pipe 1-1/4 Std.	2.27	1.66	1.38	0.140	0.130	0.620	12.8	0.184	0.222	0.543	0.368	0.305
Pipe 1 Std.	1.68	1.32	1.05	0.133	0.124	0.460	10.6	0.0830	0.126	0.423	0.166	0.177
Pipe 3/4 Std.	1.13	1.05	0.824	0.113	0.105	0.310	10.0	0.0350	0.0671	0.336	0.0700	0.0942
Pipe 1/2 Std.	0.850	0.840	0.622	0.109	0.101	0.230	8.32	0.0160	0.0388	0.264	0.0320	0.0555
Extra strong (x-strong)												
Pipe 12 x-strong	65.5	12.8	11.8	0.500	0.465	17.9	27.4	339	53.2	4.35	678	70.2
Pipe 10 x-strong	54.8	10.8	9.75	0.500	0.465	15.0	23.1	199	37.0	3.64	398	49.2
Pipe 8 x-strong	43.4	8.63	7.63	0.500	0.465	11.9	18.5	100	23.1	2.89	199	31.0

(continued)

TABLE C.7 (continued)
Pipe: Dimensions and Properties

Shape	Nominal Wt. (lb/ft)	Dimensions Outside Diameter (in.)	Inside Diameter (in.)	Nominal Wall Thickness (in.)	Design Wall Thickness (in.)	Area (in.^{2})	D/t	I (in.^{4})	S (in.^{3})	r (in.)	J (in.^{4})	Z (in.^{3})
Pipe 6 x-strong	28.6	6.63	5.76	0.432	0.403	7.88	16.4	38.3	11.6	2.20	76.6	15.6
Pipe 5 x-strong	20.8	5.56	4.81	0.375	0.349	5.72	15.9	19.5	7.02	1.85	39.0	9.50
Pipe 4 x-strong	15.0	4.50	3.83	0.337	0.315	4.14	14.3	9.12	4.05	1.48	18.2	5.53
Pipe 3-1/2 x-strong	12.5	4.00	3.36	0.318	0.296	3.44	13.5	5.94	2.97	1.31	11.9	4.07
Pipe 3 x-strong	10.3	3.50	2.90	0.300	0.280	2.83	12.5	3.70	2.11	1.14	7.40	2.91
Pipe 2-1/2 x-strong	7.67	2.88	2.32	0.276	0.257	2.11	11.2	1.83	1.27	0.930	3.66	1.77
Pipe 2 x-strong	5.03	2.38	1.94	0.218	0.204	1.39	11.6	0.827	0.696	0.771	1.65	0.964
Pipe 1-1/2 x-strong	3.63	1.90	1.50	0.200	0.186	1.00	10.2	0.372	0.392	0.610	0.744	0.549
Pipe 1-1/4 x-strong	3.00	1.66	1.28	0.191	0.178	0.830	9.33	0.231	0.278	0.528	0.462	0.393
Pipe 1 x-strong	2.17	1.32	0.957	0.179	0.166	0.600	7.92	0.101	0.154	0.410	0.202	0.221
Pipe 3/4 x-strong	1.48	1.05	0.742	0.154	0.143	0.410	7.34	0.0430	0.0818	0.325	0.0860	0.119
Pipe 1/2 x-strong	1.09	0.840	0.546	0.147	0.137	0.300	6.13	0.0190	0.0462	0.253	0.0380	0.0686
Double-extra strong (xx-strong)												
Pipe 8 xx-strong	72.5	8.63	6.88	0.875	0.816	20.0	10.6	154	35.8	2.78	308	49.9
Pipe 6 xx-strong	53.2	6.63	4.90	0.864	0.805	14.7	8.23	63.5	19.2	2.08	127	27.4
Pipe 5 xx-strong	38.6	5.56	4.06	0.750	0.699	10.7	7.96	32.2	11.6	1.74	64.4	16.7
Pipe 4 xx-strong	27.6	4.50	3.15	0.674	0.628	7.64	7.17	14.7	6.53	1.39	29.4	9.50
Pipe 3 xx-strong	18.6	3.50	2.30	0.600	0.559	5.16	6.26	5.79	3.31	1.06	11.6	4.89
Pipe 2-1/2 xx-strong	13.7	2.88	1.77	0.552	0.514	3.81	5.59	2.78	1.94	0.854	5.56	2.91
Pipe 2 xx-strong	9.04	2.38	1.50	0.436	0.406	2.51	5.85	1.27	1.07	0.711	2.54	1.60

Source: Courtesy of the American Institute of Steel Construction, Chicago, IL.

TABLE C.8
Available Strength in Axial Compression, Kips: W Shapes, $F_y = 50$ ksi

Shape	W14x						W12x					W10x				
Wt/ft	82	74	68	61	53	48	58	53	50	45	40	54	49	45	39	33
Design $\phi_c P_n$	LRFD	LRFD	LRFD	LRFD	LRFD	LRFD	LRFD	LRFD	LRFD	LRFD	LRFD	LRFD	LRFD	LRFD	LRFD	LRFD
Effective length KL (ft)																
0	1080	980	899	806	702	636	767	701	657	590	526	712	649	597	516	437
6	1020	922	844	757	633	573	722	659	595	534	475	672	612	543	469	395
7	995	901	826	740	610	552	707	644	574	516	458	658	599	525	452	381
8	970	878	804	721	585	529	689	628	551	495	439	643	585	505	435	365
9	942	853	781	700	557	504	670	610	526	472	419	625	569	483	415	348
10	912	826	755	677	528	477	649	590	499	448	397	607	551	460	395	330
11	880	797	728	652	497	449	627	569	471	422	375	586	533	435	373	311
12	846	766	700	626	465	420	603	547	443	396	351	565	513	410	351	292
13	810	734	670	599	433	391	578	525	413	370	328	543	493	384	328	272
14	774	701	639	572	401	361	553	501	384	343	304	520	471	358	305	253
15	736	667	608	543	369	332	527	477	354	317	280	496	450	332	282	233
16	698	632	576	515	338	304	500	452	326	291	257	472	428	306	260	213
17	660	598	544	486	308	276	473	427	297	265	234	448	405	281	238	195
18	621	563	512	457	278	250	446	402	270	241	212	423	383	256	216	177
19	583	528	480	428	250	224	420	378	244	217	191	399	360	233	195	159
20	546	494	448	400	226	202	393	353	220	196	172	375	338	210	176	143
22	473	428	387	345	186	167	342	306	182	162	142	327	295	174	146	118
24	403	365	329	293	157	140	293	261	153	136	120	282	254	146	122	99.5
26	343	311	281	250	133	120	249	222	130	116	102	241	216	124	104	84.8
28	296	268	242	215	115	103	215	192	112	99.8	88.0	208	186	107	90.0	73.1
30	258	234	211	187	100	89.9	187	167	98	87.0	76.6	181	162	93.4	78.4	63.7
32	227	205	185	165	88.1		165	147	86	76.4	67.3	159	143	82.1	68.9	56.0
34	201	182	164	146			130					141	126			
36	179	162	146	130			116					126	113			
38	161	146	131	117			104					113	101			
40	145	131	119	105			93.9					102	91.3			

Effective length KL (ft) with respect to least radius of gyration r_y

Source: Courtesy of the American Institute of Steel Construction, Chicago, IL.

TABLE C.9
W Shapes: Available Moment versus Unbraced Length

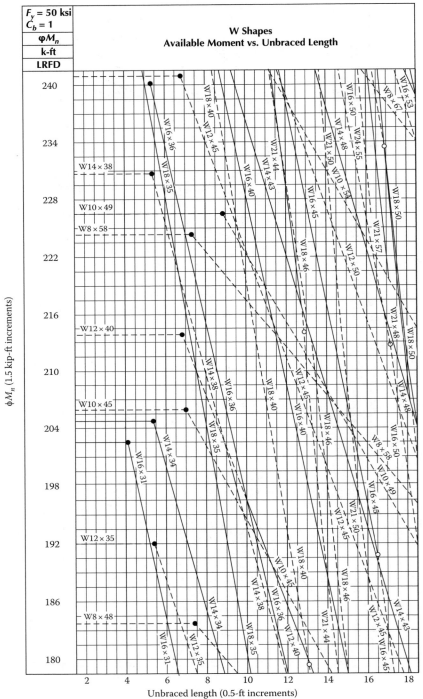

Source: Courtesy of the American Institute of Steel Construction, Chicago, IL.

TABLE C.10
LRFD

Standard Load Table for Open Web Steel Joists, K-Series

Based on a 50 ksi Maximum Yield Strength—Loads Shown in Pounds per Linear Foot (plf)

Joist Designation	8K1	10K1	12K1	12K3	12K5	14K1	14K3	14K4	14K6	16K2	16K3	16K4	16K5	16K6	16K7	16K9
Depth (in.)	8	10	12	12	12	14	14	14	14	16	16	16	16	16	16	16
Approx. Wt. (lbs./ft)	5.1	5.0	5.0	5.7	7.1	5.2	6.0	6.7	7.7	5.5	6.3	7.0	7.5	8.1	8.6	10.0
Span (ft.)																
8	**825**															
	550															
9	**825**															
	550															
10	**825**	**825**														
	480	550														
11	**798**	**825**														
	377	542														
12	**666**	**825**	**825**	**825**	**825**											
	288	455	550	550	550											
13	**565**	**718**	**825**	**825**	**825**											
	225	363	510	510	510											
14	**486**	**618**	**750**	**825**	**825**	**825**	**825**	**825**	**825**							
	179	289	425	463	463	550	550	550	550							
15	**421**	**537**	**651**	**814**	**825**	**766**	**825**	**825**	**825**							
	145	234	344	428	434	475	507	507	507							
16	**369**	**469**	**570**	**714**	**825**	**672**	**825**	**825**	**825**	**825**	**825**	**825**	**825**	**825**	**825**	**825**
	119	192	282	351	396	390	467	467	467	550	550	550	550	550	550	550
17		**415**	**504**	**630**	**825**	**592**	**742**	**825**	**825**	**768**	**825**	**825**	**825**	**825**	**825**	**825**
		159	234	291	366	324	404	443	443	488	526	526	526	526	526	526

(continued)

TABLE C.10 (continued)
LRFD

Standard Load Table for Open Web Steel Joists, K-Series

Based on a 50 ksi Maximum Yield Strength—Loads Shown in Pounds per Linear Foot (plf)

Joist Designation	8K1	10K1	12K1	12K3	12K5	14K1	14K3	14K4	14K6	16K2	16K3	16K4	16K5	16K6	16K7	16K9
Depth (in.)	8	10	12	12	12	14	14	14	14	16	16	16	16	16	16	16
Approx. Wt. (lbs./ft)	5.1	5.0	5.0	5.7	7.1	5.2	6.0	6.7	7.7	5.5	6.3	7.0	7.5	8.1	8.6	10.0
Span (ft.) ↓																
18		**369**	**448**	**561**	**760**	**528**	**661**	**795**	**825**	**684**	**762**	**825**	**825**	**825**	**825**	**825**
		134	197	245	317	272	339	397	408	409	456	490	490	490	490	490
19		**331**	**402**	**502**	**681**	**472**	**592**	**712**	**825**	**612**	**682**	**820**	**825**	**825**	**825**	**825**
		113	167	207	269	230	287	336	383	347	386	452	455	455	455	455
20		**298**	**361**	**453**	**613**	**426**	**534**	**642**	**787**	**552**	**615**	**739**	**825**	**825**	**825**	**825**
		97	142	177	230	197	246	267	347	297	330	386	426	426	426	426
21			**327**	**409**	**555**	**385**	**483**	**582**	**712**	**499**	**556**	**670**	**754**	**822**	**825**	**825**
			123	153	198	170	212	248	299	255	285	333	373	405	406	406
22			**298**	**373**	**505**	**351**	**439**	**529**	**648**	**454**	**505**	**609**	**687**	**747**	**825**	**825**
			106	132	172	147	184	215	259	222	247	289	323	361	385	385

	C1	C2	C3	C4	C5	C6	C7	C8	C9	C10	C11	C12	C13	C14
23	**271**	**340**	**462**	**321**	**402**	**483**	**592**	**415**	**462**	**556**	**627**	**682**	**760**	**825**
	93	116	150	128	160	188	226	194	216	252	282	307	339	363
24	**249**	**312**	**423**	**294**	**367**	**442**	**543**	**381**	**424**	**510**	**576**	**627**	**697**	**825**
	81	101	132	113	141	165	199	170	189	221	248	269	298	346
25				**270**	**339**	**408**	**501**	**351**	**390**	**469**	**529**	**576**	**642**	**771**
				100	124	145	175	150	167	195	219	238	263	311
26				**249**	**313**	**376**	**462**	**324**	**360**	**433**	**489**	**532**	**592**	**711**
				88	110	129	156	133	148	173	194	211	233	276
27				**231**	**289**	**349**	**427**	**300**	**334**	**402**	**453**	**493**	**549**	**658**
				79	98	115	139	119	132	155	173	188	208	246
28				**214**	**270**	**324**	**397**	**279**	**310**	**373**	**421**	**459**	**510**	**612**
				70	88	103	124	106	118	138	155	168	186	220
29								**259**	**289**	**348**	**391**	**427**	**475**	**570**
								95	106	124	139	151	167	198
30								**241**	**270**	**324**	**366**	**399**	**444**	**532**
								86	96	112	126	137	151	178
31								**226**	**252**	**304**	**342**	**373**	**415**	**498**
								78	87	101	114	124	137	161
32								**213**	**237**	**285**	**321**	**349**	**388**	**466**
								71	79	92	103	112	124	147

Source: Courtesy of the Steel Joist Institute, Forest, VA.

TABLE C.10
LRFD

Standard Load Table for Open Web Steel Joists, K-Series

Based on a 50 ksi Maximum Yield Strength—Loads Shown in Pounds per Linear Foot (plf)

Joist Designation	18K3	18K4	18K5	18K6	18K7	18K9	18K10	20K3	20K4	20K5
Depth (in.)	18	18	18	18	18	18	18	20	20	20
Approx. Wt. (lbs./ft.)	6.6	7.2	7.7	8.5	9	10.2	11.7	6.7	7.6	8.2
Span (ft.) ↓	825	825	825	825	825	825	825			
18	550	550	550	550	550	550	550			
19	771	825	825	825	825	825	825			
	494	523	523	523	523	523	523			
20	694	825	825	825	825	825	825	775	825	825
	423	490	490	490	490	490	490	517	550	550
21	630	759	825	825	825	825	825	702	825	825
	364	426	460	460	460	460	460	463	520	520
22	573	690	777	825	825	825	825	639	771	825
	316	370	414	438	438	438	438	393	461	490
23	523	630	709	774	825	825	825	583	703	793
	276	323	362	393	418	418	418	344	402	451
24	480	577	651	709	789	825	825	535	645	727
	242	284	318	345	382	396	396	302	353	396
25	441	532	600	652	727	825	825	493	594	669
	214	250	281	305	337	377	377	266	312	350
26	408	492	553	603	672	807	825	456	549	618
	190	222	249	271	299	354	361	236	277	310
27	378	454	513	558	622	747	825	421	508	573
	169	198	222	241	267	315	347	211	247	277
28	351	423	477	519	577	694	822	391	472	532
	151	177	199	216	239	282	331	189	221	248
29	327	394	444	483	538	646	766	364	439	495
	136	159	179	194	215	254	298	170	199	223
30	304	367	414	451	502	603	715	340	411	462
	123	144	161	175	194	229	269	153	179	201
31	285	343	387	421	469	564	669	318	384	433
	111	130	146	158	175	207	243	138	162	182
32	267	322	363	396	441	529	627	298	360	406
	101	118	132	144	159	188	221	126	147	165
33	252	303	342	372	414	498	589	280	339	381
	92	108	121	131	145	171	201	114	134	150
34	237	285	321	349	390	468	555	264	318	358
	84	98	110	120	132	156	184	105	122	137
35	223	268	303	330	367	441	523	249	300	339
	77	90	101	110	121	143	168	96	112	126
36	211	253	286	312	348	417	495	235	283	319
	70	82	92	101	111	132	154	88	103	115

20K6	20K7	20K9	20K10	22K4	22K5	22K6	22K7	22K9	22K10	22K11
20	20	20	20	22	22	22	22	22	22	22
8.9	9.3	10.8	12.2	8	8.8	9.2	9.7	11.3	12.6	13.8
825	825	825	825							
550	550	550	550							
825	825	825	825							
520	520	520	520							
825	825	825	825	825	825	825	825	825	825	825
490	490	490	490	548	548	548	548	548	548	548
825	825	825	825	777	825	825	825	825	825	825
468	468	468	468	491	518	518	518	518	518	518
792	825	825	825	712	804	825	825	825	825	825
430	448	448	448	431	483	495	495	495	495	495
729	811	825	825	657	739	805	825	825	825	825
380	421	426	426	381	427	464	474	474	474	474
673	750	825	825	606	682	744	825	825	825	825
337	373	405	405	338	379	411	454	454	454	454
624	694	825	825	561	633	688	768	825	825	825
301	333	389	389	301	337	367	406	432	432	432
579	645	775	825	522	588	640	712	825	825	825
269	298	353	375	270	302	328	364	413	413	413
540	601	723	825	486	547	597	664	798	825	825
242	268	317	359	242	272	295	327	387	399	399
504	561	675	799	453	511	556	619	745	825	825
218	242	286	336	219	245	266	295	349	385	385
471	525	631	748	424	478	520	580	697	825	825
198	219	259	304	198	222	241	267	316	369	369
442	492	592	702	397	448	489	544	654	775	823
179	199	235	276	180	201	219	242	287	337	355
415	463	566	660	373	421	459	511	615	729	798
163	181	214	251	164	183	199	221	261	307	334
391	435	523	621	352	397	432	481	579	687	774
149	165	195	229	149	167	182	202	239	280	314
369	411	493	585	331	373	408	454	546	648	741
137	151	179	210	137	153	167	185	219	257	292
348	388	466	553	313	354	385	429	516	612	700
125	139	164	193	126	141	153	169	201	236	269

(*continued*)

TABLE C.10 (continued)
LRFD

Standard Load Table for Open Web Steel Joists, K-Series

Based on a 50 ksi Maximum Yield Strength—Loads Shown in Pounds per Linear Foot (plf)

Joist Designation	18K3	18K4	18K5	18K6	18K7	18K9	18K10	20K3	20K4	20K5
Depth (in.)	18	18	18	18	18	18	18	20	20	20
Approx. Wt. (lbs./ft.)	6.6	7.2	7.7	8.5	9	10.2	11.7	6.7	7.6	8.2
Span (ft.) ↓										
37								222	268	303
								81	95	106
38								211	255	286
								74	87	98
39								199	241	271
								69	81	90
40								190	229	258
								64	75	84
41										
42										
43										
44										

Source: Courtesy of the Steel Joist Institute, Forest, VA.

20K6	20K7	20K9	20K10	22K4	22K5	22K6	22K7	22K9	22K10	22K11
20	20	20	20	22	22	22	22	22	22	22
8.9	9.3	10.8	12.2	8	8.8	9.2	9.7	11.3	12.6	13.8
330	367	441	523	297	334	364	406	487	579	663
115	128	151	178	116	130	141	156	186	217	247
312	348	418	496	280	316	345	384	462	549	628
106	118	139	164	107	119	130	144	170	200	228
297	330	397	471	267	300	327	364	438	520	595
98	109	129	151	98	110	120	133	157	185	211
282	313	376	447	253	285	310	346	417	495	565
91	101	119	140	91	102	111	123	146	171	195
				241	271	295	330	396	471	538
				85	95	103	114	135	159	181
				229	259	282	313	378	448	513
				79	88	96	106	126	148	168
				219	247	268	300	360	427	489
				73	82	89	99	117	138	157
				208	235	256	286	343	408	466
				68	76	83	92	109	128	146

TABLE C.11
Design Guide LRFD Weight Table for Joist Girders
Based on 50 ksi Yield Strength

Joist Girder Weight—Pounds PER Linear Foot

Factored Load on Each Panel Point—Kips

Girder Span (ft.)	Joist Spaces (ft.)	Girder Depth (in.)	6.0	9.0	12.0	15.0	18.0	21.0	24.0	27.0	30.0	36.0	42.0	48.0	54.0	60.0	66.0	72.0	78.0	84.0
20	2N @ 10.00	20	16	19	19	19	19	20	24	24	25	30	37	41	46	50	56	62	70	75
		24	16	19	19	19	19	20	21	21	25	28	32	36	41	42	49	52	53	66
		28	16	19	19	19	19	20	20	21	23	26	28	32	39	40	42	46	48	49
	3N @ 6.67	20	15	15	19	19	20	23	24	27	31	36	44	48	54	74	75	81	84	89
		24	15	16	16	16	19	20	23	26	27	33	36	45	47	53	56	68	79	82
		28	15	16	16	16	17	20	24	24	26	31	36	44	46	49	53	57	68	80
	4N @ 5.00	20	15	15	19	21	25	29	33	38	41	50	57	65	71	88	97	100	107	120
		24	15	16	17	20	23	26	29	32	35	44	50	55	62	71	85	90	100	102
		28	16	16	17	19	22	25	28	30	34	39	49	50	59	63	72	86	91	91
	5N @ 4.00	20	15	17	21	26	31	36	39	48	51	62	71	82	99	99	109	120	141	142
		24	16	16	20	23	26	30	35	39	43	53	60	68	80	91	101	103	110	120
		28	16	16	18	22	27	28	33	37	39	48	55	64	68	77	93	95	107	111
	6N @ 3.33	20	16	19	25	29	36	41	50	57	58	72	82	99	107	118	138	141	144	147
		24	16	18	22	28	31	37	43	46	53	61	70	85	102	102	111	123	119	130
		28	17	18	22	26	30	33	40	42	47	58	68	76	83	96	109	112		
	8N @ 2.50	20	19	25	32	41	51	58	65	72	82	99	118	139	142	149	153	155	166	
		24	17	22	29	36	42	50	54	61	69	86	103	107	128	124	135			
		28	18	22	29	34	40	47	54	61	67	76	88	107	112					
22	2N @ 11.00	20	21	21	21	22	22	23	24	24	25	34	39	43	49	55	62	69	76	78
		24	18	21	21	22	22	22	23	24	24	30	33	41	41	45	51	55	61	73
		28	18	21	21	21	22	22	22	23	24	27	30	33	41	42	46	48	51	58
	3N @ 7.33	20	15	18	18	19	22	24	26	29	33	42	45	53	68	70	76	84	88	94
		24	15	15	19	19	20	23	24	26	30	35	40	45	48	55	61	74	81	84
		28	15	16	16	16	19	20	23	24	27	32	36	45	47	52	54	59	74	82
	4N @ 5.50	20	15	16	19	23	26	30	36	39	44	55	62	71	82	95	96	106	119	134
		24	15	15	17	20	25	27	29	34	38	48	52	58	71	79	89	98	101	107
		28	16	16	16	19	22	25	28	32	35	40	49	54	60	72	79	87	90	97

Group	p																		
5N @ 4.40	20	15	17	24	27	34	38	42	49	55	65	75	96	98	111	126	137	116	133
	24	16	16	20	24	28	33	38	40	48	56	62	73	85	100	101	110	105	111
	28	16	16	18	22	26	30	32	38	41	51	57	65	73	86	92	102		
6N @ 3.67	20	16	21	27	33	39	49	57	65	79	97	106	118	137					
	24	16	19	23	28	32	39	51	58	66	82	98	101	109	120	142	144		
	28	16	18	22	26	30	34	39	45	47	55	73	86	92	102	113	127	105	
8N @ 2.75	20	19	27	36	43	56	64	71	80	96	106	135	138	145	149	152	142	144	148
	24	18	24	31	38	46	53	60	68	75	101	105	125	118	131	104	113	127	
	28	18	22	28	34	40	47	54	62	69	79	87	106						
3N @ 8.33	20	18	22	28	34	47	54	69	79	87	106	118	125	131	149	152	164		
	24	15	18	19	22	27	30	41	49	59	66	76	86	89	97		89	118	131
	28	15	18	19	20	25	26	32	39	43	51	59	67	71	81		81	70	76
	32	16	15	19	19	23	24	29	34	39	45	47	55	59	67	74	81		
	36	16	16	16	17	21	24	27	32	36	44	46	52	54	58	68	79		
4N @ 6.25	20	15	20	25	31	35	43	49	55	58	70	78	93	99	109	119	135	134	
	24	15	19	21	26	29	36	40	47	53	57	64	72	88	97	100	120	106	
	28	15	17	20	24	25	31	37	40	43	51	58	66	72	89	90	102	101	
	32	16	17	19	23	25	30	34	37	49	49	54	60	69	79	86	96	91	
	36	16	17	19	24	26	28	32	36	38	50	56	63	73	85	88	92		
5N @ 5.00	20	15	25	31	38	43	55	58	73	78	93	100	109	125	134	134			
	24	15	21	26	31	36	42	47	53	61	75	81	98	102	112	109			
	28	16	20	24	28	31	37	41	47	56	62	72	79	93	101	97			
	32	16	19	23	26	30	33	38	41	51	57	65	73	83	93	89			
	36	17	19	22	26	28	31	36	39	48	54	64	69	75	88	79	73		
6N @ 4.17	20	16	24	29	38	55	58	69	78	94	104	116	134	138	118	127	145	143	
	24	16	20	25	31	44	50	56	64	75	97	99	107	116	104	120	148	127	
	28	16	18	23	28	38	44	51	55	67	73	87	101	105	102	105	129	117	
	32	16	18	22	26	34	39	44	50	61	69	77	89		98	108			
	36	16	18	24	30	36	39	43	49	58	67	74	84	96					
8N @ 3.12	20	21	29	39	48	58	70	78	94	99	115	134	138		127	147	152	154	166
	24	19	26	33	41	50	57	65	75	81	99	118	116		117	129	136	148	166
	28	18	23	30	38	44	53	60	67	75	86	103	105		114	121			167
	32	18	24	28	34	39	47	54	65	71	78	87							
	36	18	22	29	34	40	46	52	61	63	76	87	101						

25

(continued)

TABLE C.11 (continued)
Design Guide LRFD Weight Table for Joist Girders
Based on 50 ksi Yield Strength

Joist Girder Weight—Pounds PER Linear Foot

Girder Span (ft.)	Joist Spaces (ft.)	Girder Depth (in.)	6.0	9.0	12.0	15.0	18.0	21.0	24.0	27.0	30.0	36.0	42.0	48.0	54.0	60.0	66.0	72.0	78.0	84.0
												Factored Load on Each Panel Point—Kips								
28	10N @ 2.50	20	26	38	49	63	78	94	100	115	134									
		24	23	33	42	54	65	75	89	99	104	130								
		28	21	30	38	48	56	64	74	84	101	109	134	147						
		32	21	28	36	43	52	62	69	76	87	107	118	130	153					
		36	22	28	37	44	52	64	71	77	85	100	116	130	151	157				
	3N @ 9.33	24	18	18	19	22	24	27	29	36	39	43	53	62	70	71	78	85	89	98
		28	18	18	19	20	22	25	26	28	31	39	43	46	55	61	66	76	83	86
		32	15	18	19	19	21	23	24	27	28	34	39	45	48	53	58	66	80	81
	4N @ 7.00	24	15	16	20	24	27	32	38	40	48	55	62	71	82	95	104	106	120	135
		28	15	15	18	21	25	28	32	36	39	49	56	64	71	79	96	97	106	107
		32	15	15	17	20	23	25	29	33	37	43	50	58	62	70	85	90	99	102
	5N @ 5.60	24	15	18	24	29	34	39	46	52	58	66	78	96	102	111	126	136		
		28	15	17	21	26	30	35	39	46	50	61	68	77	90	99	107	114	130	142
		32	16	17	20	24	27	32	37	41	44	56	62	70	80	93	102	107	112	119
	6N @ 4.67	24	16	21	28	35	41	49	55	63	70	79	96	106	134	137	138	142		
		28	15	20	24	30	36	42	50	54	58	71	82	99	107	118	111	123	144	146
		32	16	19	23	28	32	37	43	49	53	64	74	84	101	102				
	7N @ 4.00	24	18	24	32	41	49	56	64	74	79	96	110	135						
		28	17	22	27	35	43	51	57	62	69	82	99	108	129	140	143	146		
		32	16	21	27	31	38	44	52	55	63	74	85	102	108	123				
	8N @ 3.50	24	20	28	37	48	55	64	74	79	95	105	134							
		28	18	25	32	39	50	58	65	72	81	99	108	129	141					
		32	17	24	29	38	43	53	60	64	70	86	103	113	127	147	149			
	10N @ 2.80	24	24	36	46	57	70	79	96	102	117	137								
		28	23	30	41	50	60	69	82	99	100	120	141							
		32	21	30	38	46	55	66	71	80	93	109	126	147						

30	3N @ 10.00	24	18	18	21	24	27	31	35	38	40	48	58	66	71	80	92	98	117	119
		28	18	18	19	22	25	27	30	35	37	42	49	56	63	70	79	82	93	99
		32	18	18	19	20	22	26	28	31	32	39	46	51	57	64	71	73	83	84
		36	16	19	19	19	21	23	26	28	31	35	39	46	52	57	64	65	73	75
	4N @ 7.50	24	16	18	23	29	33	37	42	49	53	64	76	85	101	104	126	127	149	150
		28	15	16	21	25	30	33	37	42	45	53	61	73	81	86	103	104	126	128
		32	15	16	18	22	26	30	34	37	43	48	55	62	70	77	87	103	105	116
		36	16	16	17	22	24	27	31	34	36	43	52	59	64	74	78	88	91	105
	5N @ 6.00	24	15	15	25	30	37	43	51	58	66	73	86	96	109	125	134			
		28	15	17	23	27	32	37	44	50	57	65	75	88	97	102	112	128	138	
		32	16	17	21	24	29	35	39	45	51	58	63	77	90	100	101	107	117	133
		36	16	17	20	24	27	31	36	40	46	52	60	70	80	86	94	103	110	118
	6N @ 5.00	24	16	24	29	37	45	52	58	66	73	94	104	116	134					
		28	16	20	27	32	38	44	51	57	65	75	88	97	99	107	112	121	140	
		32	16	19	24	29	34	40	46	51	58	65	77	90	100	109	101	111	121	
		36	16	18	23	26	31	37	43	46	52	60	70	84	86	94	102	103	110	
	8N @ 3.75	24	21	32	40	51	63	73	83	99	111	124	146	148						
		28	20	30	37	44	53	61	73	80	86	104	114	126	144					
		32	18	26	34	42	49	55	63	73	79	90	104	117	121	123				
		36	17	23	32	39	46	54	61	69	71	76	89	108						
	10N @ 3.00	24	25	38	51	66	78	99	111	123	134	138	142	151	161	169				
		28	24	36	47	57	69	80	94	113	116	129	132	130	154	154				
		32	22	31	39	52	58	74	82	95	105	119		121	134					
		36	22	30	39	48	54	68	79	84	91									
32	3N @ 10.67	24	18	19	21	26	27	34	38	40	42	54	61	70	75	84	88	102	102	113
		28	16	17	18	24	26	28	31	34	37	40	43	55	60	69	70	76	85	89	93
		32	17	17	18	21	25	26	28	32	34	37	39	49	54	61	62	67	77	80	86
		36	15	17	19	20	23	26	30	30	36	36	43	50	58	65	67	70	77
	4N @ 8.00	24	18	19	23	26	32	37	40	47	55	61	72	86	94	103	114	133	134	135
		28	15	18	20	24	28	32	37	40	45	55	62	70	78	94	96	105	121	107
		32	15	15	20	22	25	29	32	37	40	49	56	64	71	83	82	97	102	102
		36	15	16	17	21	24	26	30	34	36	43	50	58	65	70	85	90	99	102

(continued)

TABLE C.11 (continued)
Design Guide LRFD Weight Table for Joist Girders
Based on 50 ksi Yield Strength

Joist Girder Weight—Pounds PER Linear Foot

Factored Load on Each Panel Point—Kips

Girder Span (ft.)	Joist Spaces (ft.)	Girder Depth (in.)	6.0	9.0	12.0	15.0	18.0	21.0	24.0	27.0	30.0	36.0	42.0	48.0	54.0	60.0	66.0	72.0	78.0	84.0
	5N @ 6.40	24	15	20	27	33	39	44	51	57	65	77	93	100	123	133				
		28	15	18	24	28	34	39	46	52	58	66	74	96	101	110	126	137		
		32	15	17	22	26	32	35	41	46	53	61	68	77	90	99	105	114	130	142
		36	16	17	21	24	27	33	37	42	47	56	62	70	79	93	102	106	117	120
	6N @ 5.33	24	17	24	31	39	47	55	61	69	76	94	103	133	134					
		28	16	21	27	35	40	48	55	60	67	79	96	105	117	137				
		32	16	20	25	30	36	42	50	54	58	71	82	99	103	118	139	142		
		36	16	19	24	28	34	38	44	49	55	66	73	84	101	102	111	123	144	146
	8N @ 4.00	24	22	32	40	54	61	72	86	93	103	133	134	137	141	147				
		28	19	27	35	45	55	63	70	80	95	105	109	120	127					
		32	18	25	32	39	50	58	65	71	81	99	103	113						
		36	18	24	31	38	43	53	59	67	71	86								
35	4N @ 8.75	28	16	19	23	27	31	36	41	46	52	60	74	79	94	100	111	117	137	138
		32	15	18	21	24	28	33	37	39	45	53	60	73	80	92	100	106	112	127
		36	15	16	20	23	27	30	33	37	41	561	55	62	74	83	94	97	107	113
		40	15	16	17	21	26	27	30	37	38	46	52	61	64	75	90	95	96	108
	5N @ 7.00	28	15	20	26	32	37	43	52	57	59	73	86	100	109	126	136			
		32	15	18	24	29	34	37	45	50	53	66	75	88	100	102	112	128	138	
		36	16	17	23	27	29	35	40	46	48	62	68	77	90	100	104	115	131	133
		40	16	17	22	25	27	33	37	43	47	56	63	70	80	95	102	107	115	125
	6N @ 5.83	28	17	24	30	37	44	52	58	65	73	93	96	117	134	139	140	142	144	146
		32	16	21	27	33	38	46	53	57	65	79	81	100	117					
		36	16	20	25	31	36	41	48	54	58	70	77	99	102	113	121	123	144	145
		40	16	20	24	28	34	38	44	49	55	64		84	101	104	115			

	7N @ 5.00	28	19	27	34	43	52	59	66	74	86	101	115	135						
		32	17	24	30	39	47	53	61	67	75	97	103	118	137					
		36	17	23	28	35	42	48	55	62	69	82	99	105	120	141	144			
		40	17	22	27	32	39	44	50	55	63	73	86	102	107	118	133	147		
	8N @ 4.38	28	21	30	39	48	59	69	78	94	98	115	136							
		32	20	27	36	42	53	61	69	79	88	101	118	138						
		36	19	26	32	39	48	55	62	71	77	99	109	121	141					
		40	18	24	30	37	44	54	60	65	73	86	102	113	127	147	149			
38	4N @ 9.50	32	16	19	21	26	31	34	39	43	48	58	67	74	87	100	101	111	127	138
		36	15	17	21	24	28	33	35	39	44	53	60	74	75	93	97	106	112	123
		40	15	16	20	23	27	30	34	37	41	51	55	62	74	83	94	98	107	109
		44	16	16	20	22	26	28	30	35	38	46	52	58	65	75	90	95	95	108
	5N @ 7.60	32	15	20	25	31	36	42	46	52	59	70	86	96	101	111	126	137		
		36	16	20	24	28	33	38	45	47	53	64	74	89	98	103	112	129	138	
		40	16	20	23	26	31	35	40	46	48	59	70	78	91	101	105	113	117	134
		44	17	20	22	25	30	33	39	41	48	56	63	75	80	93	102	107	111	118
	6N @ 6.33	32	17	24	30	35	41	49	55	62	70	86	98	105	125	136				
		36	16	21	27	33	39	47	50	57	61	75	89	100	107	118	141	142		
		40	16	21	25	31	36	40	48	55	59	71	82	99	102	109	121	143	142	
		44	17	20	24	29	33	38	44	49	55	64	77	84	102	104	115	123	145	147
	8N @ 4.75	32	20	29	38	47	56	64	74	86	95	105	135							
		36	19	28	35	42	50	57	65	76	81	101	113	138	140					
		40	19	26	32	40	48	55	62	67	78	100	103	121	142	144				
		44	20	24	30	39	47	51	57	64	71	86	102	113	127	147	149			
40	4N @ 10.00	32	17	20	23	29	37	40	47	50	56	64	73	86	103	114	126	128	149	151
		36	17	19	22	29	31	37	40	44	51	57	65	74	87	103	104	125	127	128
		40	17	18	22	25	29	33	37	40	47	52	62	73	77	87	96	104	117	127
		44	16	17	20	24	29	31	36	38	41	49	59	66	74	78	84	96	106	106
		48	17	17	20	24	25	30	32	37	39	48	53	59	67	78	78	85	99	106
	5N @ 8.00	32	15	21	26	32	38	43	52	55	62	73	86	101	109	124	134			
		36	16	20	24	30	34	39	45	53	55	66	74	88	102	102	112	128	138	
		40	16	20	24	27	32	37	41	46	51	62	68	77	90	100	105	115	130	142
		44	17	20	23	29	32	37	41	49	50	58	70	82	84	99	116	118	130	141
		48	17	20	23	26	31	34	40	41	50	57	68	75	85	95	100	119	120	132

(continued)

TABLE C.11 (continued)
Design Guide LRFD Weight Table for Joist Girders
Based on 50 ksi Yield Strength

Joist Girder Weight—Pounds PER Linear Foot

Girder Span (ft.)	Joist Spaces (ft.)	Girder Depth (in.)	Factored Load on Each Panel Point—Kips																	
			6.0	9.0	12.0	15.0	18.0	21.0	24.0	27.0	30.0	36.0	42.0	48.0	54.0	60.0	66.0	72.0	78.0	84.0
	6N @ 6.67	32	16	24	30	38	44	52	58	65	72	93	100	115	133					
		36	17	22	27	34	39	47	53	60	67	79	97	102	117	137	141			
		40	16	21	26	30	36	43	48	54	62	71	82	99	103	114	130	142		
		44	17	21	24	28	36	40	47	51	55	66	78	91	102	107	116	134	142	146
		48	17	21	24	31	36	42	46	53	57	69	79	86	100	109	132	133	135	164
	7N @ 5.71	32	18	26	33	43	52	58	66	74	86	101	115	135	136					
		36	17	24	31	39	47	53	61	67	75	97	103	117	119					
		40	17	24	29	35	43	49	55	62	69	82	99	105	111	140				
		44	20	22	28	33	39	48	55	59	64	78	92	102	122	134	143			
		48	20	23	28	36	41	48	54	61	66	80	86	108		122	136	164	167	
	8N @ 5.00	32	21	29	38	48	58	67	78	94	96	115	135	137						
		36	19	27	36	46	53	60	68	80	88	102	118	120						
		40	19	25	34	39	49	58	65	72	82	99	109	120	141					
		44	21	27	33	39	47	56	63	70	75	93	103	122	136	147				
		48	20	25	32	42	47	55	62	69	80	90	104		136	155	170			
	10N @ 4.00	32	29	39	51	64	79	92	112	123	125	149								
		36	25	36	47	60	69	81	94	103	125	150								
		40	24	36	45	56	66	75	82	96	115	129	152							
		44	23	32	41	51	60	71	82	84	99	119	143	161						
		48	23	32	41	52	58	68	76	85	94	121	134	152						

Source: Courtesy of the Steel Joist Institute, Forest, VA.

Appendix D: Concrete

TABLE D.1
Diameter, Area and Unit Weight of Steel Bars

Bar Number	3	4	5	6	7	8	9	10	11	14	18
Diameter (in.)	0.375	0.500	0.625	0.750	0.875	1.000	1.128	1.270	1.410	1.693	2.257
Area (in.^{2})	0.11	0.20	0.31	0.44	0.60	0.79	1.00	1.27	1.56	2.25	4.00
Unit weight per foot (lb)	0.376	0.668	1.043	1.502	2.044	2.670	3.400	4.303	5.313	7.65	13.60

TABLE D.2
Areas of Group of Steel Bars (in.^{2})

Number of Bars	Bar Size								
	#3	#4	#5	#6	#7	#8	#9	#10	#11
1	0.11	0.20	0.31	0.44	0.60	0.79	1.00	1.27	1.56
2	0.22	0.40	0.62	0.88	1.20	1.58	2.00	2.54	3.12
3	0.33	0.60	0.93	1.32	1.80	2.37	3.00	3.81	4.68
4	0.44	0.80	1.24	1.76	2.40	3.16	4.00	5.08	6.24
5	0.55	1.00	1.55	2.20	3.00	3.93	5.00	6.35	7.80
6	0.66	1.20	1.86	2.64	3.60	4.74	6.00	7.62	9.36
7	0.77	1.40	2.17	3.08	4.20	5.53	7.00	8.89	10.9
8	0.88	1.60	2.48	3.52	4.80	6.32	8.00	10.2	12.5
9	0.99	1.80	2.79	3.96	5.40	7.11	9.00	11.4	14.0
10	1.10	2.00	3.10	4.40	6.00	7.90	10.0	12.7	15.6
11	1.21	2.20	3.41	4.84	6.60	8.69	11.0	14.0	17.2
12	1.32	2.40	3.72	5.28	7.20	9.48	12.0	15.2	18.7
13	1.43	2.60	4.03	5.72	7.80	10.3	13.0	16.5	20.3
14	1.54	2.80	4.34	6.16	8.40	11.1	14.0	17.8	21.8
15	1.65	3.00	4.65	6.60	9.00	11.8	15.0	19.0	23.4
16	1.76	3.20	4.96	7.04	9.60	12.6	16.0	20.3	25.0
17	1.87	3.40	5.27	7.48	10.2	13.4	17.0	21.6	26.5
18	1.98	3.60	5.58	7.92	10.8	14.2	18.0	22.9	28.1
19	2.09	3.80	5.89	8.36	11.4	15.0	19.0	24.1	29.6
20	2.20	4.00	6.20	8.80	12.0	15.8	20.0	25.4	31.2

TABLE D.3
Minimum Required Beam Widths (in.)

Number of Bars in One Layer	Bar Size							
	#3 and #4	#5	#6	#7	#8	#9	#10	#11
2	6.0	6.0	6.5	6.5	7.0	7.5	8.0	8.0
3	7.5	8.0	8.0	8.5	9.0	9.5	10.5	11.0
4	9.0	9.5	10.0	10.5	11.0	12.0	13.0	14.0
5	10.5	11.0	11.5	12.5	13.0	14.0	15.5	16.5
6	12.0	12.5	13.5	14.0	15.0	16.5	18.0	19.5
7	13.5	14.5	15.0	16.0	17.0	18.5	20.5	22.5
8	15.0	16.0	17.0	18.0	19.0	21.0	23.0	25.0
9	16.5	17.5	18.5	20.0	21.0	23.0	25.5	28.0
10	18.0	19.0	20.5	21.5	23.0	25.5	28.0	31.0

Note: Tabulated values based on No. 3 stirrups, minimum clear distance of 1 in., and a 1-1/2 in cover.

TABLE D.4
Coefficient of Resistance ($\bar{K}$) versus Reinforcement Ratio (ρ) ($f'_c = 3000\,\text{psi}$; $f_y = 40{,}000\,\text{psi}$)

ρ	$\bar{K}$(ksi)	ρ	$\bar{K}$(ksi)	ρ	$\bar{K}$(ksi)	ρ	$\bar{K}$(ksi)	ρ	$\bar{K}$(ksi)	ρ	$\bar{K}$(ksi)	ε_t^a
0.0010	0.0397	0.0054	0.2069	0.0098	0.3619	0.0142	0.5047	0.0173	0.5981	0.02033	0.6836	0.00500
0.0011	0.0436	0.0055	0.2105	0.0099	0.3653	0.0143	0.5078	0.0174	0.6011	0.0204	0.6855	0.00497
0.0012	0.0476	0.0056	0.2142	0.0100	0.3686	0.0144	0.5109	0.0175	0.6040	0.0205	0.6882	0.00493
0.0013	0.0515	0.0057	0.2178	0.0101	0.3720	0.0145	0.5140	0.0176	0.6069	0.0206	0.6909	0.00489
0.0014	0.0554	0.0058	0.2214	0.0102	0.3754	0.0146	0.5171	0.0177	0.6098	0.0207	0.6936	0.00485
0.0015	0.0593	0.0059	0.2251	0.0103	0.3787	0.0147	0.5202	0.0178	0.6126	0.0208	0.6963	0.00482
0.0016	0.0632	0.0060	0.2287	0.0104	0.3821	0.0148	0.5233	0.0179	0.6155	0.0209	0.6990	0.00478
0.0017	0.0671	0.0061	0.2323	0.0105	0.3854	0.0149	0.5264	0.0180	0.6184	0.0210	0.7017	0.00474
0.0018	0.0710	0.0062	0.2359	0.0106	0.3887	0.0150	0.5294	0.0181	0.6213	0.0211	0.7044	0.00470
0.0019	0.0749	0.0063	0.2395	0.0107	0.3921	0.0151	0.5325	0.0182	0.6241	0.0212	0.7071	0.00467
0.0020	0.0788	0.0064	0.2431	0.0108	0.3954	0.0152	0.5355	0.0183	0.6270	0.0213	0.7097	0.00463
0.0021	0.0826	0.0065	0.2467	0.0109	0.3987	0.0153	0.5386	0.0184	0.6298	0.0214	0.7124	0.00460
0.0022	0.0865	0.0066	0.2503	0.0110	0.4020	0.0154	0.5416	0.0185	0.6327	0.0215	0.7150	0.00456
0.0023	0.0903	0.0067	0.2539	0.0111	0.4053	0.0155	0.5447	0.0186	0.6355	0.0216	0.7177	0.00453
0.0024	0.0942	0.0068	0.2575	0.0112	0.4086	0.0156	0.5477	0.0187	0.6383	0.0217	0.7203	0.00449
0.0025	0.0980	0.0069	0.2611	0.0113	0.4119	0.0157	0.5507	0.0188	0.6412	0.0218	0.7230	0.00446
0.0026	0.1019	0.0070	0.2646	0.0114	0.4152	0.0158	0.5537	0.0189	0.6440	0.0219	0.7256	0.00442
0.0027	0.1057	0.0071	0.2682	0.0115	0.4185	0.0159	0.5567	0.0190	0.6468	0.0220	0.7282	0.00439
0.0028	0.1095	0.0072	0.2717	0.0116	0.4218	0.0160	0.5597	0.0191	0.6496	0.0221	0.7308	0.00436
0.0029	0.1134	0.0073	0.2753	0.0117	0.4251	0.0161	0.5627	0.0192	0.6524	0.0222	0.7334	0.00432
0.0030	0.1172	0.0074	0.2788	0.0118	0.4283	0.0162	0.5657	0.0193	0.6552	0.0223	0.7360	0.00429
0.0031	0.1210	0.0075	0.2824	0.0119	0.4316	0.0163	0.5687	0.0194	0.6580	0.0224	0.7386	0.00426
0.0032	0.1248	0.0076	0.2859	0.0120	0.4348	0.0164	0.5717	0.0195	0.6608	0.0225	0.7412	0.00423
0.0033	0.1286	0.0077	0.2894	0.0121	0.4381	0.0165	0.5746	0.0196	0.6635	0.0226	0.7438	0.00419
0.0034	0.1324	0.0078	0.2929	0.0122	0.4413	0.0166	0.5776	0.0197	0.6663	0.0227	0.7464	0.00416
0.0035	0.1362	0.0079	0.2964	0.0123	0.4445	0.0167	0.5805	0.0198	0.6691	0.0228	0.7490	0.00413
0.0036	0.1399	0.0080	0.2999	0.0124	0.4478	0.0168	0.5835	0.0199	0.6718	0.0229	0.7515	0.00410
0.0037	0.1437	0.0081	0.3034	0.0125	0.4510	0.0169	0.5864	0.0200	0.6746	0.0230	0.7541	0.00407

(continued)

TABLE D.4 (continued)
Coefficient of Resistance ($\bar{K}$) versus Reinforcement Ratio (ρ) ($f_c' = 3000\,\text{psi}$; $f_y = 40,000\,\text{psi}$)

ρ	$\bar{K}$(ksi)	ρ	$\bar{K}$(ksi)	ρ	$\bar{K}$(ksi)	ρ	$\bar{K}$(ksi)	ρ	$\bar{K}$(ksi)	ρ	$\bar{K}$(ksi)	ε_t[a]
0.0038	0.1475	0.0082	0.3069	0.0126	0.4542	0.0170	0.5894	0.0201	0.6773	0.0231	0.7567	0.00404
0.0039	0.1512	0.0083	0.3104	0.0127	0.4574	0.0171	0.5923	0.0202	0.6800	0.0232	0.7592	0.00401
0.0040	0.1550	0.0084	0.3139	0.0128	0.4606	0.0172	0.5952	0.0203	0.6828	0.02323	0.7600	0.00400
0.0041	0.1587	0.0085	0.3173	0.0129	0.4638							
0.0042	0.1625	0.0086	0.3208	0.0130	0.4670							
0.0043	0.1662	0.0087	0.3243	0.0131	0.4702							
0.0044	0.1699	0.0088	0.3277	0.0132	0.4733							
0.0045	0.1736	0.0089	0.3311	0.0133	0.4765							
0.0046	0.1774	0.0090	0.3346	0.0134	0.4797							
0.0047	0.1811	0.0091	0.3380	0.0135	0.4828							
0.0048	0.1848	0.0092	0.3414	0.0136	0.4860							
0.0049	0.1885	0.0093	0.3449	0.0137	0.4891							
0.0050	0.1922	0.0094	0.3483	0.0138	0.4923							
0.0051	0.1958	0.0095	0.3517	0.0139	0.4954							
0.0052	0.1995	0.0096	0.3551	0.0140	0.4985							
0.0053	0.2032	0.0097	0.3585	0.0141	0.5016							

[a] $d = d_r$

TABLE D.5
Coefficient of Resistance ($\bar{K}$) ($f_c' = 3{,}000\,\text{psi}$, $f_y = 50{,}000\,\text{psi}$)

ρ	$\bar{K}$, ksi	ρ	$\bar{K}$, ksi	ρ	$\bar{K}$, ksi	ρ	$\bar{K}$, ksi	ρ	$\bar{K}$ (ksi)	ε_t
0.0020	0.098	0.0056	0.265	0.0092	0.418	0.0128	0.559	0.0163	0.685	0.0050
0.0021	0.103	0.0057	0.269	0.0093	0.422	0.0129	0.563	0.0164	0.688	0.0049
0.0022	0.108	0.0058	0.273	0.0094	0.427	0.0130	0.567	0.0165	0.692	0.0049
0.0023	0.112	0.0059	0.278	0.0095	0.431	0.0131	0.571	0.0166	0.695	0.0048
0.0024	0.117	0.0060	0.282	0.0096	0.435	0.0132	0.574	0.0167	0.698	0.0048
0.0025	0.122	0.0061	0.287	0.0097	0.439	0.0133	0.578	0.0168	0.702	0.0047
0.0026	0.127	0.0062	0.291	0.0098	0.443	0.0134	0.582	0.0169	0.705	0.0047
0.0027	0.131	0.0063	0.295	0.0099	0.447	0.0135	0.585	0.017	0.708	0.0047
0.0028	0.136	0.0064	0.300	0.0100	0.451	0.0136	0.589	0.0171	0.712	0.0046
0.0029	0.141	0.0065	0.304	0.0101	0.455	0.0137	0.593	0.0172	0.715	0.0046
0.0030	0.146	0.0066	0.309	0.0102	0.459	0.0138	0.596	0.0173	0.718	0.0045
0.0031	0.150	0.0067	0.313	0.0103	0.463	0.0139	0.600	0.0174	0.722	0.0045
0.0032	0.155	0.0068	0.317	0.0104	0.467	0.0140	0.604	0.0175	0.725	0.0044
0.0033	0.159	0.0069	0.322	0.0105	0.471	0.0141	0.607	0.0176	0.728	0.0044
0.0034	0.164	0.0070	0.326	0.0106	0.475	0.0142	0.611	0.0177	0.731	0.0043
0.0035	0.169	0.0071	0.330	0.0107	0.479	0.0143	0.614	0.0178	0.735	0.0043
0.0036	0.174	0.0072	0.334	0.0108	0.483	0.0144	0.618	0.0179	0.738	0.0043
0.0037	0.178	0.0073	0.339	0.0109	0.487	0.0145	0.622	0.018	0.741	0.0042
0.0038	0.183	0.0074	0.343	0.0110	0.491	0.0146	0.625	0.0181	0.744	0.0042
0.0039	0.187	0.0075	0.347	0.0111	0.494	0.0147	0.629	0.0182	0.748	0.0041
0.0040	0.192	0.0076	0.352	0.0112	0.498	0.0148	0.632	0.0183	0.751	0.0041
0.0041	0.197	0.0077	0.356	0.0113	0.502	0.0149	0.636	0.0184	0.754	0.0041
0.0042	0.201	0.0078	0.360	0.0114	0.506	0.0150	0.639	0.0185	0.757	0.0040
0.0043	0.206	0.0079	0.364	0.0115	0.510	0.0151	0.643	0.0186	0.760	0.0040
0.0044	0.210	0.0080	0.368	0.0116	0.514	0.0152	0.646			
0.0045	0.215	0.0081	0.373	0.0117	0.518	0.0153	0.650			
0.0046	0.219	0.0082	0.377	0.0118	0.521	0.0154	0.653			
0.0047	0.224	0.0083	0.381	0.0119	0.525	0.0155	0.657			
0.0048	0.229	0.0084	0.385	0.0120	0.529	0.0156	0.660			
0.0049	0.233	0.0085	0.389	0.0121	0.533	0.0157	0.664			
0.0050	0.238	0.0086	0.394	0.0122	0.537	0.0158	0.667			
0.0051	0.242	0.0087	0.398	0.0123	0.541	0.0159	0.671			
0.0052	0.247	0.0088	0.402	0.0124	0.544	0.0160	0.674			
0.0053	0.251	0.0089	0.406	0.0125	0.548	0.0161	0.677			
0.0054	0.256	0.0090	0.410	0.0126	0.552	0.0162	0.681			
0.0055	0.260	0.0091	0.414	0.0127	0.556					

TABLE D.6
Coefficient of Resistance ($\bar{K}$) versus Reinforcement Ratio (ρ)
($f_c' = 3{,}000$ psi; $f_y = 60{,}000$ psi)

ρ	$\bar{K}$ (ksi)	ρ	$\bar{K}$ (ksi)	ρ	$\bar{K}$ (ksi)	ε_t^*
0.0010	0.0593	0.0059	0.3294	0.0108	0.5657	
0.0011	0.0651	0.0060	0.3346	0.0109	0.5702	
0.0012	0.0710	0.0061	0.3397	0.0110	0.5746	
0.0013	0.0768	0.0062	0.3449	0.0111	0.5791	
0.0014	0.0826	0.0063	0.3500	0.0112	0.5835	
0.0015	0.0884	0.0064	0.3551	0.0113	0.5879	
0.0016	0.0942	0.0065	0.3602	0.0114	0.5923	
0.0017	0.1000	0.0066	0.3653	0.0115	0.5967	
0.0018	0.1057	0.0067	0.3703	0.0116	0.6011	
0.0019	0.1115	0.0068	0.3754	0.0117	0.6054	
0.0020	0.1172	0.0069	0.3804	0.0118	0.6098	
0.0021	0.1229	0.0070	0.3854	0.0119	0.6141	
0.0022	0.1286	0.0071	0.3904	0.0120	0.6184	
0.0023	0.1343	0.0072	0.3954	0.0121	0.6227	
0.0024	0.1399	0.0073	0.4004	0.0122	0.6270	
0.0025	0.1456	0.0074	0.4054	0.0123	0.6312	
0.0026	0.1512	0.0075	0.4103	0.0124	0.6355	
0.0027	0.1569	0.0076	0.4152	0.0125	0.6398	
0.0028	0.1625	0.0077	0.4202	0.0126	0.6440	
0.0029	0.1681	0.0078	0.4251	0.0127	0.6482	
0.0030	0.1736	0.0079	0.4300	0.0128	0.6524	
0.0031	0.1792	0.0080	0.4348	0.0129	0.6566	
0.0032	0.1848	0.0081	0.4397	0.0130	0.6608	
0.0033	0.1903	0.0082	0.4446	0.0131	0.6649	
0.0034	0.1958	0.0083	0.4494	0.0132	0.6691	
0.0035	0.2014	0.0084	0.4542	0.0133	0.6732	
0.0036	0.2069	0.0085	0.4590	0.0134	0.6773	
0.0037	0.2123	0.0086	0.4638	0.0135	0.6814	
0.0038	0.2178	0.0087	0.4686	0.01355	0.6835	0.00500
0.0039	0.2233	0.0088	0.4734	0.0136	0.6855	0.00497
0.0040	0.2287	0.0089	0.4781	0.0137	0.6896	0.00491
0.0041	0.2341	0.0090	0.4828	0.0138	0.6936	0.00485
0.0042	0.2396	0.0091	0.4876	0.0139	0.6977	0.00480
0.0043	0.2450	0.0092	0.4923	0.0140	0.7017	0.00474
0.0044	0.2503	0.0093	0.4970	0.0141	0.7057	0.00469
0.0045	0.2557	0.0094	0.5017	0.0142	0.7097	0.00463
0.0046	0.2611	0.0095	0.5063	0.0143	0.7137	0.00458
0.0047	0.2664	0.0096	0.5110	0.0144	0.7177	0.00453
0.0048	0.2717	0.0097	0.5156	0.0145	0.7216	0.00447
0.0049	0.2771	0.0098	0.5202	0.0146	0.7256	0.00442
0.0050	0.2824	0.0099	0.5248	0.0147	0.7295	0.00437
0.0051	0.2876	0.0100	0.5294	0.0148	0.7334	0.00432
0.0052	0.2929	0.0101	0.5340	0.0149	0.7373	0.00427
0.0053	0.2982	0.0102	0.5386	0.0150	0.7412	0.00423
0.0054	0.3034	0.0103	0.5431	0.0151	0.7451	0.00418
0.0055	0.3087	0.0104	0.5477	0.0152	0.7490	0.00413
0.0056	0.3139	0.0105	0.5522	0.0153	0.7528	0.00408
0.0057	0.3191	0.0106	0.5567	0.0154	0.7567	0.00404
0.0058	0.3243	0.0107	0.5612	0.01548	0.7597	0.00400

^a $d = d_r$.

TABLE D.7
Coefficient of Resistance ($\bar{K}$) versus Reinforcement Ratio (ρ) ($f'_c = 4{,}000$ psi; $f_y = 40{,}000$ psi)

ρ	$\bar{K}$ (ksi)	ρ	$\bar{K}$ (ksi)	ρ	$\bar{K}$ (ksi)	ρ	$\bar{K}$ (ksi)	ρ	$\bar{K}$ (ksi)	ρ	$\bar{K}$ (ksi)	ρ	$\bar{K}$ (ksi)	ε_i^a
0.0010	0.0398	0.0054	0.2091	0.0098	0.3694	0.0142	0.5206	0.0186	0.6626	0.0229	0.7927	0.0271	0.9113	0.00500
0.0011	0.0437	0.0055	0.2129	0.0099	0.3729	0.0143	0.5239	0.0187	0.6657	0.0230	0.7956	0.0272	0.9140	0.00497
0.0012	0.0477	0.0056	0.2166	0.0100	0.3765	0.0144	0.5272	0.0188	0.6688	0.0231	0.7985	0.0273	0.9167	0.00494
0.0013	0.0516	0.0057	0.2204	0.0101	0.3800	0.0145	0.5305	0.0189	0.6720	0.0232	0.8014	0.0274	0.9194	0.00491
0.0014	0.0555	0.0058	0.2241	0.0102	0.3835	0.0146	0.5338	0.0190	0.6751	0.0233	0.8043	0.0275	0.9221	0.00488
0.0015	0.0595	0.0059	0.2278	0.0103	0.3870	0.0147	0.5372	0.0191	0.6782	0.0234	0.8072	0.0276	0.9248	0.00485
0.0016	0.0634	0.0060	0.2315	0.0104	0.3906	0.0148	0.5405	0.0192	0.6813	0.0235	0.8101	0.0277	0.9275	0.00482
0.0017	0.0673	0.0061	0.2352	0.0105	0.3941	0.0149	0.5438	0.0193	0.6844	0.0236	0.8130	0.0278	0.9302	0.00480
0.0018	0.0712	0.0062	0.2390	0.0106	0.3976	0.0150	0.5471	0.0194	0.6875	0.0237	0.8159	0.0279	0.9329	0.00477
0.0019	0.0752	0.0063	0.2427	0.0107	0.4011	0.0151	0.5504	0.0195	0.6905	0.0238	0.8188	0.0280	0.9356	0.00474
0.0020	0.0791	0.0064	0.2464	0.0108	0.4046	0.0152	0.5536	0.0196	0.6936	0.0239	0.8217	0.0281	0.9383	0.00471
0.0021	0.0830	0.0065	0.2501	0.0109	0.4080	0.0153	0.5569	0.0197	0.6967	0.0240	0.8245	0.0282	0.9410	0.00469
0.0022	0.0869	0.0066	0.2538	0.0110	0.4115	0.0154	0.5602	0.0198	0.6998	0.0241	0.8274	0.0283	0.9436	0.00466
0.0023	0.0908	0.0067	0.2574	0.0111	0.4150	0.0155	0.5635	0.0199	0.7029	0.0242	0.8303	0.0284	0.9463	0.00463
0.0024	0.0946	0.0068	0.2611	0.0112	0.4185	0.0156	0.5667	0.0200	0.7059	0.0243	0.8331	0.0285	0.9490	0.00461
0.0025	0.0985	0.0069	0.2648	0.0113	0.4220	0.0157	0.5700	0.0201	0.7090	0.0244	0.8360	0.0286	0.9516	0.00458
0.0026	0.1024	0.0070	0.2685	0.0114	0.4254	0.0158	0.5733	0.0202	0.7120	0.0245	0.8388	0.0287	0.9543	0.00455
0.0027	0.1063	0.0071	0.2721	0.0115	0.4289	0.0159	0.5765	0.0203	0.7151	0.0246	0.8417	0.0288	0.9569	0.00453
0.0028	0.1102	0.0072	0.2758	0.0116	0.4323	0.0160	0.5798	0.0204	0.7181	0.0247	0.8445	0.0289	0.9596	0.00450
0.0029	0.1140	0.0073	0.2795	0.0117	0.4358	0.0161	0.5830	0.0205	0.7212	0.0248	0.8473	0.0290	0.9622	0.00447
0.0030	0.1179	0.0074	0.2831	0.0118	0.4392	0.0162	0.5863	0.0206	0.7242	0.0249	0.8502	0.0291	0.9648	0.00445
0.0031	0.1217	0.0075	0.2868	0.0119	0.4427	0.0163	0.5895	0.0207	0.7272	0.0250	0.8530	0.0292	0.9675	0.00442
0.0032	0.1256	0.0076	0.2904	0.0120	0.4461	0.0164	0.5927	0.0208	0.7302	0.0251	0.8558	0.0293	0.9701	0.00440
0.0033	0.1294	0.0077	0.2941	0.0121	0.4495	0.0165	0.5959	0.0209	0.7333	0.0252	0.8586	0.0294	0.9727	0.00437
0.0034	0.1333	0.0078	0.2977	0.0122	0.4530	0.0166	0.5992	0.0210	0.7363	0.0253	0.8615	0.0295	0.9753	0.00435
0.0035	0.1371	0.0079	0.3013	0.0123	0.4564	0.0167	0.6024	0.0211	0.7393	0.0254	0.8643	0.0296	0.9779	0.00432
0.0036	0.1410	0.0080	0.3049	0.0124	0.4598	0.0168	0.6056	0.0212	0.7423	0.0255	0.8671	0.0297	0.9805	0.00430
0.0037	0.1448	0.0081	0.3086	0.0125	0.4632	0.0169	0.6088	0.0213	0.7453	0.0256	0.8699	0.0298	0.9831	0.00427

(continued)

TABLE D.7 (continued)
Coefficient of Resistance ($\bar{K}$) versus Reinforcement Ratio (ρ) ($f'_c = 4{,}000$ psi; $f_y = 40{,}000$ psi)

ρ	$\bar{K}$ (ksi)	ρ	$\bar{K}$ (ksi)	ρ	$\bar{K}$ (ksi)	ρ	$\bar{K}$ (ksi)	ρ	$\bar{K}$ (ksi)	ρ	$\bar{K}$ (ksi)	ρ	$\bar{K}$ (ksi)	$\varepsilon_t^{\,a}$
0.0038	0.1486	0.0082	0.3122	0.0126	0.4666	0.0170	0.6120	0.0214	0.7483	0.0257	0.8727	0.0299	0.9857	0.00425
0.0039	0.1524	0.0083	0.3158	0.0127	0.4701	0.0171	0.6152	0.0215	0.7513	0.0258	0.8754	0.0300	0.9883	0.00423
0.0040	0.1562	0.0084	0.3194	0.0128	0.4735	0.0172	0.6184	0.0216	0.7543	0.0259	0.8782	0.0301	0.9909	0.00420
0.0041	0.1600	0.0085	0.3230	0.0129	0.4768	0.0173	0.6216	0.0217	0.7572	0.0260	0.8810	0.0302	0.9935	0.00418
0.0042	0.1638	0.0086	0.3266	0.0130	0.4802	0.0174	0.6248	0.0218	0.7602	0.0261	0.8838	0.0303	0.9961	0.00415
0.0043	0.1676	0.0087	0.3302	0.0131	0.4836	0.0175	0.6279	0.0219	0.7632	0.0262	0.8865	0.0304	0.9986	0.00413
0.0044	0.1714	0.0088	0.3338	0.0132	0.4870	0.0176	0.6311	0.0220	0.7662	0.0263	0.8893	0.0305	1.0012	0.00411
0.0045	0.1752	0.0089	0.3374	0.0133	0.4904	0.0177	0.6343	0.0221	0.7691	0.0264	0.8921	0.0306	1.0038	0.00408
0.0046	0.1790	0.0090	0.3409	0.0134	0.4938	0.0178	0.6375	0.0222	0.7721	0.0265	0.8948	0.0307	1.0063	0.00406
0.0047	0.1828	0.0091	0.3445	0.0135	0.4971	0.0179	0.6406	0.0223	0.7750	0.0266	0.8976	0.0308	1.0089	0.00404
0.0048	0.1866	0.0092	0.3481	0.0136	0.5005	0.0180	0.6438	0.0224	0.7780	0.0267	0.9003	0.0309	1.0114	0.00401
0.0049	0.1904	0.0093	0.3517	0.0137	0.5038	0.0181	0.6469	0.0225	0.7809	0.0268	0.9031	0.03096	1.0130	0.00400
0.0050	0.1941	0.0094	0.3552	0.0138	0.5072	0.0182	0.6501	0.0226	0.7839	0.0269	0.9058			
0.0051	0.1979	0.0095	0.3588	0.0139	0.5105	0.0183	0.6532	0.0227	0.7868	0.0270	0.9085			
0.0052	0.2016	0.0096	0.3623	0.0140	0.5139	0.0184	0.6563	0.0228	0.7897					
0.0053	0.2054	0.0097	0.3659	0.0141	0.5172	0.0185	0.6595							

a $d = d_t$.

TABLE D.8
Coefficient of Resistance ($\bar{K}$) ($f'_c = 4{,}000$ psi, $f_y = 50{,}000$ psi)

ρ	$\bar{K}$ (ksi)	ρ	$\bar{K}$ (ksi)	ρ	$\bar{K}$ (ksi)	ρ	$\bar{K}$ (ksi)	ρ	$\bar{K}$ (ksi)	ρ	$\bar{K}$ (ksi)	ε_t
0.0030	0.147	0.0061	0.291	0.0102	0.472	0.0143	0.640	0.0184	0.795	0.0216	0.908	0.0050
0.0031	0.151	0.0062	0.296	0.0103	0.476	0.0144	0.643	0.0185	0.799	0.0217	0.912	0.0050
0.0032	0.156	0.0063	0.300	0.0104	0.480	0.0145	0.648	0.0186	0.802	0.0218	0.915	0.0050
0.0033	0.161	0.0064	0.305	0.0105	0.484	0.0146	0.651	0.0187	0.806	0.0219	0.919	0.0049
0.0034	0.166	0.0065	0.309	0.0106	0.489	0.0147	0.655	0.0188	0.810	0.022	0.922	0.0049
0.0035	0.170	0.0066	0.314	0.0107	0.493	0.0148	0.659	0.0189	0.813	0.0221	0.925	0.0048
0.0036	0.175	0.0067	0.318	0.0108	0.497	0.0149	0.663	0.0190	0.817	0.0222	0.929	0.0048
0.0037	0.180	0.0068	0.323	0.0109	0.501	0.0150	0.667	0.0191	0.820	0.0223	0.932	0.0048
0.0038	0.185	0.0069	0.327	0.0110	0.505	0.0151	0.671	0.0192	0.824	0.0224	0.936	0.0047
0.0039	0.189	0.0070	0.332	0.0111	0.510	0.0152	0.675	0.0193	0.828	0.0225	0.939	0.0047
0.0040	0.194	0.0071	0.336	0.0112	0.514	0.0153	0.679	0.0194	0.831	0.0226	0.942	0.0047
0.0041	0.199	0.0072	0.341	0.0113	0.518	0.0154	0.682	0.0195	0.835	0.0227	0.946	0.0046
0.0042	0.203	0.0073	0.345	0.0114	0.522	0.0155	0.686	0.0196	0.838	0.0228	0.949	0.0046
0.0043	0.208	0.0074	0.350	0.0115	0.526	0.0156	0.690	0.0197	0.842	0.0229	0.952	0.0046
0.0044	0.213	0.0075	0.354	0.0116	0.530	0.0157	0.694	0.0198	0.845	0.023	0.956	0.0045
0.0045	0.217	0.0076	0.359	0.0117	0.534	0.0158	0.698	0.0199	0.849	0.0231	0.959	0.0045
0.0046	0.222	0.0077	0.363	0.0118	0.539	0.0159	0.702	0.0200	0.852	0.0232	0.962	0.0045
0.0047	0.227	0.0078	0.368	0.0119	0.543	0.0160	0.706	0.0201	0.856	0.0234	0.969	0.0044
0.0048	0.231	0.0079	0.372	0.0120	0.547	0.0161	0.709	0.0202	0.859	0.0235	0.972	0.0044
		0.0080	0.376	0.0121	0.551	0.0162	0.713	0.0203	0.863	0.0236	0.975	0.0043
		0.0081	0.381	0.0122	0.555	0.0163	0.717	0.0204	0.866	0.0237	0.978	0.0043
		0.0082	0.385	0.0123	0.559	0.0164	0.721	0.0205	0.870	0.0238	0.982	0.0043
		0.0083	0.389	0.0124	0.563	0.0165	0.725	0.0206	0.873	0.0239	0.985	0.0043
		0.0084	0.394	0.0125	0.567	0.0166	0.728	0.0207	0.877	0.024	0.988	0.0042
		0.0085	0.398	0.0126	0.571	0.0167	0.732	0.0208	0.880	0.0241	0.991	0.0042
		0.0086	0.403	0.0127	0.575	0.0168	0.736	0.0209	0.884	0.0242	0.995	0.0042
		0.0087	0.407	0.0128	0.580	0.0169	0.327	0.0210	0.887	0.0243	0.998	0.0041
		0.0088	0.411	0.0129	0.584	0.0170	0.743	0.0211	0.891	0.0244	1.001	0.0041
		0.0089	0.416	0.0130	0.588	0.0171	0.747	0.0212	0.894	0.0245	1.004	0.0041

(continued)

TABLE D.8 (continued)
Coefficient of Resistance ($\bar{K}$) ($f'_c = 4,000$ psi, $f_y = 50,000$ psi)

ρ	$\bar{K}$ (ksi)	ρ	$\bar{K}$ (ksi)	ρ	$\bar{K}$ (ksi)	ρ	$\bar{K}$ (ksi)	ρ	$\bar{K}$ (ksi)	ρ	$\bar{K}$ (ksi)	ε_t
0.0049	0.236	0.0090	0.420	0.0131	0.592	0.0172	0.751	0.0213	0.898	0.0246	1.008	0.0040
0.0050	0.241	0.0091	0.424	0.0132	0.596	0.0173	0.755	0.0214	0.901	0.0247	1.011	0.0040
0.0051	0.245	0.0092	0.429	0.0133	0.600	0.0174	0.758	0.0215	0.904	0.0248	1.014	0.0040
0.0052	0.250	0.0093	0.433	0.0134	0.604	0.0175	0.762					
0.0053	0.255	0.0094	0.437	0.0135	0.608	0.0176	0.766					
0.0054	0.259	0.0095	0.442	0.0136	0.612	0.0177	0.769					
0.0055	0.264	0.0096	0.446	0.0137	0.616	0.0178	0.773					
0.0056	0.268	0.0097	0.450	0.0138	0.620	0.0179	0.777					
0.0057	0.273	0.0098	0.455	0.0139	0.624	0.0180	0.780					
0.0058	0.278	0.0099	0.459	0.0140	0.628	0.0181	0.784					
0.0059	0.282	0.0100	0.463	0.0141	0.632	0.0182	0.788					
0.0060	0.287	0.0101	0.467	0.0142	0.636	0.0183	0.791					

TABLE D.9
Coefficient of Resistance ($\bar{K}$) versus Reinforcement Ratio (ρ) ($f'_c = 4{,}000$ psi; $f_y = 60{,}000$ psi)

ρ	$\bar{K}$ (ksi)	ρ	$\bar{K}$ (ksi)	ρ	$\bar{K}$ (ksi)	ρ	$\bar{K}$ (ksi)	ρ	$\bar{K}$ (ksi)	ρ	$\bar{K}$ (ksi)	ρ	$\bar{K}$ (ksi)	ε_t^a
0.0010	0.0595	0.0039	0.2259	0.0068	0.3835	0.0097	0.5322	0.0126	0.6720	0.0154	0.7985	0.01806	0.9110	0.00500
0.0011	0.0654	0.0040	0.2315	0.0069	0.3888	0.0098	0.5372	0.0127	0.6766	0.0155	0.8029	0.0181	0.9126	0.00498
0.0012	0.0712	0.0041	0.2371	0.0070	0.3941	0.0099	0.5421	0.0128	0.6813	0.0156	0.8072	0.0182	0.9167	0.00494
0.0013	0.0771	0.0042	0.2427	0.0071	0.3993	0.0100	0.5471	0.0129	0.6859	0.0157	0.8116	0.0183	0.9208	0.00490
0.0014	0.0830	0.0043	0.2482	0.0072	0.4046	0.0101	0.5520	0.0130	0.6906	0.0158	0.8159	0.0184	0.9248	0.00485
0.0015	0.0889	0.0044	0.2538	0.0073	0.4098	0.0102	0.5569	0.0131	0.6952	0.0159	0.8202	0.0185	0.9289	0.00481
0.0016	0.0946	0.0045	0.2593	0.0074	0.4150	0.0103	0.5618	0.0132	0.6998	0.0160	0.8245	0.0186	0.9329	0.00477
0.0017	0.1005	0.0046	0.2648	0.0075	0.4202	0.0104	0.5667	0.0133	0.7044	0.0161	0.8288	0.0187	0.9369	0.00473
0.0018	0.1063	0.0047	0.2703	0.0076	0.4254	0.0105	0.5716	0.0134	0.7090	0.0162	0.8331	0.0188	0.9410	0.00469
0.0019	0.1121	0.0048	0.2758	0.0077	0.4306	0.0106	0.5765	0.0135	0.7136	0.0163	0.8374	0.0189	0.9450	0.00465
0.0020	0.1179	0.0049	0.2813	0.0078	0.4358	0.0107	0.5814	0.0136	0.7181	0.0164	0.8417	0.0190	0.9490	0.00461
0.0021	0.1237	0.0050	0.2868	0.0079	0.4410	0.0108	0.5862	0.0137	0.7227	0.0165	0.8459	0.0191	0.9529	0.00457
0.0022	0.1294	0.0051	0.2922	0.0080	0.4461	0.0109	0.5911	0.0138	0.7272	0.0166	0.8502	0.0192	0.9569	0.00453
0.0023	0.1352	0.0052	0.2977	0.0081	0.4513	0.0110	0.5959	0.0139	0.7318	0.0167	0.8544	0.0193	0.9609	0.00449
0.0024	0.1410	0.0053	0.3031	0.0082	0.4564	0.0111	0.6008	0.0140	0.7363	0.0168	0.8586	0.0194	0.9648	0.00445
0.0025	0.1467	0.0054	0.3086	0.0083	0.4615	0.0112	0.6056	0.0141	0.7408	0.0169	0.8629	0.0195	0.9688	0.00441
0.0026	0.1524	0.0055	0.3140	0.0084	0.4666	0.0113	0.6104	0.0142	0.7453	0.0170	0.8671	0.0196	0.9727	0.00437
0.0027	0.1581	0.0056	0.3194	0.0085	0.4718	0.0114	0.6152	0.0143	0.7498	0.0171	0.8713	0.0197	0.9766	0.00434
0.0028	0.1638	0.0057	0.3248	0.0086	0.4768	0.0115	0.6200	0.0144	0.7543	0.0172	0.8754	0.0198	0.9805	0.00430
0.0029	0.1695	0.0058	0.3302	0.0087	0.4819	0.0116	0.6248	0.0145	0.7587	0.0173	0.8796	0.0199	0.9844	0.00426
0.0030	0.1752	0.0059	0.3356	0.0088	0.4870	0.0117	0.6296	0.0146	0.7632	0.0174	0.8838	0.0200	0.9883	0.00422
0.0031	0.1809	0.0060	0.3409	0.0089	0.4921	0.0118	0.6343	0.0147	0.7676	0.0175	0.8879	0.0201	0.9922	0.00419
0.0032	0.1866	0.0061	0.3463	0.0090	0.4971	0.0119	0.6391	0.0148	0.7721	0.0176	0.8921	0.0202	0.9961	0.00415
0.0033	0.1922	0.0062	0.3516	0.0091	0.5022	0.0120	0.6438	0.0149	0.7765	0.0177	0.8962	0.0203	0.9999	0.00412
0.0034	0.1979	0.0063	0.3570	0.0092	0.5072	0.0121	0.6485	0.0150	0.7809	0.0178	0.9003	0.0204	1.0038	0.00408
0.0035	0.2035	0.0064	0.3623	0.0093	0.5122	0.0122	0.6532	0.0151	0.7853	0.0179	0.9044	0.0205	1.0076	0.00405
0.0036	0.2091	0.0065	0.3676	0.0094	0.5172	0.0123	0.6579	0.0152	0.7897	0.0180	0.9085	0.0206	1.0114	0.00401
0.0037	0.2148	0.0066	0.3729	0.0095	0.5222	0.0124	0.6626	0.0153	0.7941			0.02063	1.0126	0.00400
0.0038	0.2204	0.0067	0.3782	0.0096	0.5272	0.0125	0.6673							

^{a} $d = d_t$.

TABLE D.10
Coefficient of Resistance ($\bar{K}$) versus Reinforcement Ratio (ρ) ($f'_c = 5{,}000$ psi; $f_y = 60{,}000$ psi)

ρ	$\bar{K}$ (ksi)	ρ	$\bar{K}$ (ksi)	ρ	$\bar{K}$ (ksi)	ρ	$\bar{K}$ (ksi)	ρ	$\bar{K}$ (ksi)	ρ	$\bar{K}$ (ksi)	ρ	$\bar{K}$ (ksi)	ε_t^a
0.0010	0.0596	0.0048	0.2782	0.0086	0.4847	0.0124	0.6789	0.0162	0.8609	0.0194	1.0047	0.02257	1.1385	0.00500
0.0011	0.0655	0.0049	0.2838	0.0087	0.4899	0.0125	0.6838	0.0163	0.8655	0.0195	1.0090	0.0226	1.1398	0.00499
0.0012	0.0714	0.0050	0.2894	0.0088	0.4952	0.0126	0.6888	0.0164	0.8701	0.0196	1.0134	0.0227	1.1438	0.00496
0.0013	0.0773	0.0051	0.2950	0.0089	0.5005	0.0127	0.6937	0.0165	0.8747	0.0197	1.0177	0.0228	1.1479	0.00492
0.0014	0.0832	0.0052	0.3005	0.0090	0.5057	0.0128	0.6986	0.0166	0.8793	0.0198	1.0220	0.0229	1.1520	0.00489
0.0015	0.0890	0.0053	0.3061	0.0091	0.5109	0.0129	0.7035	0.0167	0.8839	0.0199	1.0263	0.0230	1.1560	0.00485
0.0016	0.0949	0.0054	0.3117	0.0092	0.5162	0.0130	0.7084	0.0168	0.8885	0.0200	1.0307	0.0231	1.1601	0.00482
0.0017	0.1008	0.0055	0.3172	0.0093	0.5214	0.0131	0.7133	0.0169	0.8930	0.0201	1.0350	0.0232	1.1641	0.00479
0.0018	0.1066	0.0056	0.3227	0.0094	0.5266	0.0132	0.7182	0.0170	0.8976	0.0202	1.0393	0.0233	1.1682	0.00475
0.0019	0.1125	0.0057	0.3282	0.0095	0.5318	0.0133	0.7231	0.0171	0.9022	0.0203	1.0435	0.0234	1.1722	0.00472
0.0020	0.1183	0.0058	0.3338	0.0096	0.5370	0.0134	0.7280	0.0172	0.9067	0.0204	1.0478	0.0235	1.1762	0.00469
0.0021	0.1241	0.0059	0.3393	0.0097	0.5422	0.0135	0.7328	0.0173	0.9112	0.0205	1.0521	0.0236	1.1802	0.00465
0.0022	0.1300	0.0060	0.3448	0.0098	0.5473	0.0136	0.7377	0.0174	0.9158	0.0206	1.0563	0.0237	1.1842	0.00462
0.0023	0.1358	0.0061	0.3502	0.0099	0.5525	0.0137	0.7425	0.0175	0.9203	0.0207	1.0606	0.0238	1.1882	0.00459
0.0024	0.1416	0.0062	0.3557	0.0100	0.5576	0.0138	0.7473	0.0176	0.9248	0.0208	1.0648	0.0239	1.1922	0.00456
0.0025	0.1474	0.0063	0.3612	0.0101	0.5628	0.0139	0.7522	0.0177	0.9293	0.0209	1.0691	0.0240	1.1961	0.00453
0.0026	0.1531	0.0064	0.3667	0.0102	0.5679	0.0140	0.7570	0.0178	0.9338	0.0210	1.0733	0.0241	1.2001	0.00449
0.0027	0.1589	0.0065	0.3721	0.0103	0.5731	0.0141	0.7618	0.0179	0.9383	0.0211	1.0775	0.0242	1.2041	0.00446
0.0028	0.1647	0.0066	0.3776	0.0104	0.5782	0.0142	0.7666	0.0180	0.9428	0.0212	1.0817	0.0243	1.2080	0.00443
0.0029	0.1704	0.0067	0.3830	0.0105	0.5833	0.0143	0.7714	0.0181	0.9473	0.0213	1.0859	0.0244	1.2119	0.00440

d		d		d		d		d		d		d		
0.0030	0.1762	0.0068	0.3884	0.0106	0.5884	0.0144	0.7762	0.0182	0.9517	0.0214	1.0901	0.0245	1.2159	0.00437
0.0031	0.1819	0.0069	0.3938	0.0107	0.5935	0.0145	0.7810	0.0183	0.9562	0.0215	1.0943	0.0246	1.2198	0.00434
0.0032	0.1877	0.0070	0.3992	0.0108	0.5986	0.0146	0.7857	0.0184	0.9606	0.0216	1.0985	0.0247	1.2237	0.00431
0.0033	0.1934	0.0071	0.4047	0.0109	0.6037	0.0147	0.7905	0.0185	0.9651	0.0217	1.1026	0.0248	1.2276	0.00428
0.0034	0.1991	0.0072	0.4100	0.0110	0.6088	0.0148	0.7952	0.0186	0.9695	0.0218	1.1068	0.0249	1.2315	0.00425
0.0035	0.2048	0.0073	0.4154	0.0111	0.6138	0.0149	0.8000	0.0187	0.9739	0.0219	1.1110	0.0250	1.2354	0.00423
0.0036	0.2105	0.0074	0.4208	0.0112	0.6189	0.0150	0.8047	0.0188	0.9783	0.0220	1.1151	0.0251	1.2393	0.00420
0.0037	0.2162	0.0075	0.4262	0.0113	0.6239	0.0151	0.8094	0.0189	0.9827	0.0221	1.1192	0.0252	1.2431	0.00417
0.0038	0.2219	0.0076	0.4315	0.0114	0.6290	0.0152	0.8142	0.0190	0.9872	0.0222	1.1234	0.0253	1.2470	0.00414
0.0039	0.2276	0.0077	0.4369	0.0115	0.6340	0.0153	0.8189	0.0191	0.9916	0.0223	1.1275	0.0254	1.2509	0.00411
0.0040	0.2332	0.0078	0.4422	0.0116	0.6390	0.0154	0.8236	0.0192	0.9959	0.0224	1.1316	0.0255	1.2547	0.00408
0.0041	0.2389	0.0079	0.4476	0.0117	0.6440	0.0155	0.8283	0.0193	1.0003	0.0225	1.1357	0.0256	1.2585	0.00406
0.0042	0.2445	0.0080	0.4529	0.0118	0.6490	0.0156	0.8329					0.0257	1.2624	0.00403
0.0043	0.2502	0.0081	0.4582	0.0119	0.6540	0.0157	0.8376					0.0258	1.2662	0.00400
0.0044	0.2558	0.0082	0.4635	0.0120	0.6590	0.0158	0.8423							
0.0045	0.2614	0.0083	0.4688	0.0121	0.6640	0.0159	0.8469							
0.0046	0.2670	0.0084	0.4741	0.0122	0.6690	0.0160	0.8516							
0.0047	0.2726	0.0085	0.4794	0.0123	0.6739	0.0161	0.8562							

[a] $d = d_r$

TABLE D.11
Values of ρ Balanced, ρ for $\varepsilon_t = 0.005$, and ρ Minimum for Flexure

f_y	f_c'	3000 psi $\beta_1 = 0.85$	4000 psi $\beta_1 = 0.85$	5000 psi $\beta_1 = 0.80$	6000 psi $\beta_1 = 0.75$
Grade 40	ρ balanced	0.0371	0.0495	0.0582	0.0655
40,000 psi	ρ when $\varepsilon_t = 0.005$	0.0203	0.0271	0.0319	0.0359
	ρ min for flexure	0.0050	0.0050	0.0053	0.0058
Grade 50	ρ balanced	0.0275	0.0367	0.0432	0.0486
50,000 psi	ρ when $\varepsilon_t = 0.005$	0.0163	0.0217	0.0255	0.0287
	ρ min for flexure	0.0040	0.0040	0.0042	0.0046
Grade 60	ρ balanced	0.0214	0.0285	0.0335	0.0377
60,000 psi	ρ when $\varepsilon_t = 0.005$	0.0136	0.0181	0.0212	0.0239
	ρ min for flexure	0.0033	0.0033	0.0035	0.0039
Grade 75	ρ balanced	0.0155	0.0207	0.0243	0.0274
75,000 psi	ρ when $\varepsilon_t = 0.005$	0.0108	0.0144	0.0170	0.0191
	ρ min for flexure	0.0027	0.0027	0.0028	0.0031

TABLE D.12
Areas of Steel Bars per Foot of Slab (in.^{2})

Bar Spacing (in.)	#3	#4	#5	#6	#7	#8	#9	#10	#11
2	0.66	1.20	1.86						
2-1/2	0.53	0.96	1.49	2.11					
3	0.44	0.80	1.24	1.76	2.40	3.16	4.00		
3-1/2	0.38	0.69	1.06	1.51	2.06	2.71	3.43	4.35	
4	0.33	0.60	0.93	1.32	1.80	2.37	3.00	3.81	4.68
4-1/2	0.29	0.53	0.83	1.17	1.60	2.11	2.67	3.39	4.16
5	0.26	0.48	0.74	1.06	1.44	1.90	2.40	3.05	3.74
5-1/2	0.24	0.44	0.68	0.96	1.31	1.72	2.18	2.77	3.40
6	0.22	0.40	0.62	0.88	1.20	1.58	2.00	2.54	3.12
6-1/2	0.20	0.37	0.57	0.81	1.11	1.46	1.85	2.34	2.88
7	0.19	0.34	0.53	0.75	1.03	1.35	1.71	2.18	2.67
7-1/2	0.18	0.32	0.50	0.70	0.96	1.26	1.60	2.03	2.50
8	0.16	0.30	0.46	0.66	0.90	1.18	1.50	1.90	2.34
9	0.15	0.27	0.41	0.59	0.80	1.05	1.33	1.69	2.08
10	0.13	0.24	0.37	0.53	0.72	0.95	1.20	1.52	1.87
11	0.12	0.22	0.34	0.48	0.65	0.86	1.09	1.39	1.70
12	0.11	0.20	0.31	0.44	0.60	0.79	1.00	1.27	1.56
13	0.10	0.18	0.29	0.41	0.55	0.73	0.92	1.17	1.44
14	0.09	0.17	0.27	0.38	0.51	0.68	0.86	1.09	1.34
15	0.09	0.16	0.25	0.35	0.48	0.64	0.80	1.02	1.25
16	0.08	0.15	0.23	0.33	0.45	0.59	0.75	0.95	1.17
17	0.08	0.14	0.22	0.31	0.42	0.56	0.71	0.90	1.10
18	0.07	0.13	0.21	0.29	0.40	0.53	0.67	0.85	1.04

TABLE D.13
Size and Pitch of Spirals

Diameter of Column (in.)	Out to Out of Spiral (in.)	f'_c			
		2500	3000	4000	5000
$f_y = 40,000$					
14, 15	11,12	$\frac{3}{8}-2$	$\frac{3}{8}-1\frac{3}{4}$	$\frac{1}{2}-2\frac{1}{2}$	$\frac{1}{2}-1\frac{3}{4}$
16	13	$\frac{3}{8}-2$	$\frac{3}{8}-1\frac{3}{4}$	$\frac{1}{2}-2\frac{1}{2}$	$\frac{1}{2}-2$
17–19	14–16	$\frac{3}{8}-2\frac{1}{4}$	$\frac{3}{8}-1\frac{3}{4}$	$\frac{1}{2}-2\frac{1}{2}$	$\frac{1}{2}-2$
20–23	17–20	$\frac{3}{8}-2\frac{1}{4}$	$\frac{3}{8}-1\frac{3}{4}$	$\frac{1}{2}-2\frac{1}{2}$	$\frac{1}{2}-2$
24–30	21–27	$\frac{3}{8}-2\frac{1}{4}$	$\frac{3}{8}-2$	$\frac{1}{2}-2\frac{1}{2}$	$\frac{1}{2}-2$
$f_y = 60,000$					
14, 15	11, 12	$\frac{1}{4}-1\frac{3}{4}$	$\frac{3}{8}-2\frac{3}{4}$	$\frac{3}{8}-2$	$\frac{1}{2}-2\frac{3}{4}$
16–23	13–20	$\frac{1}{4}-1\frac{3}{4}$	$\frac{3}{8}-2\frac{3}{4}$	$\frac{3}{8}-2$	$\frac{1}{2}-3$
24–29	21–26	$\frac{1}{4}-1\frac{3}{4}$	$\frac{3}{8}-3$	$\frac{3}{8}-2\frac{1}{4}$	$\frac{1}{2}-3$
30	17	$\frac{1}{4}-1\frac{3}{4}$	$\frac{3}{8}-3$	$\frac{3}{8}-2\frac{1}{4}$	$\frac{1}{2}-3\frac{1}{4}$

TABLE D.14
Maximum Number of Bars in One Row

Recommended Spiral or Tie Bar Number	Core Size (in.) = Column Size − 2 × Cover	Circular Area (in.²)	#5	#6	#7	#8	#9	#10	#11[a]	Square Area (in.²)	#5	#6	#7	#8	#9	#10	#11[a]
3	9	63.6	8	7	7	6	—	—	—	81	8	8	8	8	4	4	4
	10	78.5	10	9	8	7	6	—	—	100	12	8	8	8	8	4	4
	11	95.0	11	10	9	8	7	6	—	121	12	12	8	8	8	8	4
	12	113.1	12	11	10	9	8	7	6	144	12	12	12	8	8	8	8
	13	132.7	13	12	11	10	8	7	6	169	16	12	12	12	8	8	8
	14	153.9	14	13	12	11	9	8	7	196	16	16	12	12	12	8	8
	15	176.7	15	14	13	12	10	9	8	225	16	16	12	12	12	8	8
4	16	201.1	16	15	14	12	11	9	8	256	20	16	16	16	12	12	8
	17	227.0	18	16	15	13	12	10	9	289	20	20	16	16	12	12	8
	18	254.5	19	17	15	14	12	11	10	324	20	20	16	16	16	12	12
	19	283.5	20	18	16	15	13	11	10	361	24	20	20	16	16	12	12
	20	314.2	21	19	17	16	14	12	11	400	24	24	20	20	16	12	12
	21	346.4	22	20	18	17	15	13	11	441	28	24	20	20	16	16	12
	22	380.1	23	21	19	18	15	14	12	484	28	24	24	20	20	16	12
5	23	415.5	24	22	21	19	16	14	13	529	28	28	24	24	20	16	16
	24	452.4	25	23	21	20	17	15	13	576	32	28	24	24	20	16	16
	25	490.9	26	24	22	20	18	16	14	625	32	28	28	24	20	20	16
	26	530.9	28	25	23	21	19	16	14	676	32	32	28	24	24	20	16
	27	572.6	29	26	24	22	19	17	15	729	36	32	28	28	24	20	16

[a] Use No. 4 tie for No. 11 or larger longitudinal reinforcement.

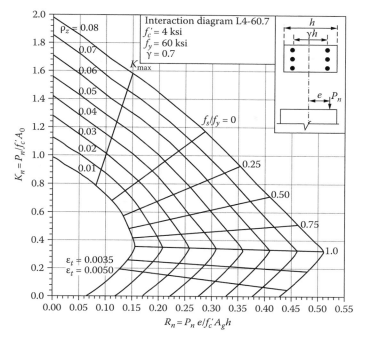

FIGURE D.15 Column interaction diagram for tied column with bars on end faces only. (Courtesy of the American Concrete Institute, Farmington Hills, MI.)

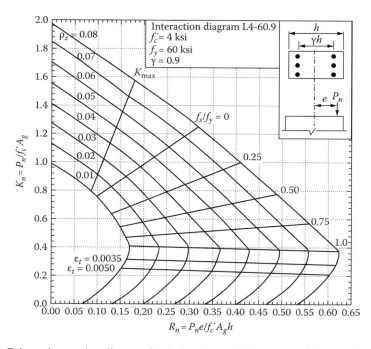

FIGURE D.16 Column interaction diagram for tied column with bars on end faces only. (Courtesy of the American Concrete Institute, Farmington Hills, MI.)

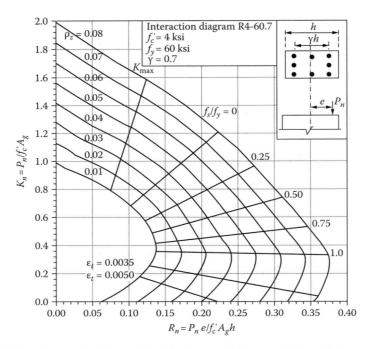

FIGURE D.17 Column interaction diagram for tied column with bars on all faces. (Courtesy of the American Concrete Institute, Farmington Hills, MI.)

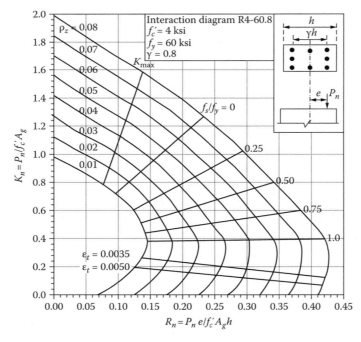

FIGURE D.18 Column interaction diagram for tied column with bars on all faces. (Courtesy of the American Concrete Institute, Farmington Hills, MI.)

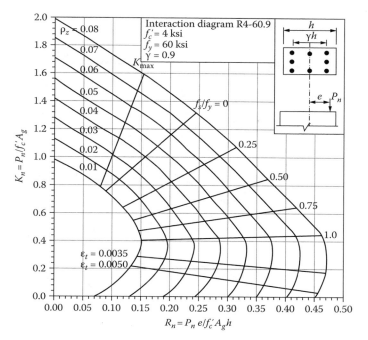

FIGURE D.19 Column interaction diagram for tied column with bars on all faces. (Courtesy of the American Concrete Institute, Farmington Hills, MI.)

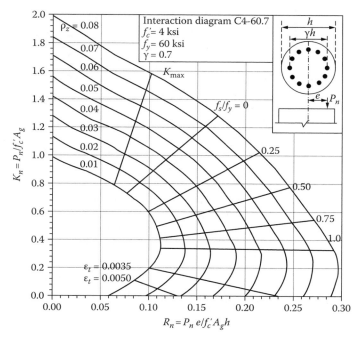

FIGURE D.20 Column interaction diagram for circular spiral column. (Courtesy of the American Concrete Institute, Farmington Hills, MI.)

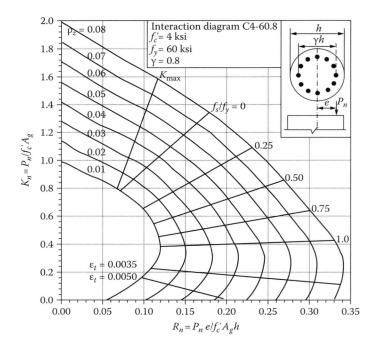

FIGURE D.21   Column interaction diagram for circular spiral column. (Courtesy of the American Concrete Institute, Farmington Hills, MI.)

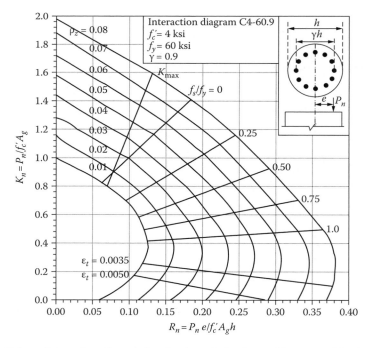

FIGURE D.22   Column interaction diagram for circular spiral column. (Courtesy of the American Concrete Institute, Farmington Hills, MI.)

References and Bibliography

Aghayere, A., Seismic load calculations per the NYS 2002 code, *Paper Presented at the American Society of Civil Engineers Rochester Chapter Meeting*, Rochester, New York, April, 2004.

Alsamsam, I. M. (ed.), *Simplified Design: Reinforced Concrete Buildings of Moderate Size and Height*, 3rd ed., Portland Cement Association, Skokie, IL, 2004.

Ambrose, J. and Tripeny, P., *Simplified Engineering for Architects and Builders*, John Wiley & Sons, Hoboken, NJ, 2006.

American Concrete Institute, *ACI Design Handbook*, SP-17, American Concrete Institute, Farmington Hills, MI, 1997.

American Concrete Institute, *Building Code Requirements for Structural Concrete with Commentary*, ACI 318-08, American Concrete Institute, Farmington Hills, MI, 2008.

American Forest and Paper Association, *Manual for Engineered Wood Construction, LRFD, Guideline: Structural Composite Lumber*, AF&PA American Wood Council, Washington, DC, 1996.

American Forest and Paper Association, *Manual for Engineered Wood Construction, LRFD, Guideline: Wood I-Joists*, AF&PA American Wood Council, Washington, DC, 1996.

American Forest and Paper Association, *Manual for Engineered Wood Construction, LRFD, Guideline: Preengineered Metal Connection*, AF&PA American Wood Council, Washington, DC, 1996.

American Forest and Paper Association, *Manual for Engineered Wood Construction, LRFD, Supplement: Structural Connection*, AF&PA American Wood Council, Washington, DC, 1996.

American Forest and Paper Association, *Manual for Engineered Wood Construction*, LRFD, *Supplement: Structural Lumber*, AF&PA American Wood Council, Washington, DC, 1996.

American Forest and Paper Association, *Manual for Engineered Wood Construction, LRFD, Supplement: Timber Poles and Piles*, AF&PA American Wood Council, Washington, DC, 1996.

American Forest and Paper Association, *Wood Frame Construction Manual for One- and Two-Family Dwellings: Commentary*, AF&PA American Wood Council, Washington, DC, 2001.

American Forest and Paper Association, *Manual for Engineered Wood Construction, ASD/LRFD*, 2005 ed., AF&PA American Wood Council, Washington, DC, 2006.

American Forest and Paper Association, *National Design Specifications for Wood Construction with Commentary and Supplement, ASD/LRFD*, 2005 ed., AF&PA American Wood Council, Washington, DC, 2006.

American Forest and Paper Association, *Solved Example Problems, ASD/LRFD*, 2005 ed., AF&PA American Wood Council, Washington, DC, 2006.

American Forest and Paper Association, *Special Design Provisions for Wind and Seismic, ASD/LRFD*, 2005 ed., AF&PA American Wood Council, Washington, DC, 2006.

American Institute of Steel Construction, *Steel Construction Manual*, 13th ed. (2005 AISC Specifications), American Institute of Steel Construction, Chicago, IL, 2006.

American Institute of Timber Construction, *Timber Construction Manual*, 5th ed., John Wiley & Sons, Hoboken, NJ, 2005.

American Society of Civil Engineers, *Minimum Design Loads for Buildings and Other Structures*, ASCE/SEI 7-05, American Society of Civil Engineers, Reston, VA, 2006.

Breyer, D. E. et al., *Design of Wood Structures, ASD/LRFD*, 6th ed., McGraw-Hill, New York, 2007.

Brockenbrough, R. L. and Merritt, F. S., *Structural Steel Designer's Handbook*, McGraw-Hill, New York, 2005.

Building Seismic Safety Council, *NEHRP Recommended Provisions: Design Examples*, FEMA 451, National Institute of Building Sciences, Washington, DC, 2006.

Ching, F. D. and Winkel, S. R., *Building Codes Illustrated: A Guide to Understanding the 2006 International Building Code*, John Wiley & Sons, Hoboken, NJ, 2007.

Concrete Reinforcing Institute, *Manual of Standard Practice*, 27th ed., Concrete Reinforcing Institute, Chicago, IL, 2003.

Fanella, D. A., *Seismic Detailing of Concrete Buildings*, 2nd ed., Portland Cement Association, Skokie, IL, 2007.

Federal Emergency Management Agency, *Seismic Load Analysis*, Federal Emergency Management Agency, Washington, DC, 2006.

Fisher, J. M. et al., Design of lateral load resisting frames using steel joists and joist girders, *Tech. Digest* 11, Steel Joist Institute, Myrtle Beach, SC, 2007.

Galambos, T. V. et al., *Basic Steel Design with LRFD*, Prentice-Hall, Englewood Cliffs, NJ, 1996.

Geschwinder, L. F., *Design Steel Your Way with the 2005 AISC Specifications*, American Institute of Steel Construction, Chicago, IL, 2006.

Ghosh, S. K., *Overview of the Wind Provisions of the 2006 International Building Code*, S. K. Ghosh Associates, Palatine, IL, 2006.

Ghosh, S. K., *Seismic Design by the 2006 International Building Code*, S. K. Ghosh Associates, Palatine, IL, 2006.

Ghosh, S. K., *Seismic Details for Reinforced Concrete Buildings*, S. K. Ghosh Associates, Palatine, IL, 2006.

International Code Council, *International Residential Code for One- and Two-Family Dwellings, 2003*, International Code Council Inc., Country Club Hills, IL, 2003.

International Code Council, *International Building Code, 2006*, International Code Council Inc., Country Club Hills, IL, 2006.

Kamara, M. E. (ed.), *Notes on ACI 318-05 with Design Applications*, Portland Cement Association, Skokie, IL, 2005.

Limbrunner, G. F. and Aghayere, A. O., *Reinforced Concrete Design*, 6th ed., Prentice-Hall, Englewood Cliffs, NJ, 2007.

Martin, L. D. and Perry, C. J. (eds.), *PCI Design Handbook*, 6th ed., Precast/Prestressed Concrete Institute, Chicago, IL, 2004.

McCormac, J. C. and Nelson, J. K., *Design of Reinforced Concrete: ACI 318-05 Code Edition*, 7th ed., John Wiley & Sons, Hoboken, NJ, 2006.

McCormac, J. C., *Structural Steel Design*, 4th ed., Prentice-Hall, Upper Saddle River, NJ, 2008.

Mehta, K. C. and Delahay, J., *Guide to the Use of the Wind Load Provisions of ASCE 7-02*, American Society of Civil Engineers, Reston, VA, 2003.

Mitchell, D. et al., *AASHTO LRFD Strut-and-Tie Model: Design Examples*, Engineering Bulletin, EB 231, Portland Cement Association, Skokie, IL, 2004.

O'Rourke, M., *Snow Loads: Guide to the Snow Load Provisions of ASCE 7-05*, American Society of Civil Engineers, Reston, VA, 2007.

Segui, W. T., *Steel Design*, 4th ed., Thomson Publishers, Toronto, Canada, 2007.

Simiu, E. and Miyata, T., *Designing of Buildings and Bridges for Wind*, John Wiley & Sons, Hoboken, NJ, 2006.

Steel Joist Institute, *Standard Specification for Joist Girders*, American National Standard SJI-JG-1.1, Revised Nov. 2003, Steel Joist Institute, Forest, VA, 2005.

Steel Joist Institute, *Standard Specification for Open Web Steel Joists, K-series*, American National Standard SJI-K-1.1, Revised Nov. 2003, Steel Joist Institute, Forest, VA, 2005.

Steel Joist Institute, *Standard Specifications*, 42nd ed., Steel Joist Institute, Forest, VA, 2005.

Underwood, R. and Chiuini, M., *Structural Design: A Practical Guide for Architects*, 2nd ed., John Wiley & Sons, Hoboken, NJ, 2007.

Whitney, C. S., *Plastic Theory of Reinforced Concrete Design*, Transactions of the American Society of Civil Engineers, vol. 68, 1942.

Winkel, S. R. et al., *Building Codes Illustrated for Elementary and Secondary Schools*, John Wiley & Sons, Hoboken, NJ, 2007.

Index